AIRO Springer Series

Volume 15

The **AIRO Springer Series** focuses on the relevance of operations research (OR) in the scientific world and in real life applications.

The series publishes peer-reviewed only works, such as contributed volumes, lectures notes, and monographs in English language resulting from workshops, conferences, courses, schools, seminars, and research activities carried out by AIRO, Associazione Italiana di Ricerca Operativa – Optimization and Decision Sciences: http://www.airo.org/index.php/it/.

The books in the series will discuss recent results and analyze new trends focusing on the following areas: Optimization and Operation Research, including Continuous, Discrete and Network Optimization, and related industrial and territorial applications. Interdisciplinary contributions, showing a fruitful collaboration of scientists with researchers from other fields to address complex applications, are welcome.

The series is aimed at providing useful reference material to students, academic and industrial researchers at an international level.

Should an author wish to submit a manuscript, please note that this can be done by directly contacting the series Editorial Board, which is in charge of the peer-review process.

THE SERIES IS INDEXED IN SCOPUS

Massimo Di Francesco · Enrico Gorgone ·
Benedetto Manca · Simone Zanda
Editors

Operations Research: Closing the Gap Between Research and Practice

International Conference on Optimization and Decision Science (ODS), Badesi, Italy, September 8–12, 2024

AIRO
ASSOCIAZIONE ITALIANA DI RICERCA OPERATIVA
OPTIMIZATION AND DECISION SCIENCE

Springer

Editors
Massimo Di Francesco
Dipartimento di Matematica e Informatica
University of Cagliari
Cagliari, Italy

Enrico Gorgone
Dipartimento di Matematica e Informatica
University of Cagliari
Cagliari, Italy

Benedetto Manca
Dipartimento di Matematica e Informatica
University of Cagliari
Cagliari, Italy

Simone Zanda
Dipartimento di Matematica e Informatica
University of Cagliari
Cagliari, Italy

ISSN 2523-7047 ISSN 2523-7055 (electronic)
AIRO Springer Series
ISBN 978-3-031-90094-5 ISBN 978-3-031-90095-2 (eBook)
https://doi.org/10.1007/978-3-031-90095-2

Mathematics Subject Classification: 90C27, 90C57, 90C15, 90C39, 90C59, 90C90

This Springer imprint is published by the registered company Springer Nature Switzerland AG
The registered company address is: Gewerbestrasse 11, 6330 Cham, Switzerland

Contents

Towards Safer Freight Rail Shunting: Integrating MILP and ML Classification Models in a Risk Management Framework

Bigi Federico, Bosi Tommaso, D'Ariano Andrea, and Viti Francesco

Abstract This paper proposes a novel risk analysis framework for the optimization of rolling stock management in rail freight shunting operations. We challenge the direct application of Machine Learning (ML) as input for operational decision-making by employing Risk Assessment strategies to evaluate how ML predictions affect the decision-making process. Our approach integrates the ML model's performance metrics into a Mixed-Integer Linear Programming (MILP) model for shunting operation. A comparative analysis based on real data from the Luxembourgish rail freight company CFL Multimodal across various destinations reveals that a risk assessment approach provides superior performance compared to the direct use of the ML input, reducing the analyzed KPIs. This study demonstrates that the use of a risk assessment framework helps mitigate potential for operational inefficiencies and unfeasibility inherent in ML-dependent models.

Keywords Rail freight operations · Shunting operations · Risk management · Machine learning · Mixed-integer linear programming · Data-driven modeling

F. Bigi (✉) · F. Viti
University of Luxembourg, Esch-Sur-Alzette, Luxembourg
e-mail: federico.bigi@uni.lu

T. Bosi · A. D'Ariano
Department of Civil, Computer Science and Aeronautical Technologies Engineering, Roma Tre University, Rome, Italy

M. Di Francesco et al. (eds.), *Operations Research: Closing the Gap Between Research and Practice*, AIRO Springer Series 15, https://doi.org/10.1007/978-3-031-90095-2_1

1 Introduction

The transportation sector, particularly road transport, significantly contributes to greenhouse gas emissions and air pollution, highlighting the need for sustainable alternatives. Rail freight transport, accounting for only 5.4% of total transport as of 2021, is recognized by the European Union as a key component in reducing its environmental impact. The EU aims to double rail freight's contribution within the next 30 years, emphasizing its role in achieving carbon neutrality [1]. However, the shift to rail freight brings challenges, particularly in the logistics and operational costs associated with operations in rail yards. Shunting, the process of rearranging wagons for train composition and maintenance, is crucial for efficient service delivery and contributes to operational delays, and cancellations, which can spread in the network [11], causing a chain effect of delays and cancellation. As rail freight aims to expand its role in sustainable transportation, optimizing shunting operations becomes critical to enhancing service reliability and minimizing environmental and economic impacts [4, 6]. In this context, new data-driven models have been developed to predict delays in the freight train sector [10]. There is a notable lack of research on the impact of rolling stock maintenance and unexpected disruptions on shunting operations, delays and cancellation. Specifically, rolling stock disruption can cause significant changes in the scheduled operations, unavailability of the fleet and shunting operations to be performed, leading to delays, cancellations, and to additional costs to be sustained by the operators. This research deals with rolling stock management through assignments to services to minimize the above-mentioned KPIs, considering shunting operations to be performed, and maintenance constraints under the form of both mileage and condition-based maintenance. We propose a risk assessment framework to implement condition-based maintenance considerations in the decision-making process for rolling stock assignment in shunting yard operations. We develop a Binary Classification ML model for rolling stock condition prediction and use the forecast, together with the performance metrics of the ML model to inform the MILP about the impact of the implementation of such prediction. In this paper, we extend the model proposed in Bosi et al. [3], Bigi et al. [2] for the Shunt-Out/Shunt-In (SISO) problem. Our methodology merges ML input, its analytics and MILP optimization with the use of risk assessment to evaluate the impact of a ML prediction on the overall operations, thus allowing the MILP model to decide whether to accept or reject the input from the ML based on the state of the system, cost of the operations, and reliability of the ML model. We incorporate reliability metrics such as True Positive Rate (TPR), True Negative Rate (TNR), False Positive Rate (FPR) and False Negative Rate (FNR) to quantify the model's performance in terms of the system vulnerability and improve the reliability of the MILP model. The remainder of the paper is organized as follows: Sect. 2 explores the literature relevant to this topic; Sect. 3 explains the methodology chosen for this paper, presenting the case study; Sect. 4 shows the result of the case study and Sect. 5 summarizes the paper and future research directions.

2 Literature Review

Shunting operation is defined by Cadarso and Marín [4] as the movement of one or more rolling stocks within a shunting yard. These operations are usually performed for a specific reason, such as fulfilling a service for an outbound train that has to be created in the departure track (demand) from the available wagons parked in the classification tracks (supply). Once this matching has been performed, for each wagon that needs to be shunted, a specific routing has to be created within the tracks of the shunting yard to connect these rolling stocks to the actual outbound train.

Thus, the shunting optimization problem is usually modeled through two interrelated and sequential sub-problems [7]: the *Rolling Stock Problem* (RSP), which addresses the planning of each wagon's service time and aims to optimize the management of rolling stocks and reduce costs; and the *Train Unit Shunting Problem* (TUSP), which deals with the routing of different rolling stocks throughout the shunting yard, parking of these in the shunting yard. The RSP is solved before tackling the TUSP in both passenger and freight train operations, as the TUSP depends on the outcomes of the RSP for essential inputs, particularly the alignment of rolling stock for inbound and outbound trains. Integrations of the two models has been proposed in Haahr and Lusby [7] through the Integrated Rolling Stock and Unit Shunting Problem (IRSUSP), which is solved through Mixed-Integer Programming (MIP). The IRSUSP has been solved also by Kamenga et al. [8, 9], which proposed the Generalized Train Unit Shunting problem (G-TUSP) and solved it through MILP formulation.

While usually the maintenance problem is solved separately, Bosi et al. [3], Bigi et al. [2] presents an MILP model for the Shunt-Out/Shunt-In (SISO) problem, which deals with the assignment of rolling stock considering shunting operation and with mileage-based maintenance integration, solving it through simulation approach for long-term assessment. The SISO model addresses two key components: Shunt-Out (SO), which involves removing wagons from an inbound train based on factors like maintenance requirements, and operational costs, and Shunt-In (SI), which focuses on selecting suitable wagons from the shunting yard for the outbound train to minimize future shunting operations, while respecting suitability constraints. Suitability is determined by the wagon mileage constraint before maintenance and type compatibility with the outbound train request. While existing literature, such as Pineda-Jaramillo and Viti [10] and Spanninger et al. [11], extensively explores disruption prediction for railway networks using data-driven modeling, to the best of the author's knowledge, condition-based maintenance prediction for rolling stock using ML and its implementation in rail freight management remains an unexplored territory. Our research proposes an improvement of the model provided in Bosi et al. [3], Bigi et al. [2] using ML as an input for condition-based information.

3 Methodology

In this section, we present an improved deterministic optimization model informed by uncertainty. The scope of this model is twofold: address the limitation identified in the earlier MILP models in Bosi et al. [3], Bigi et al. [2] and implement the ML model prediction as the information regarding the condition of the rolling stock using risk assessment. For this study, we use a Binary Classification ML model to predict the disruption of rolling stock, whose predictions are then assessed through a risk assessment approach, and the model's performance to guide the MILP decisions.

3.1 Risk Assessment

Risk assessment identifies hazards and evaluates risks in a system, leading to control measures or decisions to mitigate them. Risk is classically defined as: $R = L \cdot I$, where R represents the risk, L is the likelihood of the event occurring, I is the impact of the event should it occur. The FAIR Risk Taxonomy [5] presents a method for assessing risk impact, defined as Loss Magnitude (*LM*), outlining different losses for consideration in risk evaluations. The target of our analysis is the prediction of the condition of the rolling stock. To predict a potential disruption, we use a Binary Classification ML model. As the prediction from the model is the source of our information on the disruption, this is the *asset* that we are interested in assessing against its failure, as this could lead to a rolling stock disruption, setting off a cascade effect of other disruptive events. Therefore, L is defined as the likelihood that the ML model could provide a wrong prediction, leading to an event that could affect both rolling stocks and shunting operations. This likelihood can be computed by setting a probability threshold, which is a predetermined value usually between 0 and 1, that determines the cutoff at which a data point is classified into one of the two categories based on the predicted probability. With this, we can compute the relative performance of the classification model, namely False Positive Ratio, True Positive Ratio, True Negative Ratio, and False Negative Ratio, which defines how trustworthy is the ML model when providing predictions against the validation data. The losses analyzed in this study are relative to the potential events triggered by the failure of the ML model: the **Productivity loss** (PL), defined as the loss of availability due to disruptions occurring; the **Replacement loss** (RL) is defined as the cost for replacing the disrupted rolling stock; **Monetary loss** (MnL) is defined as fees associated with the disruptive event. With this, we can define two types of risks, the **Risk of Pointless Disruption** and the **Risk of Disruption**. The Risk of Pointless Disruption (R^{point}) represents the risk of performing unnecessary maintenance, formalized as:

$$R^{\text{point}} = \overbrace{\bar{t}_{out}^{\text{prev}} \cdot f^{\text{dest,yr}} \cdot R^{\text{dest}} \cdot Rev_{rail}}^{\text{PL}} + \overbrace{C_{\text{single}}^{\text{shunt}}}^{\text{RL}} \tag{1}$$

We associate to the *PL* the event of sending a rolling stock out for preemptive maintenance, defined as the economic impact of having the rolling stock out for maintenance for a period of time. The Risk of Disruption (R^{disr}_{grav}) defines the risk associated with encountering a disruptive event, formalized as:

$$R^{\text{disr}}_{grav} = \overbrace{p_i^{\text{mil,max}} \cdot rev^{\text{dest}}}^{\text{PL}} + \overbrace{C^{\text{corr,maint}}_{grav} + C^{\text{shunt}}_{\text{single}}}^{\text{RL}} + \overbrace{P(grav == 4) \cdot C^{\text{canc}}}^{\text{MnL}} \tag{2}$$

In our formulation, we assume different types of disruptive events, represented by $grav$. The *PL* is defined as the economic loss related to the wagon losing its potential mileage ($p_i^{\text{mil,max}}$) before the scheduled maintenance. The *RL* is represented by the cost of performing corrective maintenance, which also depends on the gr, and the cost of removing the wagon from the system. Finally, the *MnL* is specifically for our study, as the highest $grav$ of disruptive event forces us into cancellation. A conceptual framework is also presented for clarity in Fig. 1, where we can see the proposed mild integration of the ML inside the MILP.

The rolling stock arrives with an unknown status, defined in our case as the *event*. The ML model then provides its prediction on the condition of the rolling stock, which is then placed inside the MILP together with the performance metric of the ML model for the selected threshold. This allows the MILP to make more informed decisions based not only on the status of the system but also on the ML model performance. In the example provided in Fig. 1, the ML presents good metrics of Recall (True

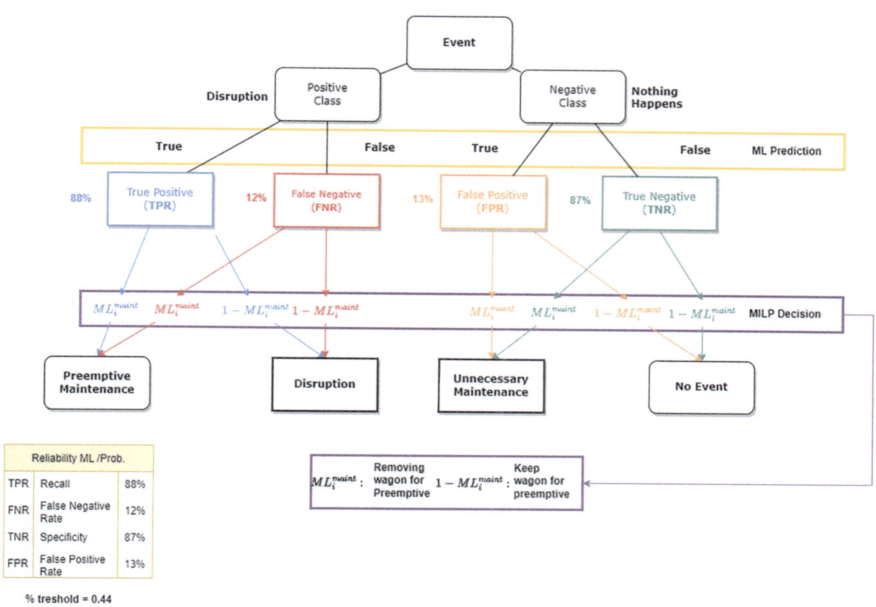

Fig. 1 Conceptual framework of the risk assessment

Positive Ratio, 88%) and Specificity (True Negative Ratio, 87%), suggesting that the MILP model will tend to trust more the ML prediction. Assuming a rolling stock condition is predicted as Non-Disrupted (False), this prediction can be either False Negative or a True Negative. Based on the performance of the ML, as well as the operations that have to be conducted in parallel to this one, the MILP will decide whether to shunt it out or not for preemptive maintenance weighting in this case the Risk of Disruption only by 12% rather than the expected probability of disruption of the wagon.

3.2 Mathematical Programming Model

The presented MILP model deals with the SISO problem, which is the problem of choosing which wagon to remove and to add based on an inbound arriving train and an expected outbound departure composition. It extends the model presented in Bosi et al. [3], Bigi et al. [2] by extending the spatial distribution of the shunting yard, thus allowing a better representation of the shunting operations, introducing more precise delays and cancellation constraints and integrating the ML input through risk assessment as explained in Sect. 3.1.

The objective function (3) maximizes the potential revenue of the train Rev_{rail} minus the costs and risk associated with the MILP decision of shunting, maintenance, and potential cancellation.

$$\max \quad Rev_{rail} - (C^{\text{maint}} + shunts \cdot C_{\text{shunt}} + Risk + Canc^{trig} \cdot C^{\text{canc}}) \tag{3}$$

The set of constraints (4) models the behavior of the model in terms of risk and revenue. They form the decision-making framework for train operations, integrating revenue considerations with risks and costs of wagon shunting and maintenance, guided by the insights from an ML model.

$$\sum_{i \in \mathcal{T}} (1 - w_i^{\text{SO}}) \cdot rev_i + \sum_{j \in TRK} \sum_{k \in PTRK} w_{i,j,k}^{\text{SI}} \cdot rev_{j,k} = Rev_{rail} \tag{4a}$$

$$\sum_{i \in \mathcal{T}} R_i^{\text{SI}} + R_i^{\text{SO}} = Risk \tag{4b}$$

$$
\begin{aligned}
& pred_i^{\mathrm{ML}} \cdot \{TPR \cdot [\\
& (\sum_{g \in \mathrm{gr}} P_g \cdot R_{i,g}^{\mathrm{disr}}) \cdot (1 - ML_i^{\mathrm{maint}}) + (C_{\mathrm{single}}^{\mathrm{shunt}} + C^{\mathrm{prev,maint}}) \cdot ML_i^{\mathrm{maint}} \\
&] + FPR \cdot [(C_{\mathrm{single}}^{\mathrm{shunt}} + C^{\mathrm{prev,maint}} + R^{\mathrm{point}}) \cdot ML_i^{\mathrm{maint}}]\} + (1 - pred_i^{\mathrm{ML}}) \cdot \\
& \{TNR \cdot (C_{\mathrm{single}}^{\mathrm{shunt}} + R^{\mathrm{point}} + C^{\mathrm{prev,maint}}) \cdot ML_i^{\mathrm{maint}} + FNR \cdot \\
& [(\sum_{g \in \mathrm{gr}} P_g \cdot R_{i,g}^{\mathrm{disr}}) \cdot (1 - ML_i^{\mathrm{maint}}) \\
& + (C_{\mathrm{single}}^{\mathrm{shunt}} + C^{\mathrm{prev,maint}}) \cdot ML_i^{\mathrm{maint}}]\} = R_i^{\mathrm{SO}} \quad \forall i \in \mathcal{T}
\end{aligned} \tag{4c}
$$

$$
\begin{aligned}
& \sum_{j \in TRK} \sum_{k \in PTRK} \{[pred_{j,k}^{\mathrm{ML}} \cdot TPR \cdot (\sum_{g \in \mathrm{gr}} P_g \cdot R_{i,j,k}^{\mathrm{disr}}) \cdot w_{i,j,k}^{\mathrm{SI}}] + \\
& + (1 - pred_{j,k}^{\mathrm{ML}}) \cdot FNR \cdot [(\sum_{g \in \mathrm{gr}} P_g \cdot R_{i,j,k}^{\mathrm{disr}}) \cdot w_{i,j,k}^{\mathrm{SI}}]\} = R_i^{\mathrm{SI}} \qquad \forall i \in \mathcal{T}
\end{aligned} \tag{4d}
$$

Equation 4a calculates the train's revenue Rev_{rail}, accounting for the wagons that are not removed by the MILP through the Shunt-Out $(1 - w_i^{\mathrm{SO}})$ and the one placed in by the Shunt-In $(w_{i,j,k}^{\mathrm{SI}})$. Equation 4b defines the overall risk $Risk$ for the Shunt-Out, R_i^{SO}, and the Shunt-In, R_i^{SI}. R_i^{SO} (Eq. 4c), defines whether to remove wagons based on ML prediction. When $pred_i^{\mathrm{ML}} = 1$, the equation weighs the $R_{i,g}^{\mathrm{disr}}$ against shunting and maintenance costs, considering the ML's TPR and FPR for the selected threshold. Similarly, when $pred_i^{\mathrm{ML}} = 0$, it compares the risk of keeping the wagon against removal costs, using TNR and FNR as weights. R_i^{SI}, described in Eq. 4d, evaluates the risk associated with wagons from the shunting yard joining the outbound train. We apply the same criteria as per R_i^{SO}, focusing on selecting wagons based on the minimum risk for their intended destination.

$$\sum_{i \in \mathcal{T}} \sum_{g \in \mathrm{gr}} (P_g \cdot C_g^{\mathrm{corr,maint}}) \cdot maint_i + C^{\mathrm{prev,maint}} \cdot ML_i^{\mathrm{maint}} = C^{\mathrm{maint}} \tag{5}$$

$$t_{\mathrm{dep}} - t_{\mathrm{op}} \leq M \cdot Canc^{trig} \tag{6}$$

$$shunts \cdot t_{\mathrm{shunt}} = t_{\mathrm{dep}} \tag{7}$$

$$shunts^{\mathrm{SI}} + shunts^{\mathrm{SO}} = shunts \tag{8}$$

$$\sum_{i \in \mathcal{T}} w_i^{\mathrm{SO}} - \sum_{i=1}^{rail-1} adj_{i,i+1} = shunts^{\mathrm{SO}} \tag{9}$$

$$2 \cdot adj_{i,i+1}^{\mathrm{SO}} \leq w_i^{\mathrm{SO}} + w_{i+1}^{\mathrm{SO}}, \quad \forall i \in [\mathcal{T}, \mathcal{T}+1] \tag{10}$$

$$maint_i + ML_i^{\mathrm{maint}} + dem_i \leq 1, \quad \forall i \in \mathcal{T} \tag{11}$$

$$dem_i \leq (1.05 - prob_i^{\mathrm{ML}}), \quad \forall i \in \mathcal{T} \tag{12}$$

$$\sum_{i \in \mathcal{T}} dem_i \leq \sum_{l \in TYP} |diff_l| \tag{13}$$

$$maint_i + ML_i^{\mathrm{maint}} + dem_i = w_i^{\mathrm{SO}}, \quad \forall i \in \mathcal{T} \tag{14}$$

$$\sum_{l \in TYP} diff_l - \sum_{i \in \mathcal{T}} (w_i^{\text{SO}} - \sum_{j \in TRK} \sum_{k \in PTRK} w_{i,j,k}^{\text{SI}}) = 0 \tag{15}$$

$$\sum_{j \in TRK} \sum_{k \in PTRK} w_i^{\text{SO}} \cdot typ_{i,j,k} + (1 - w_i^{\text{SO}}) \cdot WI_{i,l} \geq w_{i,l}^{OUT}, \quad \forall i \in \mathcal{T}, \forall l \in TYP \tag{16}$$

$$\sum_{i \in \mathcal{T}} w_{i,l}^{OUT} - WO_{i,l} = 0, \quad \forall l \in TYP \tag{17}$$

$$\sum_{l \in TYP} w_{i,l}^{OUT} = 1, \quad \forall i \in \mathcal{T} \tag{18}$$

$$\sum_{\hat{i}=1}^{\mathcal{T}-1} w_{\hat{i},j,k}^{\text{SI}} \cdot \sum_{\tilde{i}=\hat{i}+1}^{\mathcal{T}} w_{\tilde{i},j,k}^{\text{SI}} \cdot adj_{j,k} \geq adj_{j,k}^{\text{SI}}, \quad \forall j \in TRK, \forall k \in PTRK \tag{19}$$

$$\sum_{i \in \mathcal{T}} \sum_{j \in TRK} \sum_{k \in PTRK} w_{i,j,k}^{\text{SI}} - \sum_{k=1}^{PTRK-1} adj_{j,k}^{\text{SI}} = shunts^{\text{SI}} \tag{20}$$

$$\sum_{i \in \mathcal{T}} w_{i,j,k}^{\text{SI}} \leq 1, \quad \forall j \in TRK, \forall k \in PTRK \tag{21}$$

$$\sum_{j \in TRK} \sum_{k \in PTRK} w_{i,j,k}^{\text{SI}} \leq 1, \quad \forall i \in \mathcal{T} \tag{22}$$

Constraints (5)–(8) define the time to departure t_{dep}, cost of maintenance C^{maint}, and number of shunts to be performed $shunts$. Constraints (9)–(14) model the shunt-out operations, which rolling stock to remove from the inbound train w_i^{SO}, accounting for the wagon to be removed for mileage-based maintenance $maint_i$, wagon to remove for demand matching dem_i and the ones to be removed for preemptive maintenance ML_i^{maint}. Constraints (15)–(18) connect the Shunt-In and Shunt-Out variables through the type requested for the outbound train. Constraints (19)–(22) manage the Shunt-In operations, allowing only one wagon in the shunting yard to be chosen out of a Shunt-Out request.

To compare our approach to the traditional approach to ML implementation, we replace Constraint (12) with Eq. 23:

$$ML_i^{\text{maint}} = pred_i^{\text{ML,in}} \qquad \forall i \in \mathcal{T} \tag{23}$$

This allows the MILP model to treat ML_i^{maint} as input by forcing it to the prediction given by the ML $pred_i^{\text{ML,in}}$. This modification allows the use of Eqs. 4c and 4d for comparison as the risk model. For our case study, this modification will be designated as the **ML Input** scenario, while the model will be designated as **Risk Function**.

For our case study, we used a Binary Classification supervised ML model to predict rolling stock's condition. The model has been trained on real data provided by CFL Multimodal for the years 2021 and 2022. The chosen model, a Tuned Decision Tree (TDT), was tuned to optimize the F1 score, which combines Precision and Recall, and it was chosen for its higher Recall (TPR), which fits the goal of our study. For this analysis, we run scenarios on different probability thresholds, representing different risk profiles. A lower threshold (< 0.5) leads to more conservative predictions with FPR, triggering more preventive maintenance actions. This provides us with the different metrics of reliability for the ML model. The instance for the study was created using real statistics from CFL Multimodal for the year 2023, covering 7 different destinations, listed in Table 1. For the Case Study, we run the models for both the Risk Function and the ML Input scenarios for the 1-year-old fleet, and each probability threshold ranges between 0.05 and 0.95, with a cancellation threshold is 180 min and 15 min per shunting operation.

4 Results

Table 1 presents the average results obtained from running the models across all considered probability thresholds for each destination, providing a comprehensive overview of the performance for the Risk Function and ML Input scenarios. The MWC refers to *Minimum wagon change* and is defined as the minimum number of wagons that have to be changed between the inbound and outbound train. The ML Wag refers to *Wagon Flagged by ML*, representing the average number of wagons flagged as "in risk of disruption" for that specific instance. $Risk$ (Eq. 4b) is consistently lower for the Risk Function model, except for 2 destinations (Antwerp and Rostock), compared to the other model, showing a better management of the scenarios. Overall, the model exhibits improved risk performance management, averaging a reduction of approximately 5%. Cancellation rates from the Risk Function are either matching or outperforming those of the ML Input. Specifically, in cases like Trieste and Rostock, we have a decrease in the Cancellation rate of 89% and 17% respectively. Regarding the Shunts, the Risk function model consistently performs fewer shunting operations compared to the full ML as an Input model, which is also explanatory of why fewer Cancellations are seen for this instance. Finally, for the destination of Halkali, the ML Input returned Unfeasible solutions due to an elevated

Table 1 Risk function versus ML input analysis

	Instance informations		Risk function			ML input		
	MWC	ML Wag	Shunts	$Risk$	Canc. (%)	Shunts	$Risk$	Canc. (%)
Antwerp	10	0	25.17	1153286.6	100	18.5	1104504.3	100
Halkali	12	18	9.78	250061.3	0	**Unfeasible**		
Kiel	4	6.67	8.56	707510.2	22	10.72	773810.7	33
Lyon	6	0	7.33	718661.3	0	7.83	786285.0	0
Poznan	8	16.89	18.67	1344728.4	100	19	1342142.2	100
Rostock	2	11.61	11.89	644566.5	44	12.94	619173.1	61
Trieste	8	0	12	691160.4	0	12.89	701581.3	89

number of ML requests (18). Instead, the Risk Function manages not only to offer an optimal solution but also without cancellation. This underscores the resilience and reliability of the Risk Function, specifically in scenarios where ML Input may not yield usable results. Overall, the Risk Model appears to deliver a more robust and consistent assessment, particularly in terms of risk management and operational reliability.

5 Conclusion

This study presents a framework for using Risk Assessment techniques to improve shunting operations and rolling stock management in freight rail transportation. We developed a MILP with a Binary Classification ML model for the SISO problem. Our findings demonstrate that while ML can provide valuable insights, reliance on these predictions without a risk-informed buffer may lead to operational inefficiencies and even unfeasible scenarios. The robustness of the Risk Function model, which consistently shows improved performance in terms of $Risk$, highlights the necessity for a broader approach to data-driven models that values reliability and continuous data validation over inputting its results. This study supports the assumption that a comprehensive risk assessment framework, paired with an OR model, can make the models adaptable and flexible to the dynamic nature of rail operations. Further research includes the implementation of the interaction between different stakeholders for the decision-making process, a more long-term analysis for asset assessment, implementation and comparison of new data-driven models for rolling stock disruption prediction.

References

1. Agency, E.E.: Greenhouse gas emissions from transport in Europe (2021). https://www.eea.europa.eu/data-and-maps/indicators/transport-emissions-of-greenhouse-gases-7/assessment
2. Bigi, F., Bosi, T., Pineda-Jaramillo, J., Viti, F., D'Ariano, A.: Long-term fleet management for freight trains: assessing the impact of wagon maintenance through simulation of shunting policies. J. Rail Transp. Plan. & Manag. **29**, 100430 (2024)
3. Bosi, T., Bigi, F., D'Ariano, A., Viti, F., and Pineda-Jaramillo, J.: Optimal management of full train load services in the shunting yard: a comprehensive study on shunt-in shunt-out policies. Comput. & Ind. Eng. (2023)
4. Cadarso, L., Marín, Á.: Robust rolling stock in rapid transit networks. Comput. & Oper. Res. **38**(8), 1131–1142 (2011)
5. FAIR Institute. The fair institute: Quantitative information risk management. https://www.fairinstitute.org/ (2023)
6. Giacco, G.L., Carillo, D., D'Ariano, A., Pacciarelli, D., Marín, Á.G.: Short-term rail rolling stock rostering and maintenance scheduling. Transp. Res. Procedia (2014)
7. Haahr, J., Lusby, R.M.: Integrating rolling stock scheduling with train unit shunting. Eur. J. Oper. Res. (2017)

8. Kamenga, F., Pellegrini, P., Rodriguez, J., Merabet, B.: Solution algorithms for the generalized train unit shunting problem. EURO J. Transp. Logist. **10**, 100042 (2021)
9. Kamenga, F., Pellegrini, P., Rogriguez, J., Merabet, B., Houzel, B.: Train unit shunting: integrating rolling stock maintenance and capacity management in passenger railway stations. In: 8th International Conference on Railway Operations Modelling and Analysis (2019)
10. Pineda-Jaramillo, J., Viti, F.: Identifying the rail operating features associated to intermodal freight rail operation delays. Transp. Res. Part C Emerg. Technol. **147**, 103993 (2023)
11. Spanninger, T., Trivella, A., Büchel, B., Corman, F.: A review of train delay prediction approaches. J. Rail Transp. Plan. & Manag. **22**, 100312 (2022)

Optimisation of Reinforcement Learning Foreign Exchange Trading Algorithms Based on Technical Indicators

John Spencer de Wit and Jan H van Vuuren

Abstract Reinforcement learning has recently gained significant attention in the domain of financial trading due to its success in sequential decision-making problems as in the Forex market, for example. The intricate nature of learning achieved by this class of algorithms, however, often obscures the rationale behind their decision-making processes, raising concerns of opacity and unpredictability. These concerns hinder their application in high-stake domains. Explainable artificial intelligence is a relatively recent field that has emerged to overcome these shortcomings and is concerned with techniques that facilitate interpretability of the results returned by black-box models. In this vein, a novel optimisation procedure is proposed in this paper, which leverages Shapley values and the resulting increase in model interpretability, to select the most appropriate technical market indicators aimed at increasing algorithmic trading performance on the Forex market, based on deep reinforcement learning. The utility of this approach is demonstrated in the form of a case study which involves the training and evaluation of two deep reinforcement learning agents, namely a Deep Q-network agent, and a Proximal Policy Optimisation agent.

Keywords Reinforcement learning · Shapley values · Explainable artificial intelligence · Forex trading

1 Introduction

Deep Reinforcement Learning (DRL) methods have shown promise in the context of trading in financial markets in recent years [15]. The lack of interpretability of these algorithms has, however, hindered their application in high-stake domains [9]. As a result, researchers have put forward various methods for interpreting the results

J. S. de Wit (✉) · J. H. van Vuuren
Stellenbosch Unit for Operations Research in Engineering, Department of Industrial Engineering, Stellenbosch University, Matieland, South Africa
e-mail: 21586144@sun.ac.za

J. H. van Vuuren
e-mail: vuuren@sun.ac.za

M. Di Francesco et al. (eds.), *Operations Research: Closing the Gap Between Research and Practice*, AIRO Springer Series 15, https://doi.org/10.1007/978-3-031-90095-2_2

returned by DRL models. By implementing these techniques, it is possible to elucidate a DRL agent's behaviour, thereby potentially achieving explainability of RL model results which may furnish traders with the necessary insight to apply these models in high-stake domains and potentially enhance model performance [2]. Surprisingly, there is a notable absence of reported work on integrating interpretability into DRL trading algorithms applied on the *foreign exchange* (Forex) market. The objective of this paper is to bridge this gap by applying a well-known model interpretability technique to gain insight into the decision-making processes of DRL-based algorithmic trading strategies applied to the Forex market. An emphasis is placed on gaining an understanding of which technical market indicators provide the most fundamental contribution to informing a reinforcement learning agent during the agent's decision-making process. Our findings are showcased in the form of a case study, presenting both the results of the interpretability analysis and their subsequent impact on model performance for two DRL Forex trading algorithms, the *Deep Q-Network* (DQN) algorithm [11] and the *Proximal Policy Optimisation* (PPO) algorithm [20].

2 Literature Review

This section is a concise review of the literature on technical indicators employed by traders on the Forex market (in Sect. 2.1), the reinforcement learning paradigm and its main constituents (in Sect. 2.2), and the domain of explainable DRL algorithm results based on the values of technical market indicators (in Sect. 2.3).

2.1 Technical Indicators

Technical indicators, also referred to as *trading indicators*, are mathematical expressions that offer data-driven interpretations of past price movements and insight into current market conditions. These indicators are also capable of signalling specific market situations, such as overbought conditions (when an asset is overvalued due to excessive buying at unsustainably high prices) or oversold conditions (when an asset is sold at prices below its actual value). Traders use these indicators to analyse current market conditions and to identify their potential entry points into the market or exit points from the market [5].

Traders often experiment with different combinations of indicators that suit their trading styles and perceived market conditions. Trading strategies are typically built on specific combinations of indicators that confirm the trend or momentum direction of an asset's price, rather than relying on indicators which contradict each other's signals [12]. Traders often select subsets of popular and widely recognised technical indicators when making trading decisions, such as those tabulated in Table 1 [10], the selection of these subsets is, however, subjective to the trader.

Table 1 Popular technical indicators

On-balance-volume (OBV) [8]
Exponential moving average (EMA) [8]
Stochastic oscillator (STOCH) [8]
Moving average convergence divergence (MACD) [3]
Bollinger bands (BB) [3]
Relative strength index (RSI) [3]
Moving standard deviation (MSD) [7]
Average directional index (ADX) [7]
Average true range (ATR) [7]

2.2 The Paradigm of Reinforcement Learning

According to Sutton and Barto [17], reinforcement learning is focused on goal-driven learning by an agent leveraging interactions with an environment. It involves learning to map situations to actions with the aim of maximising a numerical reward signal. Apart from the agent and environment, four further subelements form part of a reinforcement learning system: A reward function, a policy, a value function, and a model of the environment [6].

The *reward function* determines the objective of a reinforcement learning problem. The environment interacts with the agent by communicating a single number to the agent, called the *reward signal*, based on the action taken by the agent and the state of the environment. The agent's sole purpose is to optimise the cumulated reward signal over a specified duration of time. In essence, the reward signal dictates which actions taken by the agent are deemed advantageous and which are considered unfavourable.

The *policy* serves as a mapping that relates observed environmental states to the actions available in a particular state. This policy can take the form of a simple lookup table or involve a computationally demanding search process. Essentially, the *policy* outlines how the agent should act when confronted with a specific state of the environment.

Whereas the reward function determines the short-term objective, the *value function* specifies what is favoured in the long term. The *value* of a state represents the amount of future reward an agent is expected to accumulate when it starts out in that specific state. Consequently, the value function takes into consideration not only the immediate reward of the current state but also the expected rewards associated with subsequent future states. More specifically, the value of a state under a specific policy is the anticipated return that the agent would receive if it were to follow that policy, starting from the state in question.

The final component of a reinforcement learning system is a *model* of the environment. The purpose of this model is to mimic the real, underlying environment, which allows for conclusions to be drawn safely about reactional agent behaviour. The model of the environment is used for planning purposes, as a means of predicting future states before they are experienced [4].

2.3 *Explainable Artificial Intelligence*

Despite significant advancements in the field of *Artificial Intelligence* (AI) across various application domains in recent years, explainability and transparency remain key obstacles hindering it from being introduced in critical domains. The concept of *Explainable AI* (XAI) was introduced to overcome these challenges by offering a means to gain insight into the inner workings of black-box AI models. XAI encompasses techniques that enable analysts and/or machines to interpret the reasoning behind AI system decisions. These techniques can be categorised as either *intrinsically* interpretable, referring to inherently interpretable models such as decision trees, or *post-hoc* interpretable, referring to approaches applied after having completed the training process [16]. XAI techniques can also be classified based on their capacity to achieve *local* interpretability or *global* interpretability. Local interpretability refers to gaining insight into the reasoning behind individual outcomes or groups of outcomes, whereas global interpretability pertains to comprehending the overall decision-making rationale of the entire model on a holistic level [13].

A notable post-hoc XAI technique involves the use of *Shapley values*. These values were first devised by Lloyd Shapley in 1953 [19] as a means to quantify the contribution of individual participants in coalitions within a cooperative game theoretic setting, as summarised briefly in Appendix A. More recently, however, Shapley values have found application in modern domains such as machine learning with the purpose of explaining black-box machine learning models [14].

3 The Forex Trading Learning Problem

This section opens in Sect. 3.1 with an articulation of the Forex trading problem as a *Markov decision process* (MDP), after which a brief discussion follows in Sect. 3.2 on the interpretation of Shapley values in the context of reinforcement learning. We finally turn our attention in Sect. 3.3 to performance metrics typically used to evaluate trading strategies.

3.1 *Reinforcement Learning Problem Description and Formulation*

The Forex trading problem is defined as deciding on the trading of one currency in exchange for another, e.g. Euro/Canadian dollar (EUR/CAD), in the hope of realising a profit. This problem can be formulated as a reinforcement learning problem by casting it as an MDP (in which a trading agent iteratively interacts with a market environment by performing a series of trading actions which allow the agent to transition through different states of the environment while receiving rewards that depend on the agent's actions). An MDP is a classical conceptualisation of a series of decision events during which actions not only influence immediate rewards, but also subsequent states and, in turn, future rewards.

The interaction between the agent and the environment is assumed to take place at a sequence of discrete, equi-temporal time steps, $t \in \{0, 1, 2, 3, \ldots, n\}$. At each time step t, the agent observes a representation of the state $S_t \in \mathcal{S}$ of the environment and, based on this state, selects an appropriate action, $A_t \in \mathcal{A}_s$. Here $\mathcal{S}$ denotes the state space of the MDP and $\mathcal{A}_s$ denotes the action space of the reinforcement agent when the environment is in state $s \in \mathcal{S}$. After an action has been selected at time step t, the agent receives a numerical reward $R_t \in \mathcal{R}$ which depends on the aforementioned action, upon which the agent finds itself in a new state $S_{t+1} \in \mathcal{S}$.

We take the time steps as days. The state space $\mathcal{S}$ therefore represents all possible perceptions by the agent of the states of the Forex market on any day and is assumed here to be based on information gathered over a sliding window dating back fifteen days, which represents a single snapshot of the grander-scale environment. Since the Forex market is exceedingly complex, it is nearly impossible to provide a full representation of its state. Instead, selected Forex market data are utilised to represent the market state. More specifically, the state space $\mathcal{S}$ is described in this paper in terms of the closing price of the Forex pair in question, together with the nine technical indicators listed in Table 1. That is, an individual state S_t is described on any day t by a set of 10-tuples containing the aforementioned ten market data values for each of the previous 15 days, equating to a state vector of 150 data instances. This presents an open problem that we aim to address in this paper, namely the curse of dimensionality that stems from the technical indicator selection process during Forex trading. Our proposed optimisation approach is aimed at reducing the dimensionality of the state vector space by providing a means of selecting the most appropriate technical indicators for a given market condition. The approach is discussed in more detail in the following section.

The action space $\mathcal{A}_s$ of the MDP formulation represents the actions available to the agent. These actions are assumed here to be opening of a long or short position. More specifically, the agent is able to open only one trade per day, while it is possible to keep trades open for multiple days. The agent learns a trading strategy which is conceptualised in the reinforcement learning context as a stochastic policy $\pi(s, a)$ that dictates an appropriate action $a \in \mathcal{A}_s$, based on a probability distribution linked to the state $s \in \mathcal{S}$ in which the agent finds the market environment.

The reward function $\mathcal{R}$ of the MDP formulation directly affects the agent's learning process by providing feedback to the agent on its actions. This feedback is the difference between the closing and opening price of the Forex pair in question when a trade is closed and opened, respectively, therefore representing the profit or loss associated with a unit of volume of the trading position. The agent receives this reward after closing each trade.

The following assumptions are made to simplify modelling of the trading process:

- No market gap or slippage is taken into account.
- Spread is not taken into account and neither are transaction costs.

- Leverage is applied with each trade (that is, funds are borrowed to amplify returns). We assume a leverage of twenty times the trade amount.
- The trade amount is directly proportional to the available portfolio balance (assumed in this paper as 8% of the portfolio balance multiplied by the leverage).

Two popular DRL algorithms are applied to solve the reinforcement learning problem associated with the MDP described above. These algorithms are the well-known *Deep Q-network* (DQN) algorithm [11] and the *Proximal Policy Optimisation* (PPO) algorithm [20].

3.2 An Interpretation of Shapley Values in Reinforcement Learning

Shapley values are commonly utilised in supervised learning to express the contributions of individual features to the predictions of a model. This paradigm of use, however, changes when Shapley values are applied to a reinforcement learning model. Instead of gaining insight into the contribution with respect to a predictive outcome, the Shapley values pertain to the policy learned by the agent in this case. More precisely, in the latter case, a Shapley value is associated with each feature in the state vector s_t and is an estimate of the contribution of that state feature to the actions a_t of an agent when following the policy $\pi(a_t, s_t)$ learned. In our case, therefore, Shapley values provide insight into the significance of the technical indicators in Table 1 when the trading strategy is learned. This, in turn, allows for the selection of the most appropriate technical indicators, thereby reducing the dimensionality of the state input vector. This mitigates, to some extent, the effects that the curse of dimensionality might have on reinforcement learning algorithms. The Shapley values are local explainers from which the explanations may be aggregated to estimate the mean contribution of each state vector feature, which is a form of global explanation.

3.3 Trading Strategy Performance Evaluation

The performance of a trading strategy, referred to here merely as *trading performance*, is usually measured by applying evaluation techniques and metrics to quantify the success of the strategy. Although the aim of a trading strategy is generally to generate high returns, this measure of success does not take into account the inherent risk associated with the strategy. Risk should ideally also be included in the evaluation process of a trading strategy.

We employed the performance metrics in Table 2 to assess the performance of the trading strategies learned by the reinforcement learning agents, with the return metric utilised as the reward function to be optimised by the agents. A notable inclusion is

Table 2 Trading strategy performance metrics

Performance metric	Description
Sharpe ratio	Represents the risk-adjusted return for a trading strategy
Returns	Positive or negative value representing the cumulative returns of a strategy
Compound annual growth rate	Annualised yearly return generated over trading period
Average drawdown	The average value of all drawdowns
Maximum drawdown	The largest loss from a peak to a trough during the trading activity
Maximum drawdown days	The maximum time spent to recover from a drawdown

the *Sharpe ratio* which is widely used in the trading domain as a measure of risk-adjusted return. This measure provides an indication as to whether a trading strategy is profitable due to well-placed trading decisions or due to risky trading positions. The three draw-down metrics also quantify risk.

In conjunction with the computation of these performance metrics, the returns generated by the agent according to the learned policy π can be plotted over time, as illustrated graphically in Appendix B. Such a visualisation may provide insight into the nature of the trading performance achieved by an agent.

4 Results and Discussion

In this section, the proposed trading performance optimisation method is demonstrated in the form of two case studies in which distinct DRL agents are trained in an unoptimised fashion by applying the DQN and PPO algorithms in the context of the EUR/CAD currency pair. The Shapley values associated with the trading policies learned are then computed and the insights thus obtained are used to guide an optimised retraining of the agents for the same currency pair by including only those features in the state vector that have a positive effect on the trading decision-making processes of the agents.

4.1 Agent Results and Discussion

Each of the DQN and PPO learning agents was allocated an initial trading budget of US \$10 000 and trained on data spanning the period 01/01/2006 to 23/12/2016 (called the *training period*) by applying the respective DRL algorithms in the context of the EUR/CAD currency pair. Thereafter, each agent was tested on unseen market data spanning the period 25/12/2016 to 31/12/2021 (called the *testing period*). Due to the stochastic nature of DRL algorithms, a single run only provides a sample instance of the performance distribution of an algorithm. In order to mitigate this bias, the training and testing procedures were repeated a total of thirty times, each with a different seed. This approach allows for a robust assessment of the performance of both DRL algorithms. The portfolio balance achieved by each (unoptimised) agent is plotted over the testing period in Figs. 2 and 4 of Appendix B. This plot represents the return distribution of the agents after thirty learning repetitions.

It is evident from these figures that the (unoptimised) DQN and PPO agents were able to increase the portfolio balance by respectively 11.54% and 15.88% on average. The overall returns generated by these agents were, however, insufficient to achieve a consistent positive trend in the portfolio balance across all iterations. Instead, some iterations of the testing period clearly resulted in negative returns. The fluctuating and unpredictable nature of the Forex market no doubt contributed to these observed trading outcomes. The PPO agent learned a policy that was superior to that learned by the DQN agent, as is evident by the greater returns generated by the former agent over the testing period.

In an endeavour to enhance the performances of these agents, an interpretability analysis was performed to identify the most influential technical indicators contributing to their respective policies learned. More specifically, the mean Shapley values associated with all nine technical indicators in Table 1 were calculated over the testing period and are illustrated graphically in Fig. 1.

From the figure, it is evident that certain technical market indicators, such as the RSI, STOCH and EMA, have a positive impact on the policies learned by both agents. Moreover, the RSI and EMA indicators have a greater positive effect on the

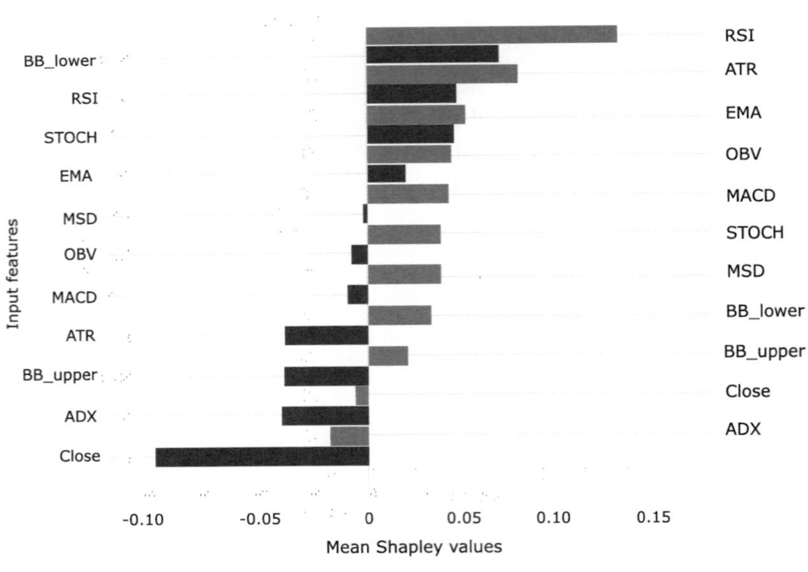

Fig. 1 The mean Shapley values associated with the policies learned by the PPO agent (blue) and the DQN agent (green) over the testing period

policy learned by the DQN agent than on that learned by the PPO agent, while the opposite is true for the STOCH technical indicator. Both agents were subsequently retrained (over the same training period) for the case where the state vector contained only those technical indicators that contributed positively to the policies learned, thus optimising the training of the agents in terms of input state vector features. These optimised agents were then tested on identical unseen data (spanning the testing period), with the resulting portfolio balances illustrated graphically in Figs. 3 and 5in Appendix B.

When evaluating these figures, it is noticeable that both optimised agents exhibited a change in decision-making when compared to that of their unoptimised counterparts, and thus a change in policy, over the testing period. The effect of this optimisation process is captured by the increased portfolio balances at the end of the testing period—representing increases in this balance of 26.04% and 40.88% for the DQN and PPO agents, respectively. The substantial performance increase observed for the optimised PPO agent is noteworthy, as illustrated by the increase in returns generated and the formation of a robust positive portfolio balance trend over all testing iterations.

4.2 *Trading Strategy Performance Evaluation*

The average trading strategy performance metrics described in Sect. 3.3 were computed for all four agents, as well as a buy-and-hold strategy which represents a risk-averse baseline trading strategy, with respect to the testing period in order to perform a numerical comparison of the qualities of the policies that these agents had learned. These performance metrics may be found in Table 3.

The basic buy-and-hold strategy was considered an appropriate strategy to benchmark against. This strategy represents the opening of a long position (to the value of \$10 000)—that is, "buying"—at the start of the testing period and keeping the position open—that is, "holding"—until the end of the testing period. This benchmark strategy was able to generate positive returns, although rather insignificant, which may be attributed to the small change in price at the start of the testing period and at

Table 3 Agent trading strategy performance metrics

Performance metric	Buy-and-hold	DQN	Optimised DQN	PPO	Optimised PPO
Sharpe	0.08	0.21	0.3	0.24	**0.51**
Returns	1.9%	11.54%	26.04%	15.88%	**40.88%**
Compound annual growth rate	0.26%	2.91%	4.81%	2.98%	**6.96%**
Average drawdown	**–1.94%**	–12.56%	–8.11%	–12.76%	–5.97%
Maximum drawdown	**–12.11%**	–32.05%	–21.46%	–17.57%	–17.92%
Maximum drawdown days	1383	1215	1029	812	**658**

the end of the testing period, while large price fluctuations were apparent during the testing period. This is evident by the strategy experiencing a maximum drawdown of –12.11% as well as a large number of drawdown days.

The optimised DQN agent exhibited a larger Sharpe ratio and a swifter recovery from drawdowns than its unoptimised counterpart. This reinforces the significance of selecting technical indicators appropriately for the purposes of informing trading decisions. In addition, the optimised DQN agent was able to learn a stabler policy, as is evident from the reduced maximum drawdown and average drawdown metrics.

There is a prominent distinction between the two PPO agents in the sense of a notable increase in the Sharpe ratio achieved by the optimised PPO agent. This metric is indicative of an increase in risk-averse decisions taken by the agent, which is further reinforced by the fact that the optimised agent experienced a decrease in average drawdown in conjunction with a significant increase in returns and compound annual growth rate. It is apparent that the optimised PPO agent not only generated higher returns, but also accomplished this in a less risky manner.

In conclusion, the optimised agents exhibited improved performance throughout the testing period when compared with their unoptimised counterparts and the risk-averse baseline strategy. The optimised PPO agent emerged as the agent with the most favourable trading strategy for the given market conditions, excelling in terms of generated returns, the Sharpe ratio metric, and exhibiting the desirable drawdown metrics.

5 Conclusion and Future Work

In this paper, we leveraged an explainability technique for the optimisation of DRL algorithms applied to the Forex market. The trading problem was first cast as an MDP, after which the popular DQN and PPO algorithms were implemented in a Forex trading context. An explainability analysis was performed thereafter to determine the most significant contributing technical indicators for the respective trading policies learned. It was shown that retraining both agents with respect to only the top contributing technical indicators led to a more favourable policy in terms of both increased returns and risk mitigation. It is therefore evident from the results presented that an increase in the explainability of reinforcement learning algorithms may lead to increased model performance in the context of Forex trading.

The work in this paper may further be expanded upon in several areas in future research. First, an increased action space may be considered which allows the agent to select from additional actions. Secondly, a wider range of technical indicators may be included in the analysis. Thirdly, a second explainability technique may be used to validate the results returned by the Shapley method. Finally, the analysis may be repeated over various combinations of timeframes and may be applied to different currency pairs in the Forex market in order to test the robustness of the optimisation approach under different market conditions.

Appendix A: Shapley Values

Consider a cooperative game theoretic setting in which M players aim to maximise returns received in terms of some objective. Index these players by the set $\mathcal{M} = \{1, \ldots, M\}$, let $\mathcal{C}$ be any subset of $\mathcal{M}$ not containing a particular player $j \in \mathcal{M}$, referred to as a *coalition*, and suppose the total expected return of the coalition is captured by a function $v(\mathcal{C})$. A Shapley value is defined as the *average marginal contribution* of player j to the coalition if the player were to be included in the coalition [1] and is computed as

$$\phi_j = \sum_{\mathcal{C} \subseteq \mathcal{M} \setminus \{j\}} \frac{|\mathcal{C}|(M - |\mathcal{C}| - 1)!}{M!} (v(\mathcal{C} \cup \{j\}) - v(\mathcal{C})). \tag{1}$$

While the intuitive definition of a Shapley value is self-explanatory, Shapley devised the aforementioned expression based on four axioms: Efficiency, symmetry, zero gain, and linearity [18]. The first axiom states that the total reward is distributed among all players, expressed mathematically by requiring that

$$\sum_{j=1}^{M} \phi_j = v(\mathcal{M}).$$

The symmetry axiom states that if i and j are two players who contribute equally to the accumulation of returns when included in any coalition, i.e. if

$$v(\mathcal{C} \cup \{i\}) = v(\mathcal{C} \cup \{j\})$$

for every subset $\mathcal{C}$ of $\mathcal{M}$ which contains neither i nor j, then their Shapley values are identical, i.e.

$$\phi_j = \phi_i.$$

The third axiom states that zero gain is assigned to players whose marginal contributions are null with respect to every coalition. That is,

$$\phi_j = 0 \text{ if } v(\mathcal{C} \cup \{j\}) = v(\mathcal{C}) \text{ for any subset } \mathcal{C} \subseteq \mathcal{M} \setminus \{j\}.$$

Finally, the axiom of linearity states if two coalition games described in terms of return functions v and w are combined, then the distributed returns correspond linearly to the returns derived from v and the gains derived from w. That is, $\phi_j(av) = a\phi_j(v)$ and $\phi_j(v + w) = \phi_j(v) + \phi_j(w)$ for every player j.

It is interesting that the Shapley value in (1) is the only set of values satisfying the aforementioned desirable axioms.

Appendix B: Returns Generated by the Trained Agents, on the Testing Environment, Over Thirty Trial Runs

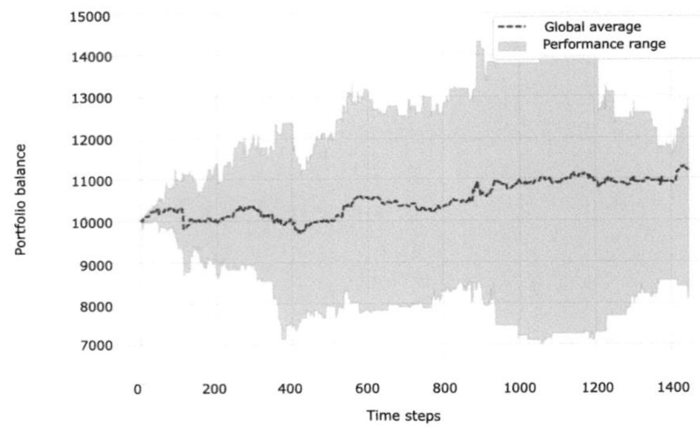

Fig. 2 Returns generated by the DQN agent, started out from a portfolio balance of \$10 000

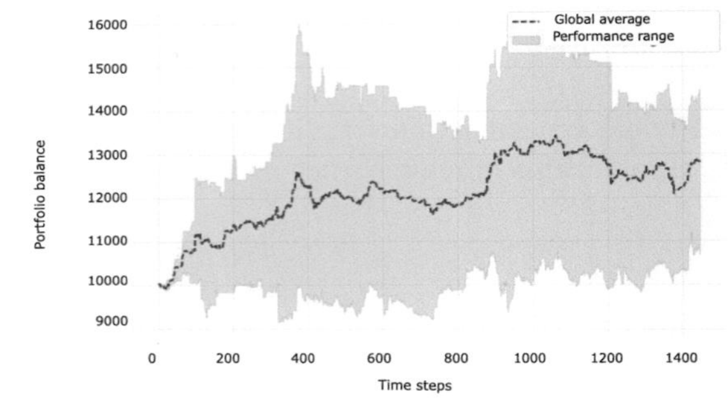

Fig. 3 Returns generated by the optimised DQN agent, started out from a portfolio balance of \$10 000

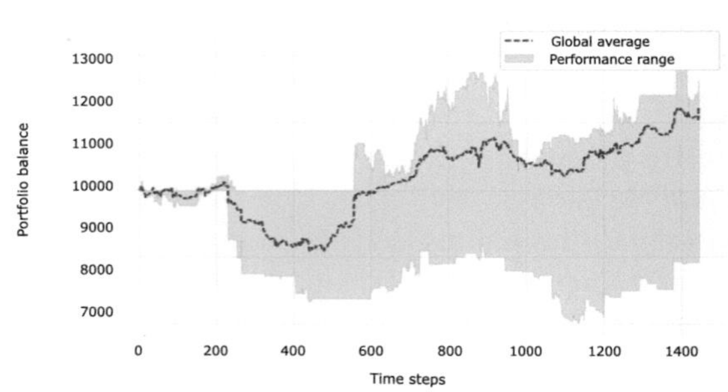

Fig. 4 Returns generated by the PPO agent, started out from a portfolio balance of \$10 000

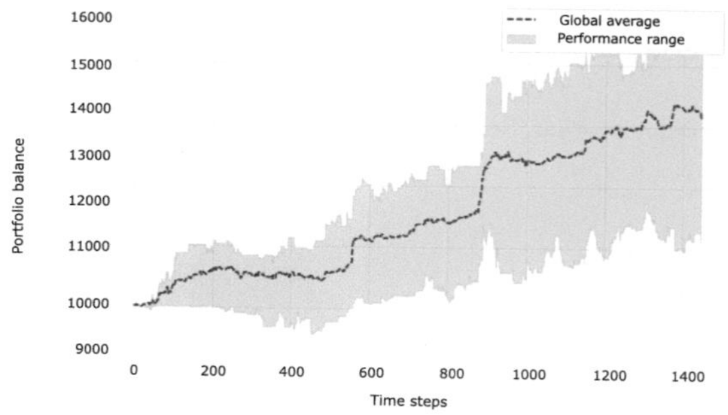

Fig. 5 Returns generated by the optimised PPO agent, started out from a portfolio balance of $10 000

References

1. Aas, K., Jullum, M., Løland, A.: Explaining individual predictions when features are dependent: More accurate approximations to Shapley values. Artif. Intell. **298**, 1–15 (2021)
2. Alharin, A., Thanh-Nam, D., Mina, S.: Reinforcement learning interpretation methods: a survey. IEEE Access **8**, 58–77 (2020). https://doi.org/10.1109/ACCESS.2020.3023394
3. Any Chart Documentation. Technical indicators mathematical description. https://encr.pw/MYo2l (2023)
4. Arulkumaran, K., Deisenroth, M.P., Brundage, M., Bharath, A.A.: Deep reinforcement learning: a brief survey. IEEE Signal Process. Mag. **34**, 26–38 (2017)
5. CMC Markets. Trading strategies every trader should know. https://rb.gy/yvenho (2022)
6. Ding, Z., Huang, Y., Yuan, H., Dong, H.: Introduction to Reinforcement Learning, pp. 47–123. Springer, Singapore (2020)
7. Financial Technical Analysis. Finta. https://pypi.org/project/finta/ (2023)
8. Forex Indicators. The best of mt4 indicators. https://forex-indicators.net/mt4-indicators (2023)
9. Glanois, C., Weng, P., Zimmer, M., Li, D., Yang, T., Hao, J., Liu, W.: A survey on interpretable reinforcement learning (2021). arXiv:2112.13112
10. Iqbal, S.: Top 10 must-have forex technical indicators that every trader should use. https://rb.gy/2nigbn (2023)
11. Jang, B., Kim, M., Harerimana, G., Kim, J.W.: Q-learning algorithms: a comprehensive classification and applications. IEEE Access **7**, 653–667 (2019)
12. Bettman, J.L., Sault, S.J., Schultz, E.L.: Fundamental and technical analysis: substitutes or complements? Int. Econ. Rev. **59**, 859–905 (2009)
13. Løver, J., Gjærum, V.B., Lekkas, A.M.: Explainable AI methods on a deep reinforcement learning agent for automatic docking. Int. Fed. Open Control PapersOnLine **54**, 146–152 (2021)
14. Lundberg, S., Lee, S.: A unified approach to interpreting model predictions. In: 31st International Conference on Neural Information Processing Systems, Red Hook (NY), pp. 4768–4777 (2017)
15. Millea, A.: Deep reinforcement learning for trading: a critical survey. Data **6**, 11–15 (2021)
16. Miller, T.: Explanation in artificial intelligence: insights from the social sciences. Artif. Intell. **267**, 1–38 (2019)
17. Piatetsky, G.: Exclusive: interview with rich sutton, the father of reinforcement learning. https://rb.gy/q3yjpq (2017)

18. Roth, A.E.: The Shapley Value: Essays in Honor of Lloyd S. Shapley. Cambridge University Press, Cambridge (1988)
19. Shapley, L.: A Value for n-Person Games. Princeton University Press, Princeton (NJ) (1953)
20. Zhang, J., Zhang, Z., Han, S., Lü, S.: Proximal policy optimization via enhanced exploration efficiency. Inf. Sci. **609**, 750–765 (2022)

On the Approximation of the Shapley Value via Machine Learning in Transportation Network Cooperative Games

Giorgio Gnecco, Yuval Hadas, Mauro Passacantando, and Marcello Sanguineti

Abstract The Shapley value, a well-established concept in cooperative game theory, serves as a metric for assessing the significance of each player in a transferable utility game. Recently, it has found application in gauging the importance of individual nodes or arcs within a network. However, in this context, the exact evaluation of the Shapley value is often computationally expensive, particularly in the case of extensive networks. This study delves into the challenge of approximating the Shapley value in a transferable utility game defined on a network, wherein the characteristics of the network are parameterized by a variable of interest (e.g., the traffic demand). We examine the smoothness of the Shapley value with respect to this parameter and leverage such smoothness to theoretically justify the adoption of machine-learning techniques for its approximate computation. Additionally, we present potential extensions for further research in this area.

This paper is dedicated to the memory of our colleague and friend Yuval Hadas, who sadly passed away after the submission of the manuscript. We will remember Yuval as an enthusiastic researcher and a very kind person, a pleasure to work with and spend free time with.

G. Gnecco (✉)
IMT Alti Studi Lucca, Lucca, Italy
e-mail: giorgio.gnecco@imtlucca.it

Y. Hadas
Bar-Ilan University, Ramat Gan, Israel
e-mail: yuval.hadas@biu.ac.il

M. Passacantando
University of Milano-Bicocca, Milan, Italy
e-mail: mauro.passacantando@unimib.it

M. Sanguineti
University of Genoa, Genoa, Italy

National Research Council of Italy, Genova, Italy

M. Sanguineti
e-mail: marcello.sanguineti@unige.it

M. Di Francesco et al. (eds.), *Operations Research: Closing the Gap Between Research and Practice*, AIRO Springer Series 15, https://doi.org/10.1007/978-3-031-90095-2_3

Keywords Transferable-utility games · Shapley value · Approximate computation · Wardrop equilibrium · Machine learning

1 Introduction

A basic question in network analysis is the following: how "important" is each node/arc in a network? Analyzing the "importance" of nodes/arcs can assist decision-makers in identifying weak components, detecting and preventing failures, controlling fire outbreaks in extensive wilderness areas, improving connectivity (for instance, in terms of travel time, costs, reliability, access, and flow), etc.

To quantify, e.g., the relative importance of nodes in a network, one possible approach uses the concept of *centrality*. The following are some classical *measures of centrality* of a node i in a network:

- *Degree centrality* relies on the concept of node degree, computed based on the adjacency matrix of the network.
- *Closeness centrality* is computed as one over the summation of the lengths of the shortest paths between node i and all the other nodes.
- *Betweenness centrality* is based on the number of shortest paths between any pair of other nodes via or not necessarily via the intermediate node of interest i.
- *Eigenvalue centrality* is based on the idea that a node i being connected to other nodes with higher scores is given more importance.

However, classical measures of centrality present several limitations:

- They do not consider how the network behaves if any subset of nodes is removed from the network.
- The *Origin-Destination (OD)* demand is not typically considered.
- Some transportation-related properties (e.g., arc capacity, route choice, and congestion) are not easily captured.

Alternative measures of node centrality based on cooperative game theory can deal with the issues reported above. Their definitions can also be extended easily to assess centrality of arcs, differently from most classical measures of node centrality (a variation of betweenness centrality holds for edges and is the *edge betweenness centrality*, see [4]). Applications range, e.g., from movement analysis [8] to traffic networks [15] and wildfire fighting [19].

In this work, we discuss advantages and limitations of one such game-theoretical measure, based on a solution concept from cooperative game theory (the Shapley value). Then, we justify theoretically its approximation by machine-learning techniques, after analyzing its smoothness with respect to a parameter of interest (the traffic demand).

2 Transportation Network Cooperative (TNc) Games

In this section, we briefly summarize a measure of centrality, based on the framework of *Transportation Network cooperative (TNc) games* (see, e.g., [7, 15, 16]), which solves the issues reported in the Introduction (although it introduces some other computational issues, as discussed later).

- In the TNc framework, a subset of the nodes (or arcs) of the network is considered as the set of players in a *Transferable Utility (TU) cooperative game model* (see the Appendix for a reminder on TU games).
- The value (utility) of each subset (coalition) of the set of all players (*grand coalition*) is defined in such a way to represent a wide range of transportation attributes, such as the total travel time.
- A suitable solution concept from cooperative game theory is applied to rank the nodes (or arcs) (e.g., the *Shapley value*, which measures the expected marginal utility provided by each player to a randomly formed coalition of players: see the Appendix for its definition). This provides a measure of the centrality of nodes (or arcs) in the network.

The key advantage of this approach is that nodes (or arcs) are ranked by considering how they contribute to the utilities of all possible subsets of nodes (or arcs).

3 Computational Issues Related to the Evaluation of the Shapley Value in TNc Games

Nevertheless, the exact evaluation of the Shapley value for TNc games poses relevant computational issues:

- In general, the computational effort increases exponentially with the number of players (although, in some special cases, this does not occur, see, e.g., [8, 12]). One possible way to reduce such a computational effort is to restrict the set of players to some special nodes or arcs of interest [5, 6].
- Such an effort is even larger when considering TNc games in which players are arcs and the utility function measures congestion [15, 16]. Indeed, in this case, the utility of each coalition is not given directly. Instead, as an additional step, one has to solve a suitable optimization problem or variational inequality for every coalition, to find its utility (see Sect. 4.1).
- The computational effort is further increased by the presence of parameters (e.g., the traffic demand).

An approximate evaluation of the Shapley value in a TNc game can be obtained using a *Monte Carlo approach*, which replaces the expectation in the definition of the Shapley value (given in the Appendix) with an empirical average. This is done, e.g., by the ApproShapley algorithm, which was originally developed in the literature to deal with the case of generic TU games, then it was extended to TNc games in [5]. Theoretical guarantees (unbiasedness and bounds on the variance of the Monte Carlo estimate of the Shapley value) can also be obtained [5].

4 Contributions of the Work

In the following, we discuss an alternative approximate evaluation of the Shapley value, based on machine learning (Sect. 4.2). It can be used when the Shapley value is a smooth function of a parameter (which is analyzed in Sect. 4.1). As discussed in Sect. 4.3, this approximation could also be combined with the Monte Carlo approach.

4.1 Analysis of Lipschitz Continuity of the Shapley Value as a Function of the Traffic Demand

For certain TNc games, the *Shapley value* is a *Lipschitz continuous function* (with Lipschitz constant L) *of a parameter* d (e.g., the *traffic demand* in TNc games modeling congestion). In particular, it is a function of bounded variation on any bounded and closed interval I (see Fig. 1 for an example).

As detailed in Sect. 4.2, such smoothness can motivate the use of machine learning for Shapley value approximation.

The utility function in the family of TNc games considered in Passacantando et al. [15] is obtained by solving several *user equilibrium traffic assignment problems* [2], one for each coalition of players, which corresponds to a suitable subnetwork of the original network.

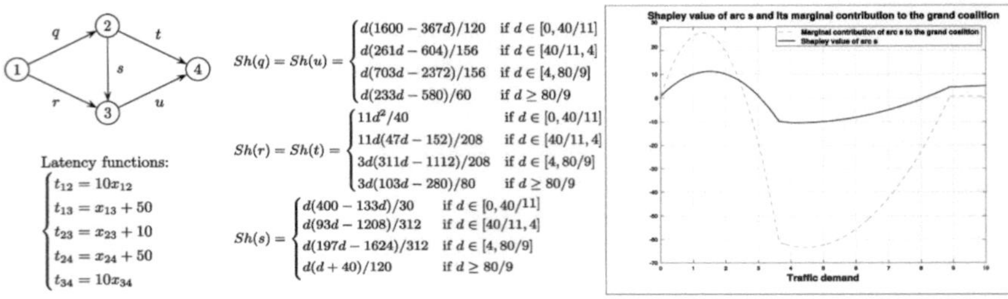

Fig. 1 ***Left***: Example of a transportation network with latency functions on the arcs (*Braess' network*) from Passacantando et al. [15]. ***Middle***: Shapley value (expressed in this case in closed form) of each arc as a function of the traffic demand d from node 1 to node 4 (see Passacantando et al. [15], and the following paragraphs for the definition of the utility function). ***Right***: Shapley value and marginal contribution to the grand coalition of the specific arc s. *Note* In this TNc game, the set of players is made by all the arcs in the network

Specifically, we consider a directed graph $G = (V, A)$, where V is the set of nodes, $A \subseteq V \times V$ is the set of arcs and $C \subseteq V \times V$ is a set of Origin–Destination (OD) pairs. For any OD pair $c \in C$, we denote by d_c the corresponding traffic demand, K_c the set of paths joining it and y_c^k the flow on a generic path $k \in K_c$. A path flow vector $y \geq 0$ is feasible if it satisfies the traffic demands, i.e., $\sum_{k\in K_c} y_c^k = d_c$ holds for any $c \in C$. For any arc $(i, j) \in A$, x_{ij} denotes the flow on (i, j), while t_{ij} is the latency (or travel time) function on the arc (i, j) which is assumed to be a function only of the flow x_{ij} on arc (i, j), and to be continuous and increasing. The travel time on each path $k \in K_c$, denoted by T_c^k, is defined as the sum of travel times on the arcs of such a path. We denote by a_{ij}^{kc} the element in position kc (with $c \in C$, $k \in K_c$) of the incidence matrix of G related to the arc $(i, j) \in A$, so that arc flows x and path flows y are linked by the following equation:

$$x_{ij} = \sum_{c\in C} \sum_{k\in K_c} a_{ij}^{kc} y_c^k, \quad \forall (i, j) \in A.$$

In this framework, we consider a TNc game where the set N of players is a subset of A. The graph associated with a generic coalition $S \subseteq N$, denoted by $G(S) := (V, S \cup (A\backslash N))$, contains the arcs of coalition S and the arcs that are not players of the game. We define $\mathcal{F}$ as the set of coalitions $S \subseteq N$ such that for each OD pair $c \in C$ there exists at least one path in $G(S)$ connecting c. For any coalition $S \in \mathcal{F}$, a feasible path flow vector $y(S)$ is called a *selfish Wardrop equilibrium* (or *user equilibrium*) in the network $G(S)$ if for any OD pair $c \in C$ and path $k \in K_c$ one has

$$T_c^k(y(S)) \begin{cases} =\lambda_c & \text{if } y_c^k > 0, \\ \geq\lambda_c & \text{if } y_c^k = 0, \end{cases}$$

where λ_c is the *equilibrium disutility* for the OD pair c.

Finding a selfish Wardrop equilibrium $y(S)$ amounts to solving the following *convex optimization problem* [13]:

$$\begin{aligned} &\text{minimize} \quad \sum_{(i,j)\in S\cup(A\backslash N)} \int_0^{x_{ij}} t_{ij}(\omega)d\omega \\ &\text{subject to:} \\ &x_{ij} = \textstyle\sum_{c\in C} \sum_{k\in K_c} a_{ij}^{kc} y_c^k, \quad \forall (i, j) \in S \cup (A\backslash N), \\ &d_c = \textstyle\sum_{k\in K_c} y_c^k, \qquad\qquad\quad \forall c \in C, \\ &x_{ij} \geq 0, \qquad\qquad\qquad\qquad\quad \forall (i, j) \in S \cup (A\backslash N), \\ &y_c^k \geq 0, \qquad\qquad\qquad\qquad\quad \forall c \in C, \forall k \in K_c \end{aligned} \tag{1}$$

The total travel time at the equilibrium is defined as:

$$TT(S) := \sum_{c\in C}\sum_{k\in K_c} T_c^k(y(S))y_c^k(S) = \sum_{(i,j)\in S\cup(A\setminus N)} t_{ij}(x(S))x_{ij}(S).$$

It is worth noting that this total travel time is well-defined since the arc equilibrium flow x exists and is unique, as being the latencies of the arcs continuous and increasing functions of the respective flows [2]. Moreover, it is worth mentioning that the convex optimization problem (1) arises as a particular case of a more general *variational inequality* approach [18], also applicable to the situation in which the latency function of each arc depends on the flows on other arcs.

We define the utility function v: $2^N \to R$ of the TNc game by considering two cases:

- If the empty coalition $\varnothing \in \mathcal{F}$, then $v(S) := TT(\varnothing) - TT(S)$ for any $S \subseteq N$.
- If the empty coalition $\varnothing \notin \mathcal{F}$, then

$$v(S) := \begin{cases} [\max_{S'\in\mathcal{F}_m} TT(S')] - TT(S) & if\ \ S \in \mathcal{F}, \\ 0 & if\ \ S \notin \mathcal{F}, \end{cases} \quad \forall S \subseteq N,$$

where $\mathcal{F}_m$ is the subset of *minimally connected* coalitions, i.e., the ones for which the removal of any player makes the resulting coalition not belong to $\mathcal{F}$.

If the latency functions t_{ij} are continuously differentiable for any arc $(i, j) \in A$, then the arc equilibrium flow $x(S)$, for any coalition $S \subseteq N$, is a Lipschitz continuous function of the vector d of traffic demands [10]. Loosely speaking, such smoothness result depends on the fact that a small variation of the demands produces a correspondingly small change in the feasible region and in the optimal solution of the optimization problem. Therefore, the total travel time at equilibrium, the utility function of each coalition, and the Shapley value of each arc are also Lipschitz continuous functions of the traffic demands.

4.2 Motivation for the Application of Supervised Machine Learning to Shapley Value Approximation

Under the previous Lipschitz continuity property (with constant L), in the univariate case (i.e., when $|C| = 1$ and $d_c = d$, likewise in the example of Fig. 1), one can approximate the Shapley value of each arc on any bounded and closed interval I by using a *one-hidden-layer feedforward neural network* with a *sigmoidal activation function* σ (i.e., a bounded, differentiable, and real function defined for all real input values, having non-negative derivative at each point, and exactly one inflection point) and M computational units, obtaining a *sup-norm approximation error* on I smaller than or equal to $\frac{c(I)\cdot L}{M}$, where $c(I) \geq 0$ depends only on the length of the interval I and on the choice of σ. Such a neural network has the following form:

$$f_M(d) = \sum_{m=1}^{M} c_m \sigma(a_m \cdot d + b_m), \tag{2}$$

where $a_m, b_m, c_m \in \mathbb{R}$ $(m = 1, \ldots, M)$ are parameters to be optimized during the training of the network.

The result above follows by an application of Theorem 1 in [3] on nonlinear approximation of univariate functions of bounded variation by this kind of neural networks.

This result suggests the application of *supervised machine-learning methods* to approximate each Shapley value function $f(d)$ by using a neural network of the form (2) and a training set made of a finite number of pairs of the form $(d_p, f(d_p))$, $p = 1, \ldots, P$.

As an example of application, Fig. 2 compares, as functions of the traffic demand d from node 1 to node 4, the Shapley value $Sh(s)$ of arc s in the example reported in Fig. 1 with its approximations obtained by training two one-hidden-layer feedforward neural networks of the form (2) on the interval $I = [0,10]$. The first neural network has $M = 3$ computational units, whereas the second neural network has $M = 12$ computational units. The square loss has been used as loss function for training, whereas the training inputs have been chosen as $d_p = (p-1)/5$, for $p = 1, \ldots, P = 51$ (which correspond with the positions of the circles in the figure). In both cases, training has been performed by batch gradient descent implemented by backpropagation, with learning rate equal to 0.01, and 10,000 training epochs. Before the application of each neural network, for scaling reasons the input d has been mapped linearly onto the interval [0, 1], and the weights of all the computational units (i.e., a_m, b_m, and c_m) have been initialized by realizations of random variables, uniformly and independently distributed on the interval $[-1,1]$. Then, the predictions coming from the two trained neural networks have been plotted on the interval I using a finer discretization with respect to the training set. The results reported in the figure show a clearly better performance of the network with the largest number of computational units.

4.3 *Possible Extensions of the Analysis*

The average error on the training set in Sect. 4.2 could be related to a corresponding *estimation error* by using either *deterministic learning theory* [1] or *statistical learning theory* [20].

The following is a possible way to combine the Monte Carlo approach briefly presented in Sect. 3 with the machine-learning approach of Sect. 4.2: one could consider the case in which a Monte Carlo approximation of the Shapley value of each arc is available on the training set, instead of the Shapley value itself. In other words, in this more general case the training set would be noisy, i.e., the training pairs would have the form $(d_p, f(d_p) + \varepsilon_p)$, $p = 1, \ldots, P$, being the ε_p realizations of independent random variables. These would be not necessarily identically distributed,

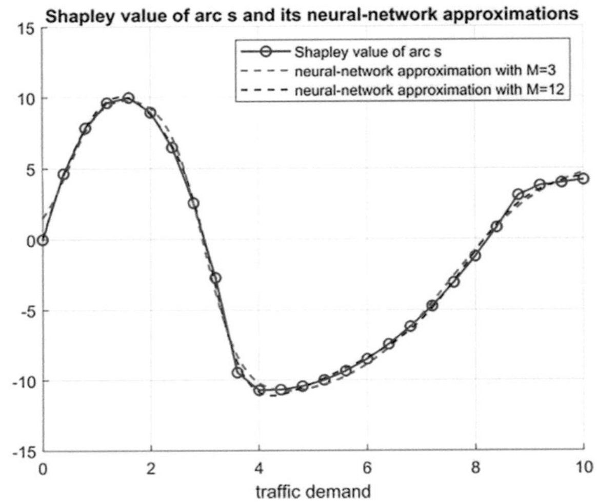

Fig. 2 Shapley value of arc s in the TNc example from Passacantando et al. [15], and its neural-network approximations obtained with $M = 3$ and $M = 12$ computational units, respectively

due to their different associated values of the parameter d. Also in this more general situation, results from statistical learning theory could be used to bound from above the estimation error. Such a bound could be combined with an upper bound on the *approximation error* of the Shapley value such as the one presented in Sect. 4.2 with the final aim to determine an upper bound on the *generalization error*, following the general statistical learning theory approach of [14].

Higher-order smoothness of the Shapley value with respect to the vector of demands could be needed to extend the bound on the sup-norm approximation error of Sect. 4.2 to the multivariate case. However, smoothness higher than Lipschitz continuity is not expected to hold in general, because a change in the vector of demands could also change the set of active constraints at optimality for some of the optimization subproblems (1), possibly resulting in loss of differentiability (likewise in the example reported in Fig. 1).

5 Conclusions

We have examined the Lipschitz continuity of the Shapley value for a class of parameter-dependent TU cooperative games on graphs. Then, we have exploited such a property to motivate the use of machine-learning techniques for the approximate computation of the Shapley value.

Variations of the results (e.g., other conditions under which the Shapley value is a smooth function of some parameters) are expected to hold for families of TU cooperative games on graphs: for instance, the cases in which the selfish Wardrop equilibrium considered in Sect. 4.1 is replaced either by the *Nash equilibrium* of a *noncooperative game model* in which the players are groups of users [9] or by the *system optimum* (see, e.g., [2]), which involves a social planner. Such cases could be analyzed under the *parametric variational inequality* approach investigated in [17]. A further interesting case concerns players that represent *joints of the human body* or *individuals in a group of people*, and the Shapley value is used as a *measure of the importance* of such players for TU cooperative game models (parametrized by time) that capture the *group motion behavior* via suitable movement features [8].

Acknowledgements The authors were partially supported by the project 2023 "Sviluppo di metodi di machine learning per la stima del valore Shapley e di sue generalizzazioni" (CUP_E53C22001930001), funded by INdAM-GNAMPA, by the PRIN PNRR 2022 project "MOTUS" (CUP: D53D23017470001), funded by the European Union—Next Generation EU program, by the PRIN 2022S8XSMY project "Optimal robust shape control for distributed parameter systems", funded by the Italian Ministry of Research, and by the Italian Ministry of Enterprises and Made in Italy (project F/310027/02/X56 "Distributed and coordinated drone swarm for fire detection and suppression activities"). M. Sanguineti was partially supported by the project of the PDGP DIT.AD021.104 "Optimization and Control Techniques" of the Institute of Marine Engineering, National Research Council of Italy, where he is research associate. G. Gnecco, M. Passacantando, and M. Sanguineti are members of INdAM-GNAMPA. M. Sanguineti is Guest Scholar at IMT—School for Advanced Studies, Lucca.

Appendix: Transferable Utility (TU) Games and Shapley Value

A *Transferable Utility (TU) cooperative game* [11] is a pair (N, v), defined as follows: $N = \{1, \dots, n\}$ is the set of players and is called the grand coalition. $S \subseteq N$ denotes a coalition. For each $S \in 2^N$, $v(S)$ denotes the utility that can be achieved by the players in S. For the empty coalition, one has $v(\varnothing) = 0$. In this class of games, there is the possibility of transferring utilities from one player to any other one without incurring any loss. If player i contributes nothing to the value of any extended coalition, then it should receive 0 in a fair allocation of the utility of the grand coalition. If player i positively contributes to at least one coalition S (and contributes non-negatively in all the other cases), then it should receive more than 0. The Shapley value represents indeed a fair way to allocate the utility of the grand coalition within the set of players (the reader is referred, e.g., to [11], for its complete axiomatic characterization), and has the following expression:

$$Sh(i) = \sum_{S \subseteq N} \left[(v(S) - v(S \setminus \{i\})) \frac{(|S| - 1)!(n - |S|)!}{n!} \right], \forall i \in N. \tag{3}$$

Equation (3) can be interpreted as the expected marginal utility of player i, when she joins the coalition formed by all the players that precede her in a random permutation of the set of players (all the permutations being equally likely).

References

1. Cervellera, C., Muselli, M.: Deterministic design for neural network learning: an approach based on discrepancy. IEEE Trans. Neural Netw. **15**, 533–544 (2004)
2. Correa, J.R., Stier-Moses, N.E.: Wardrop equilibria. In: Cochran, J.J. (ed.) Encyclopedia of Operations Research and Management Science. Wiley, Hoboken (2011)
3. Gao, B., Xu, Y.: Univariant approximation by superpositions of a sigmoidal function. J. Math. Anal. Appl. **178**, 221–226 (1993)
4. Girvan, M., Newman, M.E.: Community structure in social and biological networks. Proc. Natl. Acad. Sci. USA **99**(12), 7821–7826 (2002)
5. Gnecco, G., Hadas, Y., Sanguineti, M.: Public transport transfers assessment via transferable utility games and Shapley value approximation. Transp. A: Transp. Sci. **17**(4), 540–565 (2021)
6. Gnecco, G., Hadas, Y., Sanguineti, M.: A game-theoretic approach for reliability evaluation of public transportation transfers with stochastic travel and waiting times. EURO J. Transp. Logist. **11**, article no. 100090 (2022)
7. Hadas, Y., Gnecco, G., Sanguineti, M.: An approach to transportation network analysis via transferable utility games. Transp. Res. Part B: Methodol. **105**, 120–143 (2017)
8. Kolykhalova, K., Gnecco, G., Sanguineti, M., Volpe, G., Camurri, A.: Automated analysis of the origin of movement: an approach based on cooperative games on graphs. IEEE Trans. Human-Mach. Syst. **50**(6), 550–560 (2020)
9. Krylatov, A., Zakharov, V., Tuovinen, T.: Optimization Models and Methods for Equilibrium Traffic Assignment. Springer (2020)
10. Lu, S.: Sensitivity of static traffic user equilibria with perturbations in arc cost function and travel demand. Transp. Sci. **42**, 105–123 (2008)
11. Maschler, M., Solan, E., Zamir, S.: Game Theory. Cambridge University Press (2020)
12. Michalak, T.P., Aadithya, K.V., Szczepański, P.L., Ravindran, B., Jennings, N.R.: Efficient computation of the Shapley value for game-theoretic network centrality. J. Artif. Intell. Res. **46**, 607–650 (2013)
13. Morandi, V.: Bridging the user equilibrium and the system optimum in static traffic assignment: a review. 4OR **22**, 89–119 (2024)
14. Niyogi, P., Girosi, F.: On the relationship between generalization error, hypothesis complexity, and sample complexity for radial basis functions. Neural Comput. **8**, 819–842 (1996)
15. Passacantando, M., Gnecco, G., Hadas, Y., Sanguineti, M.: Braess' paradox: a cooperative game-theoretic point of view. Networks **78**, 264–283 (2021)
16. Passacantando, M., Gnecco, G., Hadas, Y., Sanguineti, M.: On Braess' paradox and average quality of service in transportation network cooperative games: a cooperative game-theoretic point of view. In: Cerulli, R., Dell'Amico, M., Guerriero, F., Pacciarelli, D., Sforza, A. (eds.) Optimization and Decision Science 2020, pp. 27–37. Springer (2021)
17. Passacantando, M., Raciti, F.: Lipschitz continuity results for a class of parametric variational inequalities and applications to network games. Algorithms 16, article no. 458 (2023)
18. Patriksson, M.: The Traffic Assignment Problem: Models and Methods. Dover (2015)
19. Qayyum, F., Abdel Samee, N., Alabdulhafith, M., Aziz, A., Hijjawi, M.: Shapley-Based Interpretation of Deep Learning Models for Wildfire Spread Rate Prediction. Fire Ecol. **20**(8), 1–21 (2024)
20. Vapnik, V.N.: Statistical Learning Theory. Wiley (1998)

The Cost-Benefit Trade-Off Between Problem-Specific and Algorithm-Specific Meta-learning for the Knapsack Problem

Nathan J. van der Westhuyzen and Jan H. van Vuuren

Abstract In this paper, we consider the meta-learning problem of metaheuristic optimisation performance prediction as a foundation underlying the task of algorithm selection for solving optimisation problems. The relative effectiveness of a problem-specific versus an algorithm-specific meta-learning approach is ascertained in the illustrative context of the celebrated knapsack problem. Leveraging the notion of fitness landscape analysis, our algorithm-specific approach incorporates neighbourhood structures induced by metaheuristic perturbation operators. Employing a random forest model, we conduct a comprehensive empirical analysis, contrasting algorithm-specific and problem-specific meta-learning. Two renowned metaheuristic search paradigms are considered—that of simulated annealing and that of a genetic algorithm—each with various hyperparameter and perturbation operator configurations. The numerical results provide insight into the cost-benefit trade-off between problem-specific meta-learning and algorithm-specific meta-learning. It is found that algorithm-specific information yields superior meta-learning performance, but incurs a greater computational cost during meta-feature extraction.

Keywords Meta-learning · Fitness landscape analysis · Metaheuristic optimisation

1 Introduction

According to a recent survey, hundreds of nature-inspired metaheuristics have been proposed during the last five decades [2]. Moreover, if anything, the rate at which new metaheuristics are being published seems to be on the increase.

N. J. van der Westhuyzen (✉) · J. H. van Vuuren
Stellenbosch Unit for Operations Research in Engineering, Department of Industrial Engineering, Stellenbosch University, Private bag x1, Matieland 7602, South Africa
e-mail: 21594022@sum.ac.za

J. H. van Vuuren
e-mail: vuuren@sun.ac.za

M. Di Francesco et al. (eds.), *Operations Research: Closing the Gap Between Research and Practice*, AIRO Springer Series 15, https://doi.org/10.1007/978-3-031-90095-2_4

This leaves analysts with an extensive array of approximate optimisation procedures to choose from, coupled with the potential for a greater number of hyper-parameter configurations to consider, which is crucial for uncovering high-quality solutions to optimisation problems [9]. According to the well-known phenomenon of *performance complementarity*, no algorithm dominates all other algorithms in terms of performance quality across all optimisation problems [10]. Instead, different algorithms perform best in respect of different types of optimisation problems.

The situation described above gave rise to the research field of *automated algorithm selection* (AAS) which has centred on the algorithm selection problem [18]. This problem was formulated as follows in 1976: Given a set $\mathcal{I}$ of instances of a problem P, a set of algorithms $\mathcal{A} = \{A_1, \ldots, A_n\}$, and an algorithmic performance metric $m : \mathcal{I} \times \mathcal{A} \mapsto \mathbb{R}$, construct a selector $S : \mathcal{I} \mapsto \mathcal{A}$ that maps any instance $i \in \mathcal{I}$ to an algorithm $S(i) \in \mathcal{A}$, such that $m(i, S(i))$ is maximal, for all $i \in \mathcal{I}$. It was only after several decades, however, that successful AAS frameworks became available in the literature [14, 31].

Successful AAS models have more recently leveraged the power of machine learning to make connections between problem-instance characteristics and algorithmic performance. This is facilitated by a process called *meta-learning* [20] during which so-called *meta-features* are extracted from problem instances with a view to train machine-learning models to predict metaheuristic optimisation performance when presented with unseen instances of a particular problem class.

Previous meta-learning methods have embraced a *problem-specific* (PS) approach in which meta-features are extracted from the structure of an optimisation problem instance. For example, in the context of the *knapsack problem* (KP), the average volume of items relative to the knapsack capacity may be calculated and employed as a meta-feature. Our aim in this paper is to investigate and compare this approach with an alternative—an *algorithm-specific* (AS) approach in which meta-features are extracted from a *fitness landscape analysis* (FLA). According to this approach, AS-based information may be derived from the search algorithm's ability to traverse the fitness landscape (corresponding to the problem instance under consideration). Furthermore, the local neighbourhood structures of candidate solutions induced by the perturbation operators of search algorithms are analysed and included in the meta-learning process. Several AAS models have been proposed based on this approach [3, 13], but a direct comparison between PS and AS meta-learning has yet to be made. Our objective in this paper is therefore to answer the following question, based on an empirical analysis:

What cost is associated with AS meta-learning that is agnostic to PS information typically utilised in the literature? $(\star)$

We provide an answer to this question in the context of the KP, a classical combinatorial optimisation problem, as well as an algorithmic portfolio comprising two well-known, but rather diverse, metaheuristics—*simulated annealing* (SA) [11] and *genetic algorithms* (GAs) [6]. We allow each metaheuristic to assume any one of a large number of configurations (a variety of functionality options and hyperparameter values, each inducing distinct metaheuristic search behaviour). The meta-learning model selected is the popular *random forest* model [5] which was demonstrated by Hutter et al. [8] to be the most robust machine-learning model for algorithmic performance prediction. We compare the efficacy of AS meta-learning with that of PS meta-learning, and we also consider the *best-case* scenario where AS and PS are combined.

The remainder of the paper is structured as follows. Brief overviews of related work in the fields of FLA and meta-learning performance prediction aimed at AAS are presented in Sect. 2, after which a description of our (empirical) analytic approach is presented in Sect. 3. The results of the experimental investigation are presented and discussed next, in Sect. 4, after which the paper closes in Sect. 5 with an appraisal of its contributions and suggestions for future work.

2 Related Work

Wright [30] proposed the notion of a fitness landscape to gain insight into why search algorithms perform well or poorly, and which optimisation problem instance features foster strong algorithmic performance. FLA provides a middle ground between purely algorithmic analyses and purely empirical experimentation, where anticipated algorithmic performance can be based on empirical analyses of the topological properties of the local neighbourhood structures of candidate solutions to optimisation problem instances induced by the perturbation operators of search algorithms. A vast variety of FLA measures are available in the literature, each aimed at quantifying a unique perspective of the fitness landscape. The reader is referred to the surveys by Malan and Engelbrecht [17] and Malan [16] for more information on FLA measures.

Since the inception of the algorithm selection problem [18] in 1976, the field of AAS has witnessed the proposal of a variety of effective algorithm selection frameworks and methodologies. Popular examples are the `SATzilla` model [31] and the `Autofolio` model [14]. In addition, variations on the traditional algorithm selection problem have also been studied extensively, such as the *algorithm configuration* (AC) problem [7], in which the algorithmic configuration (of a single algorithm) achieving the best performance is to be identified. More recent algorithm selection and AC models in the literature have harnessed the power of machine learning to connect distinct PS characteristics with algorithmic performance in a highly effective manner [10, 12]. These characteristics or *meta-features* form the input to a meta-learning model tailored to predict which algorithm (or configuration thereof) will yield the best optimisation performance.

The input meta-features are commonly defined in terms of PS parameters. Various feature sets have been proposed in the literature for numerous well-known optimisation problem classes—the reader is referred to the survey by Kerschke et al. [10] for further information. Alternatively, meta-features may be extracted from empirical results collected during an FLA of an optimisation problem instance equipped with the local neighbourhood structure associated with a particular metaheuristic [19]. A number of effective AAS models have been proposed based on this approach [3, 13].

3 Empirical Analysis Design

We present an empirical analysis in this paper aimed at answering the research question ($\star$). This section opens with a mathematical formulation of the KP in Sect. 3.1, and this is followed by a discussion in Sect. 3.2 on the algorithmic portfolio considered. Finally, the meta-learning procedure is described in Sect. 3.3.

3.1 The Knapsack Problem

In its classical form, the KP is a binary programming problem in which items have to be selected from a set of m items for inclusion in a knapsack so as to maximise the total benefit associated with the items selected. Each item is also associated with a volume, and the total volume of the items selected should not exceed the knapsack volume V. Denote the benefit of item i by b_i and the volume of item i by v_i. Then the objective in the KP is to

$$\begin{aligned} \text{maximise } z &= \sum_{i=1}^{m} b_i x_i \\ \text{subject to } \sum_{i=1}^{m} v_i x_i &\leq V, \\ x_i &\in \{0, 1\}, \quad i = 1, \ldots, m, \end{aligned}$$

where x_i is a binary decision variable determining which items are included in the knapsack (x_i is 1 if item i is selected, or 0 otherwise).

3.2 The Algorithmic Portfolio

The algorithmic portfolio comprises two well-known metaheuristics, namely SA and a GA, each configured from a variety of functionality options. Furthermore, various hyperparameter value combinations may be tuned for each metaheuristic. The functionality options together with hyperparameter options are summarised in Table 1. The algorithmic portfolio results in a total of 4,374 configuration variants (2,430 configuration variants for SA and 1,944 for the GA).

Table 1 The algorithm portfolio together with all functionality options and hyperparameter configurations

Metaheuristic	Functionality	Options	Hyperparameters
Simulated annealing	Perturbation operator	Bit-flip	
		Insertion	
		Probabilistic	$\gamma \in \{1.0, 0.5, 0.1\}$
	Temperature schedule	Geometric cooling	$\alpha \in \{0.85, 0.92, 0.99\}$
		Geometric reheating	$\beta \in \{1.5, 1.2, 1.05\}$
	Epoch control	Solution acceptances	$c_{\max} \in \{5, 10, 30\}$
		Solution rejections	$d \in \{1, 3, 5\}$
Genetic algorithm	Population size		$\mu \in \{20, 40, 60\}$
	Selection strategy	Tournament	$k \in \{3, 10\}$
		Rank-based	$s \in \{1.5, 2\}$
	Perturbation operator	1-point crossover	$p_c \in \{0.4, 0.7, 0.95\}$
		With bit-flip mutation	$p_m \in \{0.01, 0.05, 0.1\}$
	Replacement strategy	Steady-state	$\lambda = 1$
		Elitism	$\lambda \in \{2, 8\}$

Detailed descriptions of the basic operations of both metaheuristics may be found in Talbi [23]. The functionality options not presented in Table 1 are described briefly in this section.

Search initialisation is handled in either a *random* manner, according to a uniform distribution, or in a *hybrid* manner. The hybrid approach involves conducting a short greedy walk to uncover a (near-)locally optimal solution. The probabilistic regime for SA involves embedding a bit-flip perturbation operator within an insertion perturbation operator with probability

$$p(t) = \frac{1}{2} \exp\left(-\frac{10t}{\gamma t_{\max}}\right), \tag{1}$$

where γ denotes a decay rate, t denotes the current iteration number, and $t_{\max}$ denotes the maximum allowable number of iterations defined by the termination criteria, or else performing only the bit-flip operator. The insertion operator involves randomly removing an element at one position and relocating it to a new location.

The SA search epochs are controlled by the number of solution acceptances $c_{\max}$ and rejections $d_{\max}$, where $d_{\max} = c_{\max} + d$. Three *constraint handling techniques* (CHTs) are employed by the metaheuristics, namely *rejection*, *penalisation*, and *repair*. The rejection CHT does not allow for any solution infeasibility during the search, while the penalisation CHT involves penalising the objective function value by multiplying it by a penalty coefficient based on the degree of violation of the knapsack volume constraint. More specially, the penalisation coefficient is defined as

$$\Theta(\boldsymbol{x}) = \left(\frac{V}{\sum_{i=1}^{m} v_i x_i}\right)^2.$$

The repair CHT, on the other hand, employs a repair function to recover feasible solutions by removing items from the knapsack individually until a feasible solution is returned. Items are removed randomly according to a uniform distribution.

The metaheuristic search is terminated in a hybrid manner, by applying three termination criteria—whichever occurs first: A *static* criterion (based on a maximum allowable number of iterations), a *dynamic* criterion (based on a maximum allowable number of consecutive reheatings for SA and non-improving iterations for the GA), and a two-minute *run-time limit*.

3.3 The Meta-learning Procedure

Two random forest prediction models are trained (and tuned) on one of the three distinct forms of input information—PS, AS, and a combination of the two. First, the meta-knowledge, comprising problem instance features and algorithmic optimisation performance measures, are extracted from the problem instances. Thereafter, the collected meta-data are compiled and processed for meta-learning prediction purposes. Finally, the meta-learning prediction models are trained, tuned, and evaluated for algorithmic performance prediction. The meta-learning prediction model employed in this investigation is the `ranger` random forest model [29] (a fast-implementation variant of the random forest model). All FLA extractions, optimisation performance measurements, and meta-learning model implementations were performed on the high-performance computing cluster of Stellenbosch University (http://www.sun.ac.za/hpc).

The optimisation problem data utilised during the analysis were taken from the `MATILDA`[1] research platform [21]. Optimisation problem instance data, as well as PS meta-data, are available on this platform for a variety of problem classes. We consider 1,060 KP instances, which were sampled from a total of 5,300 KP instances[2] to alleviate the high computational burden. The set of PS meta-features was increased to include information derived from the various subclasses of the KP, such as the random generation parameters for the item profits and volumes, and to which subclasses the problem instances belong.

A diverse set of 12 FLA measures was collected empirically for use as meta-features. These include: The autocorrelation function [28], correlation length [28], entropic measures [26], an accumulated escape probability [15], Hamming distance in a level [4], a Hamming distance feasibility measure [24], a (in)feasibility pocket size measure [24], a constraint violation severity measure [24], negative slope coefficient [25], population information content [1], change rate [1], and a population evolvability metric [27].

---

[1] Available at https://matilda.unimelb.edu.au/matilda/.

[2] Random uniform sampling was employed with a sample size of 20% within each of the KP subclasses defined in [22] in order to conserve a suitable level of representation.

Table 2 The hyperparameter configuration portfolio, where $\hat{x}$ denotes the rounded-down square root of the number of input features

Hyperparameter	Model variable	Values
No. of trees	*num.trees*	100, 200, 300
No. of variable to split	*mtry*	$0.5\hat{x}$, $\hat{x}$, $1.5\hat{x}$
Minimal node size	*min.node.size*	3, 5, 7
Splitting rule	*splitrule*	'variance', 'extratrees'

The meta-learning procedure was executed in a five-fold manner, during which 30 problem instances were selected for the test set and the remainder for the training set. The meta-learning model was trained and employed to make predictions on a per-instance basis. In addition, a portfolio of hyperparameter configurations was implemented during a *grid-search* for each fold so as to tune the model hyperparameters. The `ranger` portfolio of hyperparameters together with hyperparameter value options, comprising 54 distinct configuration options, is tabulated in Table 2.

Prediction performance was evaluated in terms of three accuracy measures, namely the *mean absolute error* (MAE) and two novel accuracy measures derived for the context of AAS. We claim that traditional accuracy measures, such as the MAE utilised and others like the root mean squared error, fail to provide sufficient insight into meta-learning prediction performance in the specific context of *metaheuristic recommendation*. These measures focus on the entire algorithmic portfolio rather than specifically on the top-performing metaheuristic options, which is the primary concern during metaheuristic recommendation. For this reason, we derived two accuracy measures for assessing the true optimisation performance of the algorithmic portfolio against the estimated performance. First, the *true optimisation performance* is evaluated for those configuration variants predicted to attain the best optimisation performance and secondly, the *true algorithmic rankings* are evaluated against the predicted algorithmic rankings.

The `ranger` hyperparameter configuration with the smallest MAE for each metaheuristic was first determined. Thereafter, the meta-learning predictions were combined for evaluation purposes across the algorithmic portfolio. The derived accuracies are presented in a normalised optimisation performance format for the true performance (with 0 representing the best and 1 the worst optimisation performance in the algorithmic portfolio) and in percentage format for the true ranking (relative to the 4,374 configuration variants in the algorithmic portfolio).

4 Numerical Results

The numerical results obtained during the meta-learning investigation are presented in this section. These results are presented based on the three distinct forms of input information described in Sect. 3.3. The derived accuracy measures are employed in

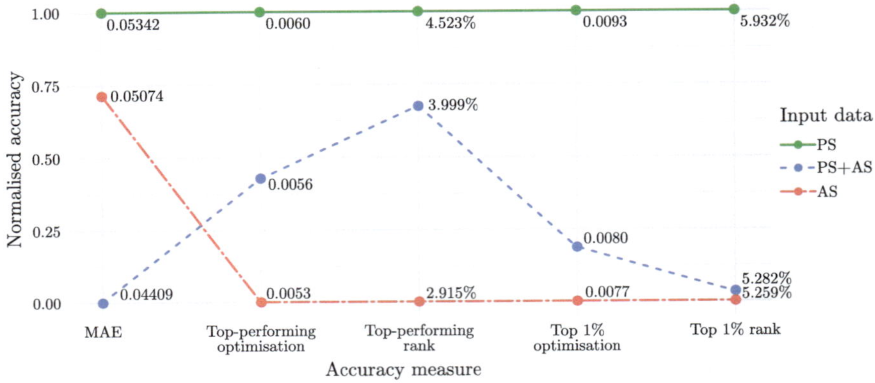

Fig. 1 Meta-learning accuracy results according to input data type, comprising the MAE for the entire algorithmic portfolio and the derived accuracy measures for the top-performing configuration variant and the top 1% of the algorithmic portfolio. True values are also displayed at each data point

two ways: Evaluating the configuration variants that are expected to perform best (the *top-performing* configuration variant), and those expected to perform within the top 1% of the algorithmic portfolio (among the top 44 of the 4,374 configuration variants). These configuration variants represent the *recommendation* that a meta-learning model would make for metaheuristic implementation. For all employed accuracy measures a lower value is indicative of superior meta-learning performance. The numerical results, with min-max normalisation applied to the vertical axis, are illustrated in Fig. 1.

From the results in Fig. 1 it is clear that the best performance with respect to MAE was achieved for the combination of PS and AS information, with an average MAE of 0.0441. This represents the gold standard where all possible input information is provided, thereby facilitating the formation of strong connections between the feature space and performance space. This standard represents the most accurate meta-learning prediction: When considering all configuration variants in the algorithmic portfolio, it is most accurate to utilise both PS and AS information. The derived accuracy measures, on the other hand, suggest that if an analyst exhibits a greater priority towards predicting only top-performing configuration variants, a meta-learning model built purely on AS information would seem to provide the most accurate results across all derived accuracy measures.

The weakest type of input information, across all accuracy measures, is that based purely on PS information, excluding all information about algorithmic search behaviour. This form of input information may be more efficient from a computational perspective, but lacks the intricacies related to how the particular metaheuristic employed functions. It should be noted, however, that high-quality meta-learning transpired across all types of input information, revealing a robust meta-learning procedure for metaheuristic optimisation performance prediction.

A cost-benefit trade-off analysis between AS and PS meta-learning may be based on the computational burden associated with meta-feature extraction, and the intended purpose of meta-learning required by the analyst. The results demonstrate that a relationship exists between the type of input information and the purpose of the meta-learning model. More specifically, if a balance between accurate meta-learning predictions and identifying top-performing configuration variants is desired, then both PS and AS information should be included, given a capacity to manage a greater computational burden. If, on the other hand, the main purpose is to identify the top-performing configuration variants and weaker configurations can be ignored, then purely AS information should be employed. Moreover, if high-quality meta-learning prediction is required and a limited computational budget is available for meta-feature extraction, then purely PS should be employed.

In comparison with the combination of PS and AS information, the MAE-cost for PS and AS meta-learning was found to be 21.1% and 15.1%, respectively. PS meta-learning, on the other hand, incurred a cost of 7.1%, while AS meta-learning yielded a benefit of 5.3% in true optimisation performance for the top-performing configuration variant. This reveals that AS meta-learning may be superior to PS meta-learning in the context of prediction accuracy, but then a greater computational capacity has to be available during meta-feature extraction.

5 Conclusion

Through our empirical analysis of metaheuristic optimisation performance prediction in the context of the KP, we have gained valuable insight into the relative effectiveness of PS and AS meta-learning approaches. The numerical results presented in this paper demonstrate that the input information should be selected based on the intended purpose of the meta-learning model. The combination of PS and AS information yields the best balance of performance across all meta-learning accuracy measures. This comprehensive input approach establishes robust connections between the feature space and performance space, making it ideal for predicting performance across an entire algorithmic portfolio.

When prioritising the prediction of top-performing configuration variants, on the other hand, a purely AS approach seems to be most effective, delivering superior results in terms of identifying the top-performing variant and the top 1% of the algorithmic portfolio. Notably, while purely PS information may offer computational efficiency, it yields a reduced capability of capturing the metaheuristic search behaviour of top-performing configuration variants. With respect to the MAE metric, PS and AS meta-learning incurred costs of 21.1% and 15.1%, respectively. With respect to true optimisation performance, PS meta-learning incurred a cost of 7.1%, while AS meta-learning yielded a benefit of 5.3%. Our study also revealed the robustness of the meta-learning procedure across all input information types, emphasising its adaptability in metaheuristic optimisation performance prediction. Ultimately, the cost-benefit trade-off between AS and PS meta-learning hinges on the computational budget available for meta-feature extraction and the specific analyst's requirements in terms of meta-learning applications.

An interesting avenue for future experimentation would be to extend the investigation reported in this paper to other optimisation problem classes, while also considering other metaheuristic search paradigms for inclusion in the algorithmic portfolio.

References

1. Aleti, A., Moser, I., Grunske, L.: Analysing the fitness landscape of search-based software testing problems. Autom. Softw. Eng. **24**, 603–621 (2017)
2. Alorf, A.: A survey of recently developed metaheuristics and their comparative analysis. Eng. Appl. Artif. Intell. **117**, 105622 (2023)
3. Beham, A., Affenzeller, M., Wagner, S.: Instance-based algorithm selection on quadratic assignment problem landscapes. In: Proceedings of the Genetic and Evolutionary Computation Conference Companion, pp. 1471–1478. ACM, Berlin (2017)
4. Belaidouni, M., Hao, J.K.: Landscapes and the maximal constraint satisfaction problem. In: Fonlupt, C., Hao, J.K., Lutton, E., Schoenauer, M., Ronald, E. (eds.) Artificial Evolution, LNCS, vol. 1829, pp. 244–255. Springer, Berlin (2000)
5. Breiman, L.: Random forests. Mach. Learn. **45**, 5–32 (2001)
6. Holland, J.H.: Adaptation in Natural and Artificial Systems: An Introductory Analysis with Applications to Biology, Control, and Artificial Intelligence. University of Michigan Press, Ann Arbor (MI) (1975)
7. Hoos, H.H.: Automated algorithm configuration and parameter tuning. In: Hamadi, Y., Monfroy, E., Saubion, F. (eds.) Autonomous Search, pp. 37–71. Springer, Berlin (2012)
8. Hutter, F., Xu, L., Hoos, H.H., Leyton-Brown, K.: Algorithm runtime prediction: methods & evaluation. Artif. Intell. **206**, 79–111 (2014)
9. Joshi, S.K., Bansal, J.C.: Parameter tuning for meta-heuristics. Knowl.-Based Syst. **189**, 105094 (2020)
10. Kerschke, P., Hoos, H.H., Neumann, F., Trautmann, H.: Automated algorithm selection: survey and perspectives. Evol. Comput. **27**(1), 3–45 (2019)
11. Kirkpatrick, S., Gelatt, C.D., Vecchi, M.P.: Optimization by simulated annealing. Science **220**(4598), 671–680 (1983)
12. Kotthoff, L.: Algorithm selection for combinatorial search problems: a survey. AI Mag. **35**(3), 48–60 (2014)
13. Liefooghe, A., Daolio, F., Verel, S., Derbel, B., Aguirre, H., Tanaka, K.: Landscape-aware performance prediction for evolutionary multiobjective optimization. IEEE Trans. Evol. Comput. **24**(6), 1063–1077 (2020)
14. Lindauer, M., Hoos, H.H., Hutter, F., Schaub, T.: AutoFolio: an automatically configured algorithm selector. J. Artif. Intell. Res. **53**, 745–778 (2015)
15. Lu, G., Li, J., Yao, X.: Fitness-probability cloud and a measure of problem hardness for evolutionary algorithms. In: Proceedings of the 11th European Conference on Evolutionary Computation in Combinatorial Optimization, pp. 108–117. Springer, Torino (2011)
16. Malan, K.M.: A survey of advances in landscape analysis for optimisation. Algorithms **14**(2), 40 (2021)
17. Malan, K.M., Engelbrecht, A.P.: A survey of techniques for characterising fitness landscapes and some possible ways forward. Inf. Sci. **241**, 148–163 (2013)
18. Rice, J.R.: The algorithm selection problem. In: Rubinoff, M., Yovits, M.C. (eds.) Advances in Computers, pp. 65–118. Academic Press, New York, NY (1976)
19. Smith-Miles, K.A.: Towards insightful algorithm selection for optimisation using meta-learning concepts. In: Proceedings of the IEEE International Joint Conference on Neural Networks, pp. 4118–4124. IEEE Press, Hong Kong (2008)

20. Smith-Miles, K.A.: Cross-disciplinary perspectives on meta-learning for algorithm selection. ACM Comput. Surv. **41**(1), 6 (2009)
21. Smith-Miles, K.A., Muñoz, M.A.: Instance space analysis for algorithm testing: methodology and software tools. ACM Comput. Surv. **55**(12), 255 (2023)
22. Smith-Miles, K.A., Christiansen, J., Muñoz, M.A.: Revisiting where are the hard knapsack problems? Via instance space analysis. Comput. Oper. Res. **128**, 105184 (2021)
23. Talbi, E.G.: Metaheuristics: From Design to Implementation. Wiley, Hoboken, NJ (2009)
24. van der Westhuyzen, N.J., van Vuuren, J.H.: Fitness landscape measures for analysing the topology of the feasible region of an optimisation problem. South African J. Ind. Eng. **34**(3), 270–285 (2023)
25. Vanneschi, L., Tomassini, M., Collard, P., Vérel, S.: Negative slope coefficient: a measure to characterize genetic programming fitness landscapes. In: Proceedings of the 9th European Conference Conference on Genetic Programming, pp. 178–189. Springer, Budapest (2006)
26. Vassilev, V.K.: An information measure of landscapes. In: Proceedings of the 7th International Conference on Genetic Algorithms, pp. 49–56. Morgan Kaufmann, East Lansing, MI (1997)
27. Wang, M., Li, B., Zhang, G., Yao, X.: Population evolvability: dynamic fitness landscape analysis for population-based metaheuristic algorithms. IEEE Trans. Evol. Comput. **22**(4), 550–563 (2018)
28. Weinberger, E.: Correlated and uncorrelated fitness landscapes and how to tell the difference. Biol. Cybern. **63**(5), 325–336 (1990)
29. Wright, M.N., Ziegler, A.: Ranger: a fast implementation of random forests for high dimensional data in C++ and R. J. Stat. Softw. **77**(1), 1–17 (2017)
30. Wright, S.: The roles of mutation, inbreeding, crossbreeding, and selection in evolution. In: Proceedings of the 6th International Congress on Genetics, pp. 356–366. Brooklyn Botanical Garden, Ithaca, NY (1932)
31. Xu, L., Hutter, F., Hoos, H.H., Leyton-Brown, K.: SATzilla: portfolio-based algorithm selection for SAT. J. Artif. Intell. Res. **32**, 565–606 (2008)

Customizing Characteristics of Multi-queue Multi-server Systems

Olga Grinchtein and Johan Karlsson

Abstract We adopt reinforcement learning approach to automatically optimize waiting time, according to constraints, in a queueing system with limited flexibility by changing priorities and/or connections between servers and queues. We use constraint solver to generate priorities of queues that satisfy constraints. Simulation of queueing system with limited flexibility is used to obtain reward.

Keywords Queueing system · Reinforcement learning · Optimization

1 Introduction

In communication systems resources are shared between the connected devices. It is not only radio resources such as base station power and access to radio channels that are shared.

Even hardware and computing resources will be shared among various functions that are to run jobs that have different capability requirements on the distributed computing resources/servers. Hence, jobs to be processed in a shared server are placed in a queue that match the capability requirements of the job. The queue is then connected with one or more servers that have the capability that match the requirement of the queue. Further, one queue may be connected to one or several servers that satisfy the capability requirements, which is one of the benefits with the distributed computing resources. In queueing system with limited flexibility there is at least one pair of a queue and a server such that jobs from the queue cannot be executed on the server, while in fully flexible system any server can serve jobs from any queue. In a system with too many computing resources/servers there will be a high cost to have those servers available but not being used. In a system with too few

O. Grinchtein (✉) · J. Karlsson
Ericsson AB, Stockholm, Sweden
e-mail: olga.grinchtein@ericsson.com

J. Karlsson
e-mail: johan.karlsson@ericsson.com

M. Di Francesco et al. (eds.), *Operations Research: Closing the Gap Between Research and Practice*, AIRO Springer Series 15, https://doi.org/10.1007/978-3-031-90095-2_5

servers available the waiting time for a job to be process will be long, and the system will perform badly.

We use ideas from reinforcement learning approach to develop optimization algorithm of multi-queue multi-server queueing system. The goal is to optimize waiting time in queueing system according to constraints by changing priorities and/or connections between servers and queues.

2 Related Work

Queueing system with limited flexibility was introduced in [7]. We are not aware of any work that optimizes waiting time in queueing system with limited flexibility by changing priorities of queues and connections between servers and queues. Afèche et al. [1] explores the problem of designing an optimal matching topology between customer classes and servers under a FCFS-ALIS service discipline. This service discipline means that when a server becomes idle it selects the customer who has been waiting the longest among those that it can serve, and a customer that can be served by multiple idle servers selects the server that has been idle the longest. Mixed-integer linear program is proposed for finding the set of matching topologies.

The adoption of reinforcement learning algorithm to optimize queueing system is not a new idea. In [5] the authors study quality of service (QoS) assurance problem for a tandem network of N queueing systems where each system is a multi-server queueing system with homogeneous servers. A service rate controller chooses a set of service rates. The idea of designing of the controller is to provide QoS guarantees on the end-to-end delay of the system. The authors formulate service control task as reinforcement learning problem. Liu et al. [4] learns the optimal control policy for queueing networks so that the average job delay is minimized by applying model-based reinforcement learning.

3 Preliminaries

3.1 Multi-queue Multi-server Queueing System

Queueing system consists of jobs, queues, servers and connections between queues and servers. Jobs have priorities and capability requirements defined by users. When a job arrives, it ends up in a queue, which capability and priority matches the capability and priority of the job. Further, there is a connection between a queue and a server if jobs from this queue can be executed on the server, thus, the required capabilities are provided by the server. Queueing system with limited flexibility [7] means that there is at least one server that cannot execute jobs from at least one of the queues due to not matching the queues capability requirement. Waiting time for a job is the difference between started time and arrival time. Started time is the time when a job started to be executed on server. Arrival time is the time when the job ends up in a queue. Mean waiting time for a queue is the sum of waiting time of jobs in the queue divided by the number of jobs. If several servers are available to execute the job, the server is selected based on predefined order. Priorities are tuned to determine a service order that minimizes the overall average waiting time in the system. If some ordering of priorities needs to be preserved, a constraint on priorities can be defined. We use non-preemptive priority policy, where a queue with highest priority is selected and job is picked up from the queue based on FIFO ordering whenever a server becomes idle. If several queues have the same priority, then a queue is selected with a job with earliest arrival time.

3.2 Simulation of Queueing System

In [9] discrete event simulation framework for queueing system with limited flexibility is described. Zhang [9] uses ideas from algorithm published in [2]. The algorithm is shown in Fig. 3-9 in [9].

We present Fig. 1 that is a version of Fig. 3-9 [9] with more details but overall functionality is the same. Let $jobList$ be a list of jobs submitted by users to the system. This list is ordered by arrival time. Each job has a priority and a capability. Let the events of the system be a list $Events$. Each event E in $Events$ has the timestamp of creation $E.T$ and corresponding job $E.Job$. When a job arrives to the system an arrival event $ArrivalEvent$ is generated and added to $Events$. When a job leaves the system departure event $DepatureEvent$ is generated and added to $Events$.The algorithm perform $2 \cdot NumberOfJobs$ iterations that is the total number of arrival and departure events. The initial step of the algorithm is to create $ArrivalEvent$ for the first job in $jobList$. At each iteration of the algorithm the next Event E from $Events$ is selected. Then algorithm checks the type of the event E.

If E is arrival event and the number of arrival events in $Events$ is less than $NumberOfJobs$ then the algorithm generates $ArrivalEvent$ for the next job in $jobList$ and adds this event to E. The algorithm finds matching queue Q for $E.Job$ and add $E.Job$ to Q. If Q contains more than one job or there is no free server supporting Q then the next iteration is performed. A server is free if it is not executing any job. If Q contains only one job and there is free server $Server$ supporting Q then departure event is created for $E.Job$ and added to E, $E.Job$ is removed from Q, and $E.Job$ is assigned to $Server$.

If E is departure event then the server $Server$ that executed $E.Job$ should be released. Then the algorithm finds list of queues $qList$ such that each queue in $qList$ is supported by $Server$. If each queue in $qList$ has no jobs then the next iteration is performed, otherwise a job is selected from queue Q' according to $timePolicy$, where Q' in $qList'$ and $qList'$ contains only queues from $qList$, and all queues in $qList'$ have at least one job. Then the job is removed from Q', the job is assigned to

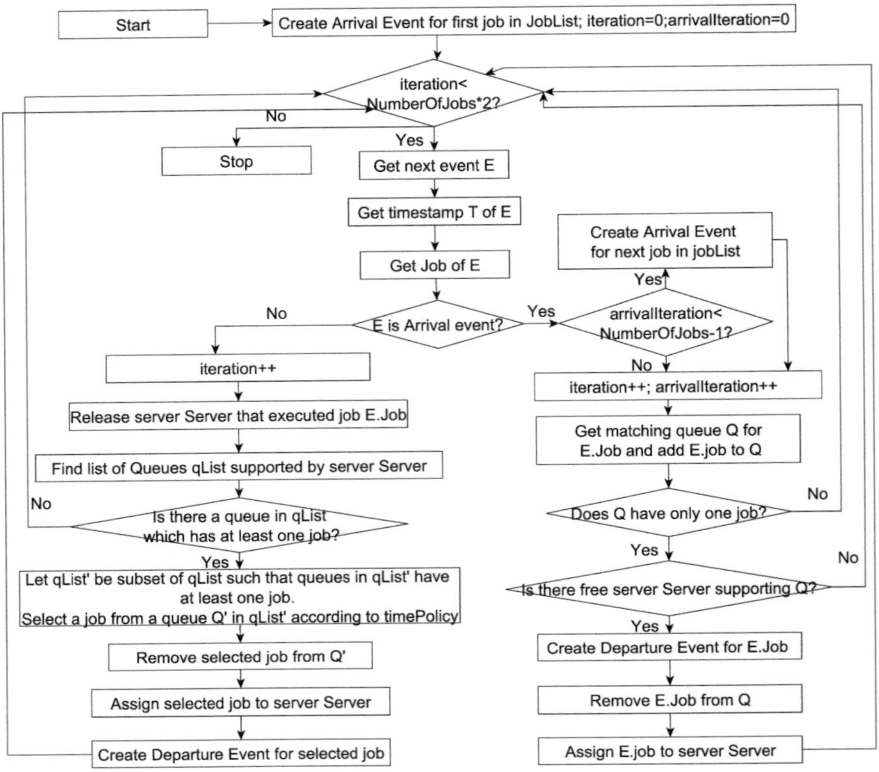

Fig. 1 Discrete event simulation algorithm of multi-queue multi-server system

server *Server*, departure event is created for selected job and the algorithm moves to next iteration. The *timePolicy* in our case is non-preemptive priority policy, Q' is a queue with highest priority in $qList'$ and job in Q' is selected based on FIFO principle.

3.3 Reinforcement Learning

Reinforcement learning problem can be defined as an agent that makes decisions by interacting with environment to achieve a goal. In Fig. 2 are shown elements of reinforcement learning problem.

The Agent-Environment Interface is described in [6]. The agent interacts with environment at discrete time stamps $t = 0, 1, 2, \ldots$. Let S be the set of possible states, s_t be some representation of the environment state that the agent receives at step t and $a_t \in A(s_t)$ be an action that agent selects, where $A(s_t)$ is the set of actions available in state s_t. At step $t+1$ the agent receives the numerical reward, r_{t+1}, and finds itself in a new state s_{t+1}.

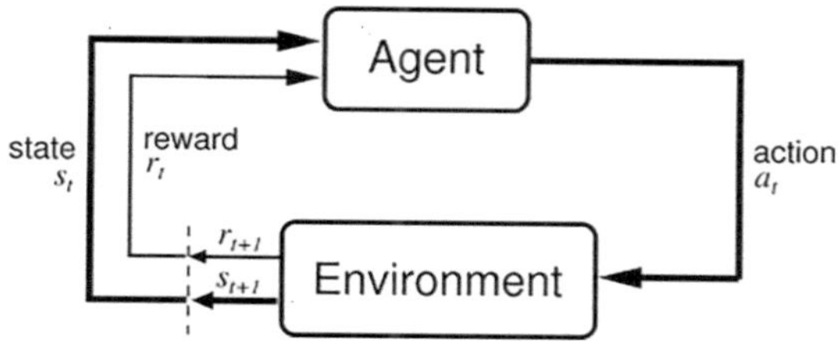

Fig. 2 The agent-environment interaction in reinforcement learning [6]

4 Priority and Topology-Based Optimization

4.1 State

Let n be the number of queues, k be number of servers, and $prio_i$ be priority of the $queue_i$, where $1 \leq i \leq n$ and $1 \leq j \leq k$. The agent will maintain a state, which is a vector of priorities of queues $(prio_1, prio_2, \ldots, prio_n)$ and triples $(queue_i, server_j, c)$, where

- c is equal to 2 if a connection between a queue and a server should not be changed. If there is no connection between a queue and a server, then it should not be added. If there is a connection between a queue and a server, then it should not be removed.
- c is equal to 1 if there is a connection between a queue and a server and this connection can be removed.
- c is equal to 0 if there is no connection between a queue and a server, but this connection can be added.

We use cs to specify constraints on topology of a queueing system. If $prio_i < prio_j$, then $queue_i$ has higher priority than $queue_j$. If the queueing system consists of 3 queues and 2 servers, then an example of a state is

$$\begin{aligned}&((1, 1, 3),\\ &\qquad (queue_1, server_1, 0), (queue_1, server_2, 1), (queue_2, server_1, 2),\\ &\qquad (queue_2, server_2, 1), (queue_3, server_1, 2), (queue_3, server_2, 2)).\end{aligned}$$

In this example

- $queue_1$ has priority 1, $queue_2$ has priority 1 and $queue_3$ has priority 3,
- there is no connection between $queue_1$ and $server_1$ but it can be added,
- there is a connection between $queue_1$ and $server_2$ and it can be removed,
- there is a connection between $queue_2$ and $server_2$ and it can be removed,
- there is a connection between $queue_2$ and $server_1$ and it should not be removed,
- there is no connection between $queue_3$ and $server_1$ and it should not be added, and
- there is a connection between $queue_3$ and $server_2$ and it should not be removed.

4.2 Action

The actions are to increase or decrease the priority of a queue by one, and remove or add a connection of the queue to a server. For a triple $(queue_i, server_j, c)$ the action can be set c to 0 if c is equal to 1, or set c to 1 if c is equal to 0. The priority change is done only if it satisfies constraints. The constraints have the form $x < y$, $x \leq y$ or $x = y$.

For example, if there are 6 queues, a constraint could be $prio_1 < prio_2 \wedge prio_1 \leq prio_6 \wedge prio_4 < prio_2 \wedge prio_2 = prio_5 \wedge prio_3 = prio_4$, where $prio_i$ is the priority of ith queue, $1 \leq prio_i \leq 6$ and $1 \leq i \leq 6$.

4.3 Environment

The environment is a simulation framework as described in Sect. 3.2. When the action creates a new queueing system by changing the priority of a queue or a connection between a queue and a server, the new queueing system is simulated and reward is calculated.

4.4 Reward

Reward at step $t = 1, 2, \ldots$ is calculated based on output from simulation framework that is mean waiting time of queues. In this paper Reward is sum of differences between mean waiting time of queues in initial queueing system and mean waiting time of queues after action is taken.

$$Reward = \sum_{1 \leq i \leq n} (\overline{Wq_i^0} - \overline{Wq_i^t}),$$

where $\overline{Wq_i^t}$ is mean waiting time of $queue_i$ after priority or topology change in queueing system, and $\overline{Wq_i^0}$ is initial mean waiting time of $queue_i$.

4.5 Summary of the Algorithm

The algorithm is presented in Fig. 3. The user defines the number of iterations N and/or in combination with the value of reward $RewardGoal$ that should be achieved. The algorithm terminates after N iterations or after less than N iterations if the reward is greater or equal to $RewardGoal$. In each iteration maximum reward after taking actions is calculated. The next state is defined by the action yielding the highest reward in the latest iteration. If maximum reward in current iteration

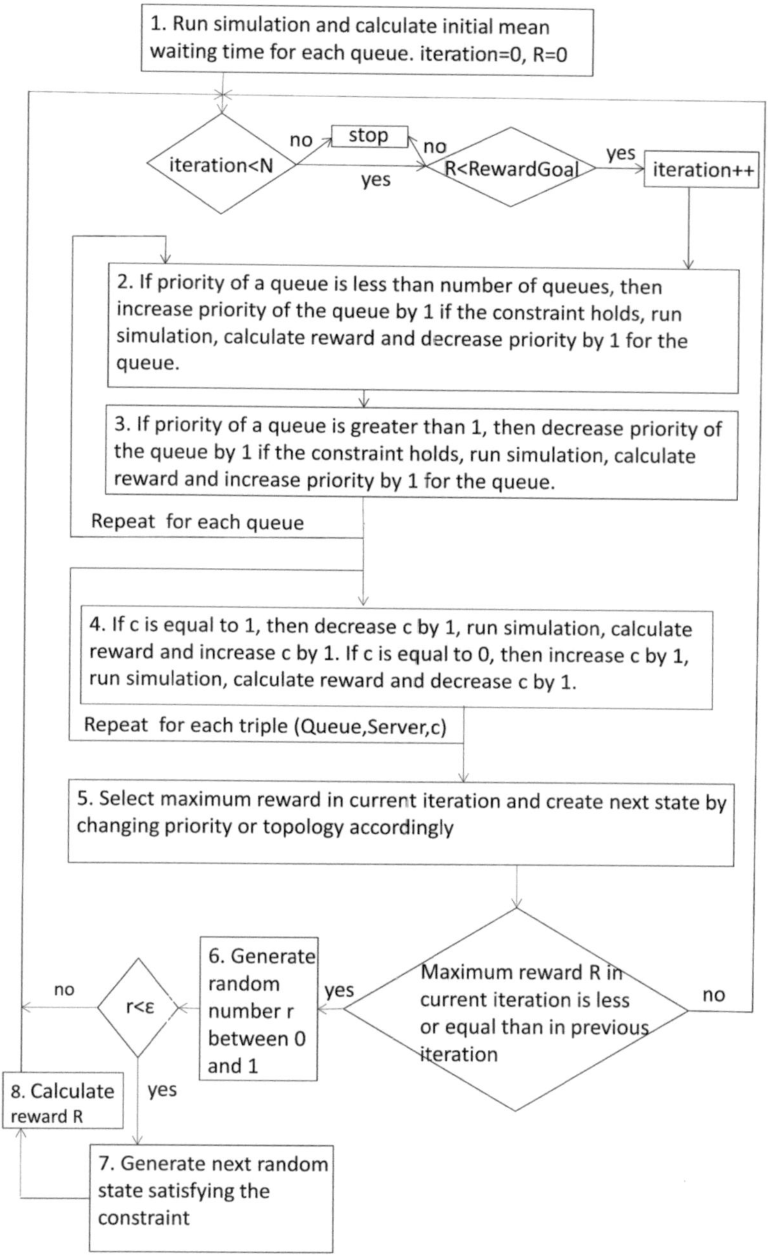

Fig. 3 The algorithm

is less than maximum reward in previous iteration, then uniform random number r between 0 and 1 is generated. If r is less than predefined ϵ, then next random state satisfying constraints is generated. The approach of random state generation is based on idea from epsilon-greedy action selection algorithm [8]. Constraint solver can be used to generate random priorities satisfying constraints. We use JaCoP constraint solver [3]. Steps 2–4 in Fig. 3 can be parallelized, but in our implementation there is no parallelization. In Figs. 4, 5 and 6 is shown an example of initial system with 10 iterations of algorithm. The iterations 5–7, where was priority change, are ommited due to lack of space. Arrival time and service time were extracted from real sequence of jobs recorded from an operational system, but priorities, capabilities, and connections between queues and servers were changed to better exemplify the approach. We set $\epsilon = 0.8$ for all experiments we perform. There are 5 queues and 8 servers, where $queue_i$ has $capability_i$ and $prio_i$, $i = 1, \ldots, 5$. The constraint on priorities is $prio_1 < prio_5 \wedge prio_4 < prio_5$. There are also constraints on topology. $queue_1$ can be connected to $server_3$ and $server_4$, connections from $server_4$ to $queue_2$, from $server_3$ to $queue_3$ and from $server_4$ to $queue_3$ can be removed, and all other connections should not be changed. The reward increases from Iteration 1 to Iteration 8. In Iteration 9 the reward is the same as in Iteration 8. There is random step after Iteration 9.

5 The Elimination of Duplicated Iterations

We define two states to be equivalent if their rewards are equal. Let $(prio_1, prio_2, \ldots, prio_n)$ be vector if priorities of the first state and $(prio'_1, prio'_2, \ldots, prio'_n)$ be vector of priorities of the second state. Let first state and second state are differed only by values in vectors of priorities. These two states are equivalent if for $i = 1, \ldots, n$ and $j = 1, \ldots, n$ if $prio_i < prio_j$ then $prio'_i < prio'_j$ and if $prio_i = prio_j$ then $prio'_i = prio'_j$.

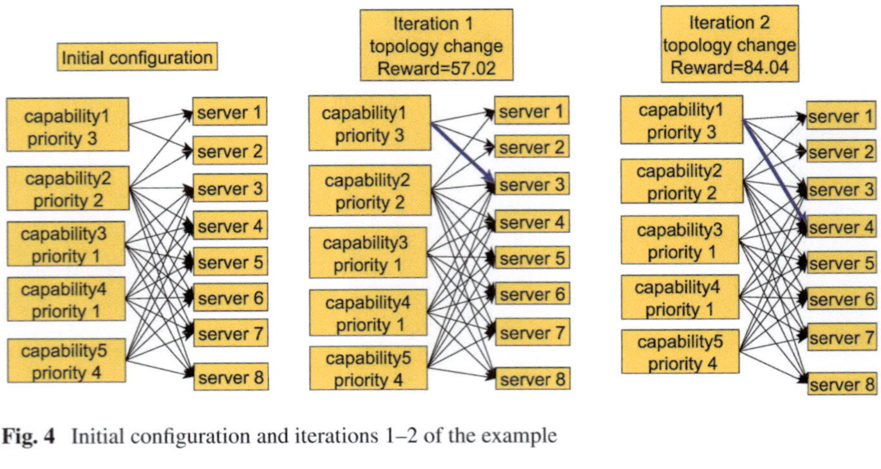

Fig. 4 Initial configuration and iterations 1–2 of the example

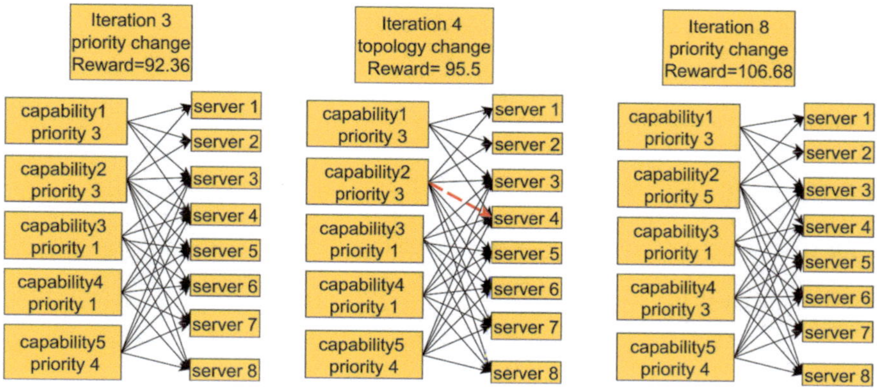

Fig. 5 Iteration 3, iteration 4 and iteration 8 of the example

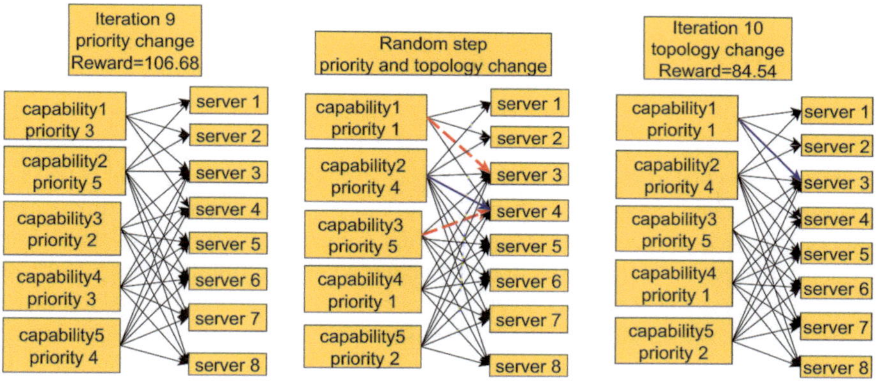

Fig. 6 Iterations 9–10 and random step of the example

An example of two equivalent states is two states that differ only in vector of priorities, and vector of priorities of the first state is $(1, 2, 2)$ and vector of priorities of the second state is $(1, 3, 3)$. In order to avoid to do simulations of queueing systems that have equivalent states, we maintain a set of states S for which simulation have already been performed. Let a state be a vector of priorities of queues $(prio_1, prio_2, \ldots, prio_i, \ldots, prio_n)$ and triples $(queue_i, server_j, c_k)$, where $k = 1, \ldots, m$. In order to simplify lookup in S, we convert each state to a string of the form

$$i_1, i_2, \ldots, i_n, leq_1 leq_2 \ldots leq_{n-1} c_{j_1} c_{j_2} \ldots c_{j_{m'}},$$

where

- for all $1 \leq i \leq n$ there is $1 \leq z \leq n$ such that $i_z = i$,
- $leq \in \{<, =\}$,
- for all $1 \leq z \leq n-1$, $prio_{i_z} < prio_{i_{z+1}}$ and $leq_{n+i_z} =' <'$, or $prio_{i_z} = prio_{i_{z+1}}$ and $leq_{n+i_z} =' ='$,
- for all $1 \leq l \leq n$, for all $1 \leq p \leq n, l \neq p$ if $prio_l = prio_p$ and $i_z = l$ and $i_{z+1} = p$, $1 \leq z \leq n$ then $l < p$,
- $1 \leq m' \leq m$, $j_1 < j_2 < \cdots < j_{m'}$, and
- for all $1 \leq z \leq m'$ there is $1 \leq z' \leq m$ such that $j_z = z'$, $c_{z'} < 2$ and $c_{j_z} = c_{z'}$.

For example, a state

$$((2, 3, 1, 1),$$
$$(queue_1, server_1, 1), (queue_1, server_2, 0), (queue_2, server_1, 2),$$
$$(queue_2, server_2, 0), (queue_3, server_1, 0), (queue_3, server_2, 2),$$
$$(queue_4, server_1, 2), (queue_4, server_2, 1))$$

is converted to a string

$$3, 4, 1, 2, =<< 10001.$$

Then we modify slightly the algorithm shown in Fig. 3. Each time a random state is generated, we convert the state to the string and check if this string is in S. If yes, new random state is generated, otherwise this string is added to S. Each time a new state is generated, it is converted to a string and if this string is in S, then simulation is not performed, but if this string does not belong to S, then simulation is performed and the string is added to S. However, this approach does not eliminate all duplicate states. It can be some hidden duplication. Assume that a queueing system consists of two queues and one server, and one queue has jobs if and only if the other queue is empty. Then the priority of these two queues do not have any impact on waiting time in the queueing system. We ran algorithm on real data, but with modified capabilities, that consisted of 17 queues, 8 servers and about 9100 jobs. Two connections between a queue and servers could be added and two connections between another queue and servers could be removed. 500 iterations took 12 min. The highest reward value 121 min was achieved at 461th iteration. The sum of mean waiting time of queues in initial queueing system is 272 min and mean of mean waiting time of queues in initial queueing system is 16 min. Therefore by changing priorities and the topology of the queueing system, mean of mean waiting time of queues can be reduced by 44%. The experiments were done on the computer equipped with 32 GB RAM and an Intel(R) Xeon(R) Gold 6132 (2.60 GHz).

6 Conclusion

We introduced an approach to improve waiting time in queueing system by changing priorities of queues and architecture of queueing system. The approach was applied on real data and the experimental results show that waiting time can be reduced significantly. However, the state space is huge, and therefore it could be interesting to explore further state space reduction algorithm for this problem. Simulations in the iteration can be run in parallel that can also reduce execution time. On the other hand in some cases the significant reduction of waiting time can happen in first few iterations, and there is no need to run 500 iterations.

Acknowledgements The authors would like to thank Hans Hannu and Viktoria Fodor for providing useful suggestions.

References

1. Afèche, P., Caldentey, R., Gupta, V.: On the optimal design of a bipartite matching queueing system. Oper. Res. **70**(1) (2022)
2. Boon, M.: Lecture Notes and Source Code in the Course 2WB05 Simulation at Eindhoven Institute of Technology. https://www.win.tue.nl/~/2WB05/#lecturenotes. Accessed 20 Nov 2023
3. Kuchcinski, K., Szymanek, R.: JaCoP—Java Constraint Programming Solver. In: Abstract from CP Solvers: Modeling, Applications, Integration, and Standardization, Co-located with the 19th International Conference on Principles and Practice of Constraint Programming (2013)
4. Liu, B., Xie, Q., Modiano, E.H.: RL-QN: a reinforcement learning framework for optimal control of queueing systems. ACM Trans. Model. Perform. Eval. Comput. Syst. **7**(1), 2:1–2:35 (2022)
5. Raeis, M., Tizghadam, A., Leon-Garcia, A.: Queue-learning: a reinforcement learning approach for providing quality of service. In: Thirty-Fifth AAAI Conference on Artificial Intelligence (2021)
6. Sutton, R.S., Barto, A.G.: Reinforcement Learning: An Introduction. MIT Press, Cambridge, MA (1998)
7. Tsitsiklis, J.N., Xu, K.: Queueing system topologies with limited flexibility. In: Harchol-Balter, M., Douceur, J.R., Xu, J. (eds.) International Conference on Measurement and Modeling of Computer Systems, SIGMETRICS '13. ACM (2013)
8. Watkins, C.J.C.H.: Learning from Delayed Rewards. Ph.D. thesis, King's College, Oxford (1989)
9. Zhang, Y.: Simulation and Analysis of Queueing System. Master's thesis, KTH Royal Insititute of Technology (2019)

Delay Constrained Routing: the Multi-flow Single-Path Case

Antonio Frangioni, Laura Galli, and Luca Mencarelli

Abstract Delay Constrained Routing (DCR) is an optimization problem arising in computer networks with stringent Quality of Service (QoS) scheduling requirements. IP packets are subject to several types of delays when traversing the network; we consider the routing problem where the QoS requirement consists in guaranteeing a controlled flow-specific *worst-case end-to-end delay* of any packet of each given flow. Depending on the scheduling algorithm employed in the routers, worst-case delay formulæ can be devised of the bandwidth (a.k.a. *reserved rate*) allocated for the flow over each arc of its (unique) origin-destination path. The (single-path) multi-flow DCR problem can then be modelled as a multicommodity-type problem with nonlinear constraints depending on the reserved rate on the chosen arcs, and it is clearly a hard problem since it generalises the nonsplittable multicommodity flow one. Even the single-flow DCR problem is hard, since it generalises the constrained shortest path one, but it can be solved efficiently enough for realistic-sized instances thanks to previously developed modelling and algorithmic approaches. We show how to leverage the latter to construct a Lagrangian approach and explore the quality of the obtained information on a set of realistic DCR instances.

Keywords Lagrangian relaxation · Heuristics · Mixed-integer non-linear programs

---

A. Frangioni · L. Galli · L. Mencarelli (✉)
Dipartimento di Informatica, Università di Pisa, Pisa, Italy
e-mail: luca.mencarelli@unipi.it

A. Frangioni
e-mail: antonio.frangioni@unipi.it

L. Galli
e-mail: l.galli@unibo.it

M. Di Francesco et al. (eds.), *Operations Research: Closing the Gap Between Research and Practice*, AIRO Springer Series 15, https://doi.org/10.1007/978-3-031-90095-2_6

1 Introduction

The IP protocol is pervasive in nowadays' computer networks, but it had originally not been designed for delay-sensitive applications (teleconferencing, online games, telemedicine, etc.). IP packets are subject to several types of delays, and the overall delay that a packet incurs when traversing the network (from origin to destination) is known as *end-to-end* delay. Bounding the end-to-end delay of an IP flow is a crucial component of Quality of Service (QoS) guarantees for many of today's real-life applications in computer networks.

When a packet reaches a network node—typically, a router—it must be processed in order to be forwarded to a queue of packets associated to the node itself; this is known as *node processing delay*, and its worst-case can be bounded by a constant for each node. Similarly, the worst-case *link delay* due to transmission and physical propagation effects over a link is a constant. Yet, the time that a packet spends in the queue before the router sends it to the link is not a constant, as it depends on the queue itself and on the scheduling protocol.

Computing the exact delay of a stream of packets (a.k.a. *packet flow*) is usually not a viable option. Indeed, while constant delays only depend on the network element considered, queuing delays are dynamic and depend on the scheduling algorithm implemented by the network routers. However, if the arrival curve of the flow can be properly bounded (which is not an unreasonable assumption), a *worst-case* analysis can be performed and the corresponding *worst-case end-to-end delay* (WCD) can be computed by a closed formula. By bounding the WCD under a given threshold (a.k.a. *flow deadline*) we obtain a QoS controlled end-to-end delay. The WCD of a flow depends on the path (assumed unique in this work, as in most of the literature) along which the packets are routed, and on the portion of bandwidth allocated to that specific flow (a.k.a. *reserved rate*) on each of the arcs of the path. Therefore, the Delay Constrained Routing Problem (DCR) can be formulated as an optimization problem involving two types of decisions: computing paths and reserving resources along them, subject to WCD constraints. The former is a discrete decision, while the latter is a continuous decision; due to the nonlinearity of the WCD constraints this therefore results in a Mixed-Integer NonLinear Problem (MINLP). A typical objective function minimizes the total bandwidth allocated for all flows along the paths, in order to contain the load of the network and leave as much as possible available resources for newly arriving flows to be accommodated at a later time.

A well-known result in computer network performance analysis allows to compute a WCD bound for each arc (a.k.a. *latency*) if the scheduling of packets is performed according to the Generalized Processor Sharing (GPS) [14] paradigm, which requires the flow arrival rate at the source to be properly constrained. GPS is an abstract model for packet scheduling and there exist several practical implementations. Among them, the Strictly Rate-Proportional (SRP) scheduling model yields a worst-case latency closed formula inversely proportional to the reserved rate. Hence DCR can be mathematically formulated assuming SRP schedulers throughout the network. Other GPS-type scheduling protocols give rise to more or less analogous formulæ [7], and therefore this work could be extended to basically all known practical routers. However, an advantage of SRP, from the optimization standpoint, is that each flow is "well isolated" from all others: the WCD formulæ only depend on its path and reserved rates, and therefore the other flows only influence the given one by consuming available bandwidth on the links. Other GPS-type scheduling protocols may rather require the introduction of extra *admission control constraints* [6] due to other aspects of one flow impacting on the WCDs of the others except its reserved rates. These constraints can be formulated as linear ones and therefore the approach of this paper could be extended in a straightforward way to other GPS-type schedulers, but this is avoided for the sake of clarity of presentation; besides, SRP is the most common protocol in practical implementations.

The seminal paper [3] defined the DCR problem for the special case of Single-Flow Single-Path (SFSP), i.e., routing of one single flow (on one single path). A number of efficient heuristic approaches were proposed, together with Dynamic Programming ones. Furthermore, a novel MINLP model was proposed that, exploiting the *Perspective Reformulation* (PR) approach [9] gives rise to tight formulations that can even be recast as a Mixed-Integer Second-Order Cone Program (MISOCP) and efficiently solved with general-purpose optimization software on realistic instances. The approach was shown, by means of simulations, to consistently yield a positive impact on networks' operations in which the routing of each single new flow to be admitted is computed by solving the corresponding SFSP-DCR problem [5]. SFSP-DCR can indeed be solved within the short time required by an actual field application of the technique.

However, this greedy approach is not guaranteed to yield the minimum possible resource consumption once a number of flows are admitted to the network. It would therefore make operational sense to compute—likely not in an online fashion but when appropriate conditions arise, like the network occupation growing or some amount of time/changes having occurred—a *global* solution to the multi-flow DCR problem to re-arrange the current set of flows in order to further reduce the resource consumption. Such a Multi-Flow Single-Path DCR problem (MFSP-DCR) is clearly way harder than SFSP-DCR, but on the other hand it can conceivably be solved without the strict time limits that have to be imposed when deciding whether or not (and how) to accept a newly incoming flow. Yet, reasonable solution times will be required for the practical applicability of the technique. We start by presenting a MINLP model and solving it with general-purpose optimization software on realistic IP network instances of the literature. Since the computing times, as it could be expected, quickly increase with the size of the problem, we explore the quality of a Lagrangian relaxation approach that exploits the previously obtained results on SFSP-DCR. Preliminary experiments provide encouraging results, showing that the Lagrangian relaxation consistently provides much tighter bounds than the continuous relaxation—even if the PR approach is used—in relatively short computing time. This result may form the basis of an effective *Lagrangian-based* branch-and-bound approach for the MFSP-DCR problem. The results have been obtained within the open-source `SMS++` modelling system [13], exploiting its powerful components for the implementation of structure-aware approaches.

2 Problem Definition

In this section we define the MFSP-DCR problem, which consists of the network model and WCD delay model.

The computer network is represented by a directed graph $G = (N, A)$. With N we denote the set of nodes representing routers, switches, etc.; each node $i \in N$ is characterized by a fixed *node delay* n_i. With A we denote the set of arcs representing the network links. Each link $(i, j) \in A$ is characterized by a fixed *link delay* l_{ij}, a *physical link speed* w_{ij}, and a *reservable capacity* $c_{ij} \leq w_{ij}$. Usually the problem is solved for "high value" flows (e.g., audio/video calls, gaming sessions), but some capacity may have to be left free for "low value" ones, which justifies why one cannot reserve the whole arc capacity.

In the network there is a set K of (unsplittable) flows, each $k \in K$ characterized by an origin $s^k \in N$, a destination $d^k \in N \setminus \{s^k\}$ and an *arrival curve* $A^k(t) : \mathbb{R}_+ \to \mathbb{R}_+$. The concept of arrival curve is used in traffic engineering to provide guarantees to data flows. In particular, we need to limit the traffic sent by sources by specifying at most how many more bits of a given flow can be expected to have entered the origin s^k (and, therefore, have been dispatched from there to reach the destination d^k) w.r.t. those arrived already after the next t time instants. In other words, if the *arrival function* $F^k(t)$ measures how many bits of flow k have entered the origin s^k at time t, we have $F^k(\bar{t} + t) - F^k(\bar{t}) \leq A^k(t)$ for all $\bar{t}$ and $t \geq 0$. For our purposes we assume the arrival curves to be entirely specified by the two parameters σ^k (burst) and ρ^k (rate) of a *leaky-bucket traffic shaper*, so that $A^k(t) = \sigma^k + t\rho^k$. In our models we consider the general case where each flow has *individual reservable capacity* $c^k_{ij} \leq c_{ij}$ and *link reservation cost* f^k_{ij} (i.e., the cost of reserving one unit of capacity on (i, j)) on each arc; very possibly one could assume $c^k_{ij} = c_{ij} = w_{ij}$ and $f^k_{ij} = f_{ij}$, but such an assumption would have no significant impact on the structure of our model. Each flow $k \in K$ has its own *deadline* δ^k, which bounds from above its WCD. Any routing of the flow satisfying this condition is a feasible *delay-constrained routing*.

The WCD of a flow $k \in K$ from the source s^k to the destination d^k depends on several factors. Here we summarize the key elements: (i) the selected *path* $p \in P^k$ in G, where $P^k = P(s^k, d^k)$ denotes the set of all simple paths between s^k and d^k in G; (ii) for each arc (i, j) of the chosen path p, the *reserved rate* $0 \leq r^k_{ij} \leq c^k_{ij}$ for the flow k along the arc; (iii) the specific characteristics of the software/hardware systems at the nodes dictating how the flows entering and leaving the nodes are managed (intra-node scheduling of different flows, queues and buffer depths, etc.). The latter point requires a nontrivial analysis via *network calculus* [11]. In all cases of interest here, the delay is *finite* only if the minimum reserved rate among all arcs of the path is at least as large as the rate of the path, i.e., $r^k_{ij} \geq \rho^k \; \forall (i, j) \in p$. If this condition is satisfied, the general form of the delay for a given routing path p is

$$\frac{\sigma^k}{\min\{ r^k_{ij} \; : \; (i, j) \in p\}} + \sum_{(i,j)\in p} \left(\theta^k_{ij} + l_{ij} + n_i \right) \tag{1}$$

where θ_{ij}^k is the latency, i.e., the delay experienced by the flow k on path p by traversing the arc (i, j) that is due to the scheduling protocol. As in [3, 5], we assume a Strictly Rate-Proportional (SRP) scheduler, that bears its name to the inverse relationship to the reserved rate r_{ij}^k. The corresponding latency expression is

$$\theta_{ij}^k = \frac{L}{r_{ij}^k} + \frac{L}{w_{ij}}, \tag{2}$$

where L is the *maximum transmit unit* (MTU), which is assumed known and constant. Crucially, the latency function (2) is a convex function of r_{ij}^k (if $r_{ij}^k \geq 0$). As already mentioned, (2) only depends on the reserved rates of the given flow k; this allows one to ignore feasibility issues related to the other flows and avoid admission control constraints [6, 7]. Since the component $L \,/\, w_{ij}$ in (2) is an additive constant term, to ease notation we aggregate the constant delay components of an arc (i, j) and define $\bar{l}_{ij} = L \,/\, w_{ij} + l_{ij} + n_i$. Hence, in the rest of our treatment, $\theta_{ij}^k = L \,/\, r_{ij}^k$ represents the variable part of the arc latency of a flow k, which depends on the reserved rate r_{ij}^k.

3 Problem Formulation

We now present a MINLP formulation for the MFSP-DCR problem, which naturally generalizes the one described in [3] for the single-flow case.

$$\min \textstyle\sum_{k\in K} \sum_{a\in A} f_a^k r_a^k \tag{3}$$

$$\sum_{a\in\delta^-(i)} x_a^k - \sum_{a\in\delta^+(i)} x_a^k = \begin{cases} -1 & \text{if } i = s^k \\ 1 & \text{if } i = d^k \\ 0 & \text{otherwise} \end{cases} \qquad i \in N,\ k \in K \tag{4}$$

$$\frac{\sigma^k}{r_{\min}^k} + \sum_{a\in A} \left(\frac{L(x_a^k)^2}{r_a^k} + \bar{l}_a x_a^k \right) \leq \delta^k \qquad k \in K \tag{5}$$

$$0 \leq r_a^k \leq c_a^k x_a^k \qquad a \in A,\ k \in K \tag{6}$$

$$\rho \leq r_{\min}^k \leq r_a^k + c_{max}(1 - x_a^k) \qquad a \in A,\ k \in K \tag{7}$$

$$\textstyle\sum_{k\in K} r_a^k \leq c_a \qquad a \in A \tag{8}$$

$$x_a^k \in \{0, 1\} \qquad a \in A,\ k \in K \tag{9}$$

$$r_{\min}^k \in [\,\rho^k, c_{max}\,] \qquad k \in K \tag{10}$$

For each flow $k \in K$ we use three types of variables: (i) binary *arc-flow* variables x_a^k indicating whether arc a belongs to the path of flow k; (ii) continuous *reserved rate* variables r_a^k representing the amount of capacity reserved on arc a for flow k; (iii) continuous *auxiliary* variables $r_{\min}^k$ capturing the minimum reserved rate along the path of flow k. Constraints (4) are the standard flow conservation constraints, with $\delta^-(i)$ and $\delta^+(i)$ denoting respectively the backward star and forward star of node i. With $c_{max} = \max_{a\in A}\{c_a\}$, constraints (6) and (7) ensure that $r_a^k = 0$ if $x_a^k = 0$, and $\rho^k \leq r^k_{\min} \leq r^k_a$ if $x^k_a = 1$. Finally, the non-linear constraint (5) models the QoS requirement in terms of WCD. The fundamental modelling trick is the use of the Perspective Reformulation technique to write the (nonconvex) delay term

$$\theta^k_a(r^k_a, x^k_a) = \begin{cases} L / r^k_a & \text{if } x^k_a = 1 \ [\Longrightarrow r^k_a \geq \rho > 0] \\ 0 & \text{if } x^k_a = 0 \ [\Longrightarrow r^k_a = 0] \end{cases}$$

as $L(x^k_a)^2 / r^k_a$. This can be shown to be the convex envelope of $\theta^k_a(r^k_a, x^k_a)$, i.e., the tightest possible convex function approximating it from below. In fact, $L(x^k_a)^2 / r^k_a$—and, separately, $\sigma / r_{\min}$—is convex and can be expressed via rotated SOCP constraints, so the corresponding MISOCP model can be solved by current off-the-shelf general-purpose solvers like `Cplex` and `Gurobi`. However, this is not the only possible approach. In fact, [9] rather defines *Perspective Cuts* (P/C), a family of (ideally, infinitely many) easy-to-separate linear inequalities that can be dynamically added to the formulation in the Branch-and-Cut fashion and that provide the same bound; we refer the interested reader to the original paper for the easy details. In fact, it has been proven that this approach can be superior in practice to the direct MISOCP formulation [10] (even if one forces the MISOCP solvers to use cutting plane techniques that are conceptually very similar to P/C). This furthermore allows to use the technique in any MILP solver that provides user cut capabilities (basically, all of them) if, like in our case, the PR is applied to all the nonlinear components in the model.

4 Lagrangian Relaxation

The model (3)–(10), in either the MISOCP formulation or with the addition of P/C, can be directly solved by current state-of-the-art general-purpose solvers. However, as our computational results will show, as the number of flows grow the computational burden quickly becomes excessive; thus, alternative approaches are required. An easy observation is that, like for standard multicommodity flow problems [8], the mutual capacity constraints (8) are the only ones that link together the different flows. Hence, by removing them, the corresponding relaxed problem decomposes into $|K|$ independent SFSP-DCR problems, one for each flow $k \in K$, that can be efficiently solved heuristically or exactly [3]. We then relax in a *Lagrangian* way these constraints with Lagrangian multipliers $\lambda_a \geq 0$, $a \in A$, obtaining the Lagrangian function

$$\phi(\lambda) = \min_{(x,r)} \left\{ \sum_{k \in K} \sum_{a \in A} \tilde{f}^k_a r^k_a \ : \ (4)\text{–}(7) \, , \ (9)\text{–}(10) \right\} - \sum_{a \in A} \lambda_a c_a \tag{11}$$

with $\tilde{f}^k_a = f^k_a + \lambda_a$. The corresponding Lagrangian dual problem

$$LDCR \quad \max\{ \, \phi(\lambda) \ : \ \lambda \geq 0 \, \} \tag{12}$$

can be efficiently solved in practice, owing in particular to the fact that $|A|$—and therefore the dimension of the multipliers space—is not too large in practice, e.g. by means of *bundle methods* [2]. This provides both lower bounds on the optimal value, which can be (much) tighter than that of the standard continuous relaxation since the Lagrangian subproblems do not have the integrality property [1], as well as a host of other primal and dual information that can be used to implement either heuristic or exact approaches to the original problem (cf. e.g. [1, 12]). The fundamental factor dictating the chances of success of such a Lagrangian approach is on one hand the quality of the obtained bound, and on the other hand the time required to obtain it, considering that a (potentially, large) number of iterations are required to find the optimal dual multipliers λ^*, each one of which requires the solution of $|K|$ SFSP-DCR problems. In the next section we will describe experiments aimed at gauging these chances of success.

5 Computational Experiments

In this section we report the details of the computational settings and the results of our experimental analysis.

5.1 *Implementation Details*

All our experiments have been performed using the open-source modelling system `SMS++` [13]. The main goal of our analysis is to compare the continuous relaxation of (3)–(10) with the Lagrangian dual (12), considering both the quality of the bound and the running time.

The former is solved using the general-purpose solver `Cplex`, interfaced via the `CPXMILPSolver` component of the `MILPSolver` sub-project of `SMS++`; other solvers could be used, such as `Gurobi`, `SCIP` and `HiGHS`, at the only cost of changing a few lines in a configuration file due to the modularity of the `SMS++` framework. In particular, the solution method requires the dynamic separation of P/C, which, in `SMS++`, is demanded to the `Block` object implementing the formulation (described in more details below) and it is therefore completely solver-agnostic.

The solution of the Lagrangian dual is rather made possible—and easy—by the `LagrangianDualSolver` sub-project of `SMS++`, which provides the same-named `Solver`. The `LagrangianDualSolver` can be registered to any `Block` with the appropriate structure (a set of sub-`Block` containing all the formulation save for the linear linking constraints, that are expressed in the "father") and automatically generates a "Lagrangian Dual `Block`" that represents the Lagrangian Dual problem w.r.t. the linking constraints. This problem can then be solved by means of any appropriate Nondifferentiable Optimization algorithm; `SMS++`

offers the BundleSolver implementation of state-of-the-art bundle-type methods, comprising crucial details like disaggregated master problems and incremental approaches. The ParallelBundleSolver derived class further exploits the common case where, like in our problem, the Lagrangian function computation requires the solution of several independent subproblems by performing it in parallel on different threads. Crucially, upon termination LagrangianDualSolver automatically provides both the optimal dual solution λ^* as well as the optimal primal solution of the "convexified relaxation" that is the primal equivalent of the Lagrangian dual [1], exactly in the same way as the continuous relaxation provides both a primal and a dual optimal solution (but hopefully "better" since the convexified relaxation is tighter, which can have a significant impact on Lagrangian approaches [12]).

To exploit these SMS++ facilities it is, however, necessary to formulate the problem in terms of a SMS++ Block object. The Block concept is meant to represent (a part of) an optimization problem with a well-defined structure, such as, in our case, the MFSP-DCR one. Crucially, a Block can have multiple sub-Block, each one representing a part of the problem (and so on recursively); in our case the sub-Block correspond to the SFSP-DCR problems related to the individual flows. For this reason we have implemented two classes: SingleFlowDCRBlock and MultiFlowDCRBlock, which define the Block concept for the solution of DCR problems relative to single- and multi-flows single-path cases, respectively. These classes take care of loading the DCR instances from a file in DIMACS standard format containing the graph topology and data, and a specific .dcr file for all other data. SingleFlowDCRBlock generates all the variables, static and dynamic constraints, and objective for the single-flow DCR problem, while MultiFlowDCRBlock generates all the variables, and constraints concerning each single-flow DCR problem by exploiting the SingleFlowDCRBlock functionalities, and then adding the mutual capacity constraints.

The Lagrangian approach requires to repeatedly call the SFSPDCRBlock to solve the corresponding subproblems. In the current experiments this is obtained by just using Cplex on the MINLP formulation of the problem (with P/C), exactly in the same way as the whole formulation is solved. Notably, SMS++ would make it trivial to rather use a specialised Solver that exploits the specific structure of the sub-Block, as shown in [4]. Thus, the performances of the Lagrangian approach could be further improved. However, we elected not to do so also for being maximally fair with the approach using the standard continuous relaxation. Similarly, we did not use ParallelBundleSolver to parallelise the Lagrangian function computation, despite the fact that this would have further reduced the running time. All the results have been obtained on a MacBook Pro with Apple M2 and 8 Gb RAM, under macOS Ventura 13.6.1. All the codes were compiled with clang 15.0.0, and the Cplex version is 22.11. The current develop branch of SMS++ and all its sub-projects has been used: we intend to make all source codes available shortly in the SMS++ repository.

5.2 Instances

We used instances from the data set described in [3], which consists of both real-world and synthetic network topologies. In particular, we considered some `SNDlib` topologies, while the synthetic network topology was generated using the `Waxman` model, with $n = 100$ and density $n/m = 0.4$.

In order to generate instances for the DCR problem, one needs both the information on the topology and all the required network parameters, including the traffic matrices (i.e., the flows). Therefore, the above mentioned topologies were combined with the data generated via the `FNSS` tool [15], which allows one to generate realistic computer network instances. Since the link capacities were originally generated to allow one flow at a time (i.e., SFSP-DCR), in order to obtain feasible instances for the MFSP-DCR case we considered subsets of flows with different cardinality and multiplied the original capacities by the number of flows. This guarantees feasibility of all our MFSP-DCR instances. In our preliminary experiments we consider the following 4 topologies: `abilene` (11 nodes, 28 links, 110 flows), `Garr199904` (23 nodes, 50 links, 506 flows), `Garr200109` (22 nodes, 48 links, 462 flows), and `w1-100-04` (100 nodes, 414 links, 664 flows). For each topology, we considered 5 subsets of flows of different sizes, denoted by column "NC" in Table 1, where each row corresponds to an instance defined by the topology, the corresponding network parameters, and the subset of flows (which in turn determines the link capacities).

5.3 Computational Results

The first group of four columns in Table 1 reports the quality of the bounds. In order to solve *exactly* the MINLP model (3)–(10), we used `Cplex` with a time limit of 1800 seconds. Column "gapE" reports the `Cplex` gap (i.e., the exact method), which can be larger than 0 if the time limit was reached before proving optimality. In columns "p-gapL" and "d-gapL" we report the primal and dual gap, respectively, of the Lagrangian bound $LDCR$ (12). To compute the *primal gap* we compare the Lagrangian bound with respect to the best feasible solution found (UB) by `Cplex` within the time limit: $(UB - LDCR)/LDCR$. The *dual gap* compares the Lagrangian bound with respect to the best lower bound (LB) found by `Cplex` within the time limit: $(LDCR - LB)/LB$. Finally, column "p-gapC" reports the primal gap corresponding to the continuous relaxation of (3)–(10) obtained by using `Cplex`. Gaps smaller than `1e-6` (the optimality tolerance for `Cplex`) are clearly meaningless and are therefore tabulated as blanks. The last three columns report the computing times in seconds, where "tL" is that for solving the Lagrangian dual using `BundleSolver`, "tC" is that for solving the continuous relaxation, and "tE" that for solving the MINLP model to integer optimality, in both cases using `Cplex`; a blank entry means the 1800 time limit is exceeded.

Table 1 Computational experiments

	NC	gapE	*p*-gapL	*d*-gapL	*p*-gapC	tE	tL	tC
abilene	12				5.01e+0	0.12	0.04	0.00
	23			6.04e-5	5.32e+0	0.59	0.09	0.00
	56		1.69e-6	9.59e-5	5.61e+0	31.76	3.38	0.01
	78		4.74e-6	7.25e-5	5.97e+0	63.46	3.95	0.01
	100	7.46e-3	2.19e-6	7.46e-3	5.75e+0		5.89	0.02
Garr199904	51				4.60e-1	0.14	0.08	0.01
	151		5.92e-6	8.20e-5	2.63e-1	0.56	0.24	0.03
	251		9.54e-6	6.95e-5	2.13e-1	2.08	0.38	0.05
	351		9.33e-6	6.49e-5	1.69e-1	3.29	0.53	0.07
	451		1.20e-5	3.26e-5	1.60e-1	54.61	0.69	0.11
Garr200109	47			1.15e-6	3.33e-1	5.43	0.07	0.01
	139		1.65e-6	7.39e-6	1.74e-1	1.07	0.21	0.03
	231		9.99e-6	5.80e-5	1.52e-1	2.02	0.34	0.05
	323		8.84e-6	5.53e-5	1.14e-1	3.30	0.47	0.07
	461		2.65e-6	1.98e-5	1.25e-1	94.64	1.70	0.11
w1-100-04	51	3.07e-2	1.36e-3	2.93e-2	5.59e+0		10.79	0.14
	101	1.11e-1	3.79e-2	7.03e-2	5.84e+0		19.45	0.33
	151	1.38e-1	2.22e-2	1.14e-1	5.60e+0		30.42	0.52
	201	1.64e-1	3.70e-2	1.22e-1	5.64e+0		40.32	0.77
	251	1.66e-1	4.65e-2	1.15e-1	5.78e+0		50.33	0.97

Table 1 shows that the Lagrangian approach consistently provides bounds that are orders of magnitude smaller with respect to the continuous relaxation, while the computing times can be significantly improved by replacing `Cplex` with a bespoke approach to solve the SFSP-DCR subproblems at each iteration of the Lagrangian.

References

1. Frangioni, A.: About Lagrangian methods in integer optimization. Ann. Oper. Res. **139**, 163–193 (2005)
2. Frangioni A.: Standard bundle methods: untrusted models and duality. In: Bagirov, A.M., Gaudioso, M., Karmitsa, N., Mäkelä, M., Taheri, S. (eds.) Numerical Nonsmooth Optimization: State of the Art Algorithms, pp. 61–116. Springer (2020)
3. Frangioni, A., Galli, L., Scutellà, M.G.: Delay-constrained shortest paths: approximation algorithms and second-order cone models. J. Optim. Theory Appl. **164**, 1051–1077 (2015)
4. A. Frangioni, L. Galli, E. Sorbera: Lagrangian approaches for QoS scheduling in computer networks. In: Optimization in Green Sustainability and Ecological Transition. AIRO Springer Series
5. Frangioni, A., Galli, L., Stea, G.: Optimal joint path computation and rate allocation for real-time traffic. Comput. J. **58**, 1416–1430 (2015)

6. Frangioni, A., Galli, L., Stea, G.: QoS routing with worst-case delay constraints: models, algorithms and performance analysis. Comput. Commun. **103**, 104–115 (2017)
7. Frangioni, A., Galli, L., Stea, G.: Accurate scheduling models and admission control. Comput. & Oper. Res. **81**, 67–77 (2017)
8. Frangioni, A., Gallo, G.: A bundle type dual-ascent approach to linear multicommodity min cost flow problems. INFORMS J. Comput. **11**(4), 370–393 (1999)
9. Frangioni, A., Gentile, C.: Perspective cuts for a class of convex 0–1 mixed integer programs. Math. Program. **106**, 225–236 (2006)
10. Frangioni, A., Gentile, C.: A computational comparison of reformulations of the perspective relaxation: SOCP vs cutting planes. Oper. Res. Lett. **37**(3), 206–210 (2009)
11. Lenzini, L., Martorini, L., Mingozzi, E., Stea, G.: Tight end-to-end per-flow delay bounds in FIFO multiplexing sink-tree networks. Perform. Eval. **63**, 956–987 (2006)
12. Reolon, S.M., Finardi, E.C., Frangioni, A.: Solving stochastic hydrothermal unit commitment with a new primal recovery technique based on Lagrangian solutions. Int. J. Electr. Power Energy Syst. **127**, 106661 (2021)
13. SMS++, a structured modelling system. https://gitlab.com/smspp/smspp-project
14. Parekh, K., Gallager, R.G.: A generalized processor sharing approach to flow control in integrated services networks: the single node case. IEEE/ACM Trans. Netw. **1**, 344–357 (1993)
15. Saino, L., Cocora, C., Pavlou, G.: A toolchain for simplifying network simulation setup. In: Proceedings of the 6th International ICST Conference on Simulation Tools and Techniques (2013)

A Robustness Analysis of Trading Strategy Ensembling Based on Trader Preferences

Dirk Johan Coetzee Koegelenberg and Jan H van Vuuren

Abstract *Foreign exchange* (Forex) trading algorithms require trading strategies involving the autonomous identification and exploitation of trading opportunities. Trading strategies are generally implemented in isolation, raising the question of whether diverse trading strategies can be combined in order to improve trading performance. It remains unclear, however, whether such a trading ensemble approach is capable of adapting to different trading styles based on predefined trading goals. Our contribution in this paper is, therefore, to evaluate the robustness of the notion of Forex trading ensembling in respect of different trading preferences when ensembling different trading strategies and when creating trading portfolios. An empirical analysis is conducted on four different Forex currency pairs during which five different trading strategies are ensembled and asset portfolios are created from the perspective of risk-averse and risk-seeking trading styles. The results indicate that an ensemble design is indeed capable of adjusting trading decisions based on these two defined trading styles. Ensembling from a risk-averse perspective reduces risk exposure while risk-seeking ensembling favours return maximisation with less emphasis on trading risk.

Keywords Algorithmic trading · Risk utility · Portfolio management

1 Introduction

While algorithmic trading allows financial traders to execute trades without user interference, the use of such algorithms would seem to contradict the well-known *efficient market hypothesis*. This hypothesis states that new market information is

D. J. C. Koegelenberg (✉) · J. H. van Vuuren
Department of Industrial Engineering, Stellenbosch University, Matieland, South Africa
e-mail: 21598053@sun.ac.za

J. H. van Vuuren
e-mail: vuuren@sun.ac.za

M. Di Francesco et al. (eds.), *Operations Research: Closing the Gap Between Research and Practice*, AIRO Springer Series 15, https://doi.org/10.1007/978-3-031-90095-2_7

reflected instantaneously in market price behaviour, providing no real opportunity for traders to benefit [15]. Trading algorithms nevertheless utilise automated trading strategies which are rules that dictate when to enter and exit financial markets with the objective of maximising profitability [9]. These trading strategies provide traders with a competitive edge and are generally kept secret (to the detriment of the academic literature) as any publication of a promising trading strategy may result in its mass market adoption, resulting in a forfeiture of any competitive advantage. As a result, traders often rely on simple technical indicator-based trading strategies which have largely been criticised for their simplicity and lacklustre trading performance. Although these claims have merit, to some degree, we believe that these trading strategies are nevertheless capable of yielding periods of profitability if applied to suitable market conditions.

The challenge for traders, therefore, lies in attempting preemptively to identify profitable trading periods for any given trading strategy. If such periods can be identified successfully, the temporal fluctuation in performance of the trading strategy may lead to extended intervals of trading inactivity. This performance fluctuation raises the question of whether different types of trading strategies, which are diverse in design, can effectively be combined in order for one trading strategy to compensate for another's inactivity. Such a multi-strategy "ensembling" approach assumes compliance with the popular catchphrase "horses for courses" and may potentially increase overall trading performance. Although the basic principles underlying algorithmic ensembling may be found in the literature on financial market time series forecasting methodologies, trading strategy ensembling remains largely unexplored. In particular, the question of the extent to which trading strategy ensembling methodologies are capable of adapting to different trading preferences also remains unanswered. The primary contribution of this paper is, therefore, to address this gap in the literature.

We contend that the trading ensembling process is essentially a two-layer approach which is first executed on different trading strategies (on an intra-asset level) and also between different assets (on an inter-asset level). Although the latter level is the well-studied problem of *portfolio management* (which is addressed abundantly in the literature), the effects of ensembling different trading strategies according to different trading preferences, and how this affects portfolio management, remain unstudied, and also form the focus in this paper.

The remainder of the paper is partitioned into the following sections. First, a concise literature overview is conducted in Sect. 2. This discussion is followed in Sect. 3 by the presentation of a two-layer ensembling methodology design. The effect of different trading preferences on ensembling behaviour is empirically evaluated thereafter. The experimental setup of this empirical analysis is described in Sect. 4. The discussion continues in Sect. 5 with the presentation and evaluation of the numerical results. The paper closes in Sect. 6 with a brief summary and the recommendation of suitable follow-up work.

2 Literature Review

This section is devoted to a very brief overview of the literature pertaining to fundamental notions underlying the work presented in this paper. More specifically, a discussion is presented in Sect. 2.1 on how algorithmic ensembling is carried out in the context of financial markets, after which the basic design of trading strategies is reviewed in Sect. 2.2. The notion of diversifying portfolio risk *via* portfolio management is finally described in Sect. 2.3. Emphasis is placed on discussing influential papers, but as the literature on trading strategy ensembling is scarce, topic-relevant papers are also considered appropriate.

2.1 Ensembling in Algorithmic Trading

The notion of *ensembling* is concerned with combining the generalisation capabilities of different expert systems in an attempt to improve upon overall prediction performance results. Ensembling in the context of the financial market data is predominantly found in the literature on time series forecasting. In this area, traditional ensemble methods, such as bagging, boosting and stacked generalisation, are typically employed to combine the predictions of different forecasting methods with a view to obtain improved forecasting performance [7]. Petropoulos et al. [18], for example, utilised heterogeneous forecasting methods which were combined according to a stacked generalisation ensembling approach that proved to be profitable when compared with suitable benchmarks. Paiva et al. [17], however, claimed that financial market forecasting techniques should ideally be fused with more traditional financial risk management strategies. When tested, the latter approach resulted in promising trading performance.

In contrast, the literature on ensembling trading strategies is not nearly as large as that on financial time series forecasting ensembling. A popular and emerging field that embraces the nuances of trading ensembling, however, is reinforcement learning. Instead of providing a holistic method of ensembling strategies, reinforcement learning primarily relies on employing different agents tasked with learning and adapting based on previous trading actions. More in line with the remit of this paper, Chiam et al. [6] adopted an evolutionary approach towards assigning importance scores to different trading strategies based on a multi-objective risk-return trade-off. The trained in-sample ensemble strategy parameters did not, however, guarantee correlated out-of-sample results and did not dynamically adjust to new out-of-sample information. Koegelenberg and van Vuuren [13], on the other hand, adopted a multi-criteria decision-making approach towards creating an *automated trading strategy decision assistance tool* (ATSDAT). They claimed that the methodology is trading strategy-agnostic and can be tuned to match different trading styles, but these claims were not proven formally.

2.2 The Basics of Trading Strategies

A *trading strategy* may simply be defined as a set of rules which generate one or more signals that indicate when to buy, hold, or sell an asset [9]. Any practical trading strategy, however, has to account for both the buy (open) and sell (close) signals at the start and end of a time bar [12]. Determining when to open a position is referred to as an *entry strategy*. Moreover, an open strategy should also determine whether either a *long* position or a *short* position is to be opened. Long positions suit trading conditions involving assets that appreciate in value. An asset is, therefore, bought at a low price and sold at a higher price, after a holding period. A short position, on the other hand, is opened when an asset is depreciating in value. The asset is subsequently sold at a high price with the aim of reclaiming the asset at a lower price in the future [2].

Conversely, *exit strategies* are strategies that identify exit points for open positions. Profitable trades are exited once the trading strategy deems the trade to have brought acceptable returns relative to current market conditions. Conversely, unprofitable trades are also exited once a trade is labelled unsuccessful and there is no indication of a profitable turn-around [12].

2.3 Portfolio Management

In Markowitz's Nobel prize-winning work on *modern portfolio theory* (MPT), his reasoning for the formulation of the portfolio selection problem was based on the phenomenon of portfolio diversification [16]. First, he successfully discredited the universally adopted principle of selecting assets which provide the maximum expected returns. The rationale behind this reasoning was that if this principle were to be followed, a trader's portfolio would only consist of one asset, which directly contradicts the notion of diversification. Such rules that promote portfolio standardisation should, therefore, be rejected [8]. MPT is concerned with determining a suitably weighted combination of selected assets (referred to as a portfolio) which maximises a trader's utility, by determining an appropriate trade-off between maximising expected portfolio returns and minimising risk. This trade-off is important as greater returns increase a trader's utility, but these returns are paired with increased levels of risk which conversely decrease a trader's utility. By adjusting the portfolio weightings, *efficient portfolios* may be uncovered as those which provide maximum returns for a specific degree of risk. The most suitable portfolio may then be selected from the set of efficient portfolios as the one that provides the trader with the greatest satisfaction.

3 Trading Ensemble Design

The intra-asset trading strategy ensemble approach adopted in this paper is a minor adaptation of the ATSDAT proposed by Koegelenberg and van Vuuren [13]. A brief description of our ATSDAT modification is presented here. The reader is referred to the original paper for a more in-depth description.

The ATSDAT is a dynamically adjusting trading strategy selection tool which assigns preference scores to a collection of trading strategies based on historical trading performance. In order to obtain comparable trading strategy performance results, returns generated by all individual strategies are normalised to daily returns according to the expression $n_{xy} = \frac{r_{xy}}{t_{xy}/1440}$, where r_{xy} is the x-th return generated by strategy y, and t_{xy} is the length of the trade, measured in minutes. The resulting normalised return series are all evaluated in terms of different risk and return-related performance measures. Trading strategies are selected according to a TOPSIS multiple criteria decision-making approach [11], with the trading strategies serving as the alternatives and the performance measures serving as the selection criteria. In addition, a risk-free investment also forms part of the alternative pool so as to accommodate the potential situation where all the strategies are underperforming and no trading should be conducted. Each of the different criteria is also assigned a user-specified importance weighting which is utilised as a method for defining trading preferences. A risk-seeking individual may, for example, prefer that trading strategy selection should be weighted heavily towards trading return performance with less emphasis placed on trading risk. Based on the historical trading performance and the criteria importance weightings, selection scores are generated for all alternatives which are combined to obtain a selection probability by evaluating each strategy's contribution to the total selection score of all the strategies combined. Thereafter, a random selection process is performed, but in this paper we deviate from this selection methodology by rather adopting a fixed user-specified threshold which has to be reached in order for a strategy to be selected. This adaptation provides more control over the trading strategy selection sensitivity and leads to more consistent and repeatable results. To account for this slight change, we refer to this methodology as the *adjusted* (*adj*)-ATSDAT.

For the inter-asset ensembling procedure, a *cardinality constrained portfolio management* (CCPM) approach is adopted. The reason for this choice is two-fold. First, we propose utilising the CCPM in order to control the lot sizes of trades (in contrast with measuring returns in terms of a centralised portfolio balance). Note that Forex brokers limit the minimum lot size allowed during trading, which may be addressed by assigning feasible lot sizes to those assets selected for trading. Portfolio management *via* a lot sizing approach acts as a method for comparing the potential prospects of trading different assets relative to the returns generated in the base currency of the currency pairs, and rewarding those assets which seem favourable with greater position sizes. This lot sizing approach should, therefore, not be viewed as a method for controlling and measuring portfolio returns according to a centralised portfolio balance. Secondly, the number of assets selected for trading may also affect the risk tolerance of the trader. A risk-averse individual may, for example, wish to choose selectively those assets which provide the best likelihood of executing profitable trades, while a risk-seeking individual may prefer to trade in all assets in order to maximise potential profitability. Keeping these two considerations in mind, the CCPM may be formulated based on a design proposed by Chang et al. [5] as the mixed-integer programming problem

$$\min_{\boldsymbol{w},z} \quad \sum_{i=1}^{N}\sum_{j=1}^{N} w_i w_j \sigma_{ij},$$

subject to the constraints

$$\begin{aligned}
\sum_{i=1}^{N} w_i \mu_i &\geq R^*, \\
\sum_{i=1}^{N} w_i &\leq 1, \\
\sum_{i=1}^{N} z_i &\geq K, \\
\epsilon z_i \leq w_i &\leq \delta z_i, \quad i = 1, \ldots, N, \\
z_i &\in \{0, 1\}, \ i = 1, \ldots, N,
\end{aligned}$$

where w_i is the weight assigned to the i-th asset, N is the size of the portfolio, σ_{ij} is the downside risk covariance (semi-covariance) as defined by Estrada [10], μ_i is the expected return of the i-th asset, R^* is some target return value for the portfolio of assets, K is the minimum number of assets selected for trading, and ϵ and δ are upper and lower limits on the portfolio weights, respectively. Notice the inequality constraint for R^* which allows for some degree of error tolerance, as the target return may not be reached exactly.

4 Experiment Setup

In order to demonstrate the combined robustness of the *adj*-ATSDAT and CCPM approach towards changing trading preferences, an empirical analysis is conducted in which the trading behaviours of both a risk-averse individual and a risk-seeking individual are considered. For the purposes of this study, we define a risk-averse trader as an individual who aims to minimise return volatility and reduce market exposure, while a risk-seeking individual aims to maximise profit at any cost. The analysis is partitioned into two parts. The first is concerned with the results of the *adj*-ATSDAT for different Forex currency pairs while we combine the *adj*-ATSDAT results of different assets into a single portfolio *via* the CCPM approach in the second part.

We consider five heterogeneous intraday trading strategies which may briefly be described as follows:

Dual CCI: A strategy which uses the commodity channel index technical indicator [14] to identify trending, overbought, and oversold market conditions.

Short-term scalping: A scalping strategy which utilises the combined consensus of Bollinger bands [3], simple moving averages, the relative strength index [23], and the moving average convergence and divergence technical indicators [1].

Bollinger squeeze: A Bollinger band breakout strategy which filters breakout trading signals *via* a long-term trend analysis and the time-segmented volume technical indicator [21].

Schaff trend cycle: A strategy which uses the Schaff trend cycle indicator [22] as well as long-term and short-term exponential moving averages to identify market turning points.

Triple supertrend: A strategy which evaluates the market trend over three different periods and which is assisted by the relative strength index, time-segmented volume, and exponential moving average technical indicators.

Trading is conducted retrospectively over the volatile first half of the 2019 calendar year in respect of the AUD/CAD, AUD/NZD, EUR/USD, and USD/CHF currency pairs, based on 15-minute bar data. All parameters of the individual trading strategies and *adj*-ATSDAT are tuned on (unseen) 2018 calendar year data. The trading balance for each asset consists of 100 units of the quote currency which are traded at a leverage of 100:1. Furthermore, a fixed lot size scaling brokerage fee of 0.0003 is considered and the risk-free investment is assumed to be a *certificate of deposit* which generates a modest 0.68% annual return (conforming to average returns at the time). It is also assumed that each asset is allowed to trade at a lot size of 0.025, resulting in a total lot size of 0.1 risked at any point in time across all assets.

The different trading preferences are incorporated by adjusting the criteria importance weightings of the TOPSIS method utilised in the adj-ATSDAT. These criteria are the average returns (returns-orientated), downside risk (risk-orientated), the return range (risk/returns-orientated), the Sharpe ratio [20] (risk-orientated), the hit rate (risk/return-orientated), and the average trade duration (risk-orientated). The relative importance of these criteria for both trading preferences are computed *via* an *analytical hierarchy process* [19] comparison matrix by outweighing the relative importance of one criterion by another from the perspective of risk-seeking and risk-averse trading individuals. TOPSIS criteria importance weights of $w_a = (0.08, 0.29, 0.10, 0.15, 0.32, 0.06)$ for the risk-averse trader and $w_s = (0.44, 0.04, 0.11, 0.10, 0.28, 0.03)$ for the risk-seeking trader are obtained as a result. Moreover, the *adj*-ATSDAT selection proportion threshold is fixed at 0.25, which means a maximum of two trading strategies are allowed to trade simultaneously at any point in time.

In the second part of the empirical analysis, the entire process is repeated, but this time the lot sizes are controlled by the CCPM approach. The parameters of the CCPM approach are selected as $K = 2$ (limiting the total amount of market exposure and losses caused by under-performing portfolio assets) for the risk-averse trader and as $K = 4$ (maximising the potential for generating large returns) in the case of the risk-seeking trader. Moreover, the portfolio weights are bound to the lot size range of 0.01 (generally, the smallest lot size accepted by Forex brokers) to 0.025 (the previously defined maximum lot size). Portfolios are re-evaluated daily, and the portfolio yielding the best Sortino ratio [4] is selected.

5 Results and Discussion

The trading performance results associated with the risk-averse and risk-seeking (presented in brackets) *adj*-ATSDAT implementations are summarised in Table 1 for all four currency pairs. On face value, the results seem to suggest that the *adj*-ATSDAT is robust to different trading preferences. This claim is, first of all, supported by the number of trades executed in the risk-averse and risk-seeking scenarios, as may be seen in Table 1. Directly comparing the number of trade executed in the two scenarios, it is evident that the risk-averse implementation was significantly more conservative in executing trades while the risk-seeking implementation adopted a more opportunistic approach, executing up to 50 trades more than that of the risk-averse scenario.

The number of trades executed also assists in describing the *adj*-ATSDAT trading behaviour relative to the remaining performance measures. First, it is important to note that the risk-seeking trading approach did not guarantee greater returns. On the contrary, the risk-seeking approach's opportunistic and risky behaviour in executing numerous trades resulted in more unprofitable trades, lowering the hit rate percentage, relative to the risk-averse scenario, and drastically affecting the returns generated. The conservative approach of the risk-averse implementation, on the other hand, was more assured when executing trades, leading to less but improved trading decisions which ultimately generated greater returns of up to 38.14%. The different trading behaviours in both scenarios are also noticeable when evaluating the annualised volatility percentage figures in Table 1, which indicate that the returns generated by the risk-seeking approach were far more volatile than those in its risk-averse counterpart. A risky trading style was, however, warranted in the case of the USD/CHF currency pair which generated significant returns, but at the cost of higher return volatility. As this behaviour is only observed for one of the four currency pairs, it supports the argument that risk-seeking behaviour may lead to greater returns, but this result is not guaranteed. Moreover, the adj-ATSDAT method requires moderate

Table 1 The trading performance results associated with the risk-averse (risk-seeking) implementations of the *adj*-ATSDAT for each respective currency pair

	Return (%)	Annualised volatility (%)	Hit rate (%)	Number of trades
AUD/CAD	18.34 (0.94)	20.91 (23.34)	78.18 (68.00)	65 (110)
AUD/NZD	38.14 (27.73)	16.08 (18.18)	84.90 (80.00)	75 (80)
EUR/USD	−24.19 (−37.18)	26.34 (44.13)	58.82 (51.85)	80 (130)
USD/CHF	14.33 (36.79)	19.67 (27.06)	72.00 (63.74)	58 (107)

computational power making the adopted method scalable to larger trading strategy portfolios. The *adj*-ATSDAT is also robust to changing trading strategy performance. This robustness improves upon traditional ensemble methods which may require periodical retraining so as to account for changing strategy dynamics which, in turn, increases computational cost.

The results emanating from the second part of the empirical analysis, when adopting the CCPM approach for both the risk-averse and risk-seeking traders, are illustrated graphically in Fig. 1a and b, respectively. In the case of the risk-averse approach, illustrated in Fig. 1a, trading was mainly conducted in respect of the AUD/NZD and USD/CHF currency pairs. It is worth noting that these were the two assets that provided the least annualised volatility figures during trading strategy ensembling, showcasing that emphasis was placed on lowering portfolio risk.

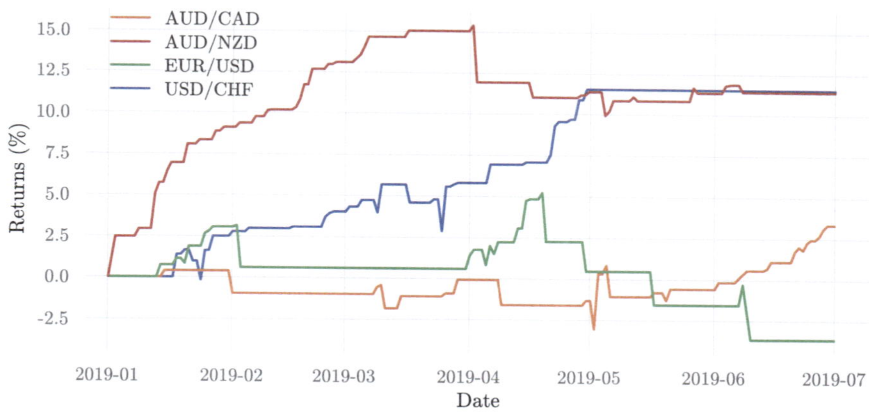

(a) Risk-averse approach for $K = 2$.

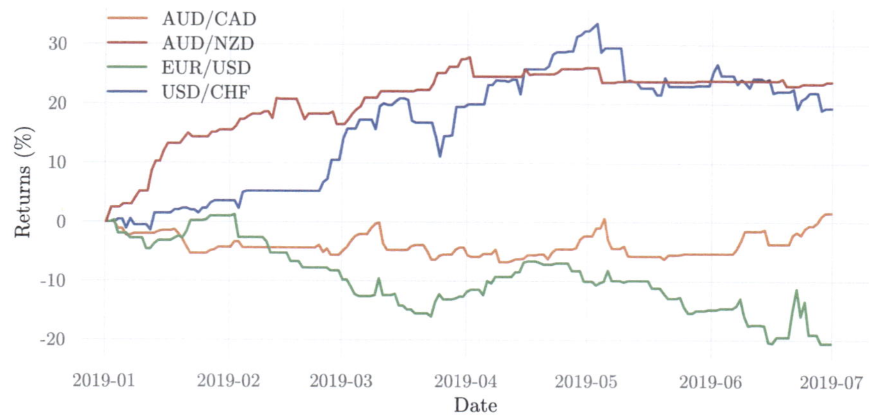

(b) Risk-seeking approach for $K = 4$.

Fig. 1 The portfolio returns percentages for **a** the risk-averse approach and **b** the risk-seeking approach, both from January 2019 to July 2019

Furthermore, although more than two assets could be traded at any point in time, the CCPM optimisation always resulted in only two assets being selected. At the same time, the minimum lot size was assigned to those selected assets (sometimes being more lenient and assigning an asset a larger position size) which drastically reduced the profitability of the strategies but improved the return variances to the risk-averse individual's satisfaction. The most significant contribution of the CCPM method was, however, its ability to limit the losses incurred by trades involving the EUR/USD currency pair from −24.19% to −2.94%.

The risk-seeking implementation, on the other hand, provided a portfolio with significantly different characteristics, but more in line with risk-seeking behaviour, as may be seen in Fig. 1b. The portfolio configuration naturally resulted in more volatile return curves as trading was allowed in all assets at all times. Unlike in the risk-averse scenario, however, larger (but not necessarily the maximum) lot sizes were more freely assigned to those assets with the best trading performance and, therefore, still allowing for significant returns to be generated. The extent of losing trades was, therefore, better controlled, as best showcased by the EUR/USD currency pair where the loss made was reduced from −37.18% to −20.53%.

6 Conclusion

In this paper, the robustness of trading strategy ensembling was evaluated for different trading preferences. More specifically, a two-layer ensembling approach was considered according to which trading preferences are specified at an intra-asset level and an inter-asset level. These two layers were populated with an existing trading strategy ensemble process and a CCPM method, respectively, which allowed for flexibility in specifying different trading preferences. The robustness of the adopted ensemble design was evaluated in the form of an empirical analysis during which risk-averse and risk-seeking trading scenarios were assessed. The results indicated that the ensemble design was able to adapt according to the predefined trading preferences and exhibited trading behaviour in line with risk-averse and risk-seeking individuals.

Related future work may include analyses of trading strategy ensembling in which it is possible to update trading preferences dynamically. This consideration lends itself to the application of prospect theory, according to which an individual's behaviour changes based on his or her current wealth status.

References

1. Appel, G.: Technical Analysis: Power Tools for Active Investors. FT Press, Upper Saddle River (NJ) (2005)
2. Archer, M.D.: Getting Started in Forex Trading Strategies. John Wiley & Sons, Hoboken (2007)

3. Bollinger, J.: Bollinger on Bollinger Bands. McGraw-Hill, New York (NY) (2001)
4. Booth, G.G., Broussard, J.P.: The Sortino ratio and extreme value theory: an application to asset allocation. In: Longin, F. (ed.) Extreme Events in Finance: A Handbook of Extreme Value Theory and Its Applications, pp. 443–464. John Wiley & Sons, Hoboken (NJ) (2016). https://doi.org/10.1002/9781118650318.ch17
5. Chang, T.-J., Meade, N., Beasley, J.E., Sharaiha, Y.M.: Heuristics for cardinality constrained portfolio optimisation. Comput. & Oper. Res. **27**(13), 1271–1302 (2000). https://doi.org/10.1016/S0305-05489900074-X
6. Chaim, S.C., Tan, K.C., Al Mamun, A.: Investigating technical trading strategy via an multi-objective evolutionary platform. Expert. Syst. Appl. **36**(7), 10408–10423 (2009). https://doi.org/10.1016/j.eswa.2009.01.058
7. Cheng, C., Xu, W., Wang, J.: A comparison of ensemble methods in financial market prediction. In: 5th International Joint Conference on Computational Sciences and Optimization, pp. 755–759, IEEE Press, Harbin (2012). https://doi.org/10.1109/CSO.2012.171
8. Constantinides, G.M., Malliaris, A.G.: Portfolio theory. In: Jarrow, R.A., Maksimovic, V., Ziemba, W.T. (eds.) Handbooks in Operations Research and Management Science, vol. 9, pp. 1–30. North-Holland, Amsterdam (1995). https://doi.org/10.1016/S0927-0507(05)80045-3
9. De Santis, A., Dellepiane, U., Lucidi, S., Renzi, S.: A derivative-free optimization approach for the autotuning of a Forex trading strategy. Optim. Lett. **15**, 1649–1664 (2021). https://doi.org/10.1007/s11590-020-01546-7
10. Estrada, J.: Systematic risk in emerging markets: the D-CAPM. Emerg. Mark. Rev. **3**(4), 365–379 (2002). https://doi.org/10.1016/S1566-0141(02)00042-0
11. Hwang, C.-L., Yoon, K.: Multiple Attribute Decision Making: Methods and Application. Springer, Heidelberg (1981)
12. Katz, J.O., McCormick, D.L.: The Encyclopedia of Trading Strategies. McGraw-Hill, New York (2000)
13. Koegelenberg, D., van Vuuren, J.H.: Heterogeneous trading strategy ensembling for intraday trading algorithms. S. Afr. J. Ind. Eng. **34**(3), 156–169 (2023). https://doi.org/10.1016/10.7166/34-3-2951
14. Lambert, D.R.: Commodity channel index: tool for trading cyclic trends. Tech. Anal. Stock. Commod. **1**(5), 120–122 (1980)
15. Malkiel, B.G.: A Random Walk Down Wall Street: The Time-tested Strategy for Successful Investing. WW Norton & Company, New York (2012)
16. Markowitz, H.M.: Portfolio selection. J. Financ. **7**(1), 77–91 (1952). https://doi.org/10.1016/S1566-0141(02)00042-0
17. Paiva, F.D., Cardoso, R., Hanaoka, G.P., Duarte, W.M.: Decision-making for financial trading: a fusion approach of machine learning and portfolio selection. Expert. Syst. Appl. **115**, 635–655 (2019). https://doi.org/10.1016/j.eswa.2018.08.003
18. Petropoulos, A., Chatzis, S.C., Siakoulis, V., Vlachogiannakis, N.: A stacked generalization system for automated Forex portfolio trading. Expert. Syst. Appl. **90**, 290–302 (2017). https://doi.org/10.1016/j.eswa.2017.08.011
19. Saaty, R.W.: The analytic hierarchy process–what it is and how it is used. Math. Model. **9**(3–5), 161–176 (1987). https://doi.org/10.1016/0270-0255(87)90473-8
20. Sharpe, W.F.: Capital asset prices: a theory of market equilibrium under conditions of risk. J. Financ. **19**(3), 425–442 (1964). https://doi.org/10.1111/j.1540-6261.1964.tb02865.x
21. TC2000: Time segmented volume. https://help.tc2000.com/m/69445/l/755873-time-segmented-volume (2020)
22. Twomey, B.: Schaff trend: a faster and more accurate indicator. https://www.investopedia.com/articles/forex/10/schaff-trend-cycle-indicator.asp (2023)
23. Wilder, J.W.: New Concepts in Technical Trading Systems. Trend Research, Mcleansville (NC) (1978)

Intervention Planning for a Home Health-and-Social-Care Institute in Switzerland

Roxanne Tison and Nicolas Zufferey

Abstract We consider an intervention planning problem proposed by AVASAD, a home health-and-social-care institute in Switzerland. The main decisions are to assign interventions to caregivers, decide the start time of each intervention, and determine the caregiver routes. Various constraints are considered, including skills, working hours, patient availabilities, travel times, and imposed start times for some interventions. Six objective functions are ranked according to a strict priority imposed by AVASAD (e.g., maximize the care continuity and the number of interventions assigned to caregivers with the recommended skills, minimize the travel and idle times, maximize the assignment of interventions at the patient preferred times). Various insertion algorithms are proposed and combined, and their results are favorably compared to the schedules employed by AVASAD.

Keywords Intervention planning · Homecare scheduling · Insertion heuristics · Lexicographic optimization

1 Introduction

The project is motivated by AVASAD, the main home-care player in the Canton of Vaud in Switzerland. The activities of AVASAD consist of home health-and-social-care services. AVASAD has nearly 5,000 caregivers who provide services to about 40,000 patients. This results in the coordination of roughly 90,000 interventions on a weekly basis. AVASAD is geographically segmented into approximately 50 structures, wherein each structure independently plans its interventions while following the below-described objectives and constraints. This project focuses on the intervention planning in one representative structure of AVASAD, with the goal to extend it

R. Tison (✉)
Afflux, EPFL innovation park, La Forge Batiment C, Lausanne, Switzerland
e-mail: roxanne.tison@afflux.ch

N. Zufferey
GSEM, University of Geneva, Geneva, Switzerland
e-mail: n.zufferey@unige.ch

M. Di Francesco et al. (eds.), *Operations Research: Closing the Gap Between Research and Practice*, AIRO Springer Series 15, https://doi.org/10.1007/978-3-031-90095-2_8

globally after this initial study. The planning horizon is a working day. Throughout the day, caregivers must perform interventions (i.e., treatments, care, and procedures) at patient locations. The considered real instances are characterized by roughly 300 interventions, 120 patients, and 30 caregivers with 15 possible skills. Consequently, in line with other works in the field, exact methods are likely to be unsuitable for real instances, necessitating the utilization of heuristic methods. As highlighted in recent studies: (a) “solutions provided by the MIP + CPLEX do not possess enough quality for practical implementations” [12]; (b) “there is not a compact model and a practically implementable exact algorithm in the literature to solve such a complicated problem” (while referring to a problem that includes care continuity) [11]. Therefore, we propose here ad-hoc insertion heuristics. The reader interested in mathematical models for related problems is referred to [3, 4].

As many real-world problems, the considered problem, denoted here as (P), contains various objectives and constraints. (P) belongs to a well-known research stream: home health-care routing and scheduling (HHCRS). The reader interested in recent literature reviews is referred to [4, 5]. Due to its various possible constraints and objectives, no universally accepted standard version of a HHCRS problem exists. Because of their complexity, such problems are typically approached as a rich vehicle routing problem [6]. While metaheuristics have proven efficiency in solving single-objective problems like the Traveling Salesman Problem [9], addressing the growing complexity that arises with multiple vehicles, objectives and constraints, is not straightforward [2]. (P) is characterized by two decision-variable types, six objective functions ranked lexicographically, and seven constraint types. To face this complexity, in line with other works [15], we propose a quick and efficient solution method, based on insertion moves that are guided with priority rules.

The contributions of this paper are twofold. First, with the help of AVASAD, we design a problem (P) that perfectly covers their needs. (P) is a new HHCRS problem due to its unique combination of six objective functions and constraints of various nature. Second, we propose and combine insertion algorithms that can generate successful results for real-life instances, while taking into account the lexicography of the objectives and the computational time limit of 10 min. The insertion algorithms developed are based on simple, flexible concepts that can be efficiently explained. Indeed, from a practical standpoint, decision makers are only likely to accept solutions they can understand.

The remaining part of this paper is organized as follows. Section 2 describes problem (P), as well as the features that make it unique and challenging in the HHCRS research field. Section 3 proposes our solution method. Section 4 presents our results. Finally, Sect. 5 concludes the paper and suggests future research directions.

2 Presentation of Problem (P)

The planning horizon is a day. The demand encompasses hundreds of interventions, where each intervention requires exactly one caregiver to take place. Each intervention is characterized by a patient, a *usual* caregiver list, a location, a type (A or B), a *recommended* skill, an *authorized* skill list, a required or desired start time, and a duration (which varies depending on the service or care provided). Type-A interventions (e.g., mental care, medication, monitor blood glucose) are more important to assign than type-B interventions (e.g., give a shower, cleaning). The caregiver list C_i associated with each intervention i contains *usual* caregivers, i.e., caregivers who have already treated the involved patient within the last three months (this list may thus evolve over time). In each list, the caregivers are ranked in descending order of frequency of intervention (over the last three months). The imposed *authorized* skill list S_i associated with each intervention i contains all the *authorized* substituting skills, ranked from the most adapted to the least. All the problem-related parameters of this study are imposed by AVASAD.

Each intervention must be planned in the caregiver working hours (constraint 1) and the patient availability periods (constraint 2), with a 5-minute tolerance parameter. The travel times and the interventions cannot overlap in time for more than 5 (parameter) minutes (constraint 3). The same patient may require multiple visits throughout the day to address various needs. S/he can receive certain interventions concurrently (e.g., medication delivery plus housekeeping), whereas other interventions cannot overlap in time (constraint 4, e.g., give a shower and nursing care).

Regarding the assignment of interventions to caregivers, two goals must be considered. First, in line with other studies where care continuity is crucial [8, 10], objective 1 consists in maximizing the number of interventions involving a patient and one *usual* caregivers. More precisely, for each patient, we consider the visit frequency of her/his caregivers: the higher a caregiver frequency, the better that caregiver is for the considered patient. For instance, a caregiver who has visited a patient 14 times in the last three months is more valued than a caregiver with 6 visits. This care continuity addresses important social aspects, thereby enhancing the overall care experience. This objective also contributes to work efficiency, as a caregiver who is familiar with a patient location can better navigate through it. Assigning a non-usual caregiver to an intervention is allowed but not rewarded in this first objective. Second, the caregiver skills play a pivotal role. More precisely, each intervention requires a specific skill (among the 15 considered skills). Objective 2 aims at maximizing the assignment of interventions to caregivers with the *recommended* skill (e.g., nursing care activities should be assigned to nurses, shower assistance should be assigned to healthcare assistants). This prioritization also favors job satisfaction for caregivers who can perform tasks aligned with their expertise. However, when it is not possible to assign an intervention to a caregiver with the *recommended* skill, it can be delegated to a caregiver with an *authorized* (usually "higher") skill. Each caregiver assigned to an intervention must have either the *recommended* skill or an *authorized* skill (constraint 5). For each skill, a hierarchical list of *authorized* substituting skills is given. For example, a healthcare assistant can perform a shower, but if no healthcare assistant is available, this intervention can be assigned to a nurse assistant. If neither healthcare nor nurse assistants are available, the task can be delegated to a nurse (this assignment is feasible but not desirable). However, if no caregiver with an *authorized* skill is available, the intervention cannot be assigned (e.g., a healthcare assistant cannot perform a nursing care). Objective 3 consists in maximizing the number of assigned interventions, with a strict priority for the type-A interventions.

Travel times between interventions are integrated into the caregiver working schedules, with buffer periods (parameters) for parking and walking between the car and the patient location. As for many extensions of the Vehicle Routing Problem, we aim at minimizing the overall travel times (objective 4) and idle times (excluding travel times) between interventions (objective 5). Another important aspect is to schedule each intervention at the time preferred by the patient (objective 6). This notion cannot be treated as a constraint since there is a high demand for interventions at the same time. For instance, scheduling all showers early in the morning, as per patient preferences, is not feasible. The type-A interventions must start precisely at the given required time (constraint 6), often for medical reasons. In contrast, the start of each type-B intervention tolerates a specific flexibility margin with respect to a given desired time (constraint 7). Considering the large amount of data (available + missing), these flexibility margins were determined by learning from the historical solutions previously employed by AVASAD for the same intervention types.

The optimization problem (i.e., decision variables, constraints, and objective functions) can be presented as follows.

Decision Variables: intervention set assigned to each caregiver; start time for each intervention.

Constraints

1. A caregiver is restricted to performing interventions only within her/his designated working hours, with a tolerance of up to 5 min (parameter).
2. A patient is eligible to receive an intervention only within her/his availability period, with a tolerance of 5 min (parameter).
3. The travel time between two consecutive interventions i and j must start at least after the end of i and finish before the start of j, with a tolerance of 5 min (parameter).
4. For each patient, some interventions cannot overlap in time.
5. An intervention must be assigned to a caregiver with a *recommended* skill or an *authorized* skill.
6. The start of a type-A intervention must be planned at its required time.
7. The start of a type-B intervention must be planned at its desired time $\pm$ its associated flexibility margin.

Objective functions. In line with various other situations in practice [1, 7, 13, 17, 18], a lexicographic approach is employed to rank the objectives once and for all (i.e., no improvement on a lower-level objective can compensate a degradation of a higher-level objective). The below prioritization is imposed by the strategy of AVASAD.

1. Maximize the number of interventions involving a patient and an associated *usual* caregiver, giving strict priority to caregivers who have a higher visit frequency with the involved patient (i.e., optimize the patient-caregiver relationship).
2. Maximize the number of interventions for which a caregiver with the *recommended* skill is assigned (in contrast with *authorized* skill).
3. Maximize the number of interventions, with a strict priority for type A.
4. Minimize the average travel time between interventions (in minutes).
5. Minimize the average idle time between interventions (in minutes).
6. Minimize the use of flexibility margins for type-B interventions (i.e., the average difference between the planned start times and the desired start times, in minutes).

Before this project, AVASAD considered the optimization of a single function with many components (having different weights) that somewhat covered the above objectives and constraints. That approach was unsuccessful. Indeed, determining the weight of each component is a sensitive task that can be challenged by many decision makers. Moreover, the obtained solutions were often unfeasible with respect to caregiver and patient availabilities. In contrast, in our project, we can make a clear distinction between objectives and constraints, and a straightforward hierarchy among objectives. The resulting optimization problem (P) was thus easily accepted by the involved practitioners, and feasible solutions can be easily obtained.

3 Methods

The proposed solution method consists of applying sequentially Algorithms 1 and 2. The former inserts interventions in the working schedule of the usual caregivers (caregiver by caregiver), whereas the latter tries to insert the remaining interventions in the schedule of the caregivers (intervention by intervention), favoring the use of the *recommended* skills over the *authorized* skills. At the end, the intervention set that cannot be scheduled is penalized through objective functions 1, 2 and 3. Only feasible solutions are considered in the proposed methods. Indeed, the consideration of non-feasible solutions would require more complex and time-consuming algorithms, which have the ability to repair solutions [16].

The pseudocode of Algorithm 1 is self-explanatory. It relies on repeating the following two main steps. First, select a caregiver c whose is the preferred one for some interventions. Second, insert interventions in the schedule of c with the use of various insertion procedures that will be explained below. These insertion procedures are Algorithm A (it tries to schedule an intervention without the use of its flexibility margin), Algorithm B (it tries to schedule an intervention with the possible use of its flexibility margin, which is only possible for type-B interventions), and Algorithm 3 (it tries to schedule a given set of interventions). Of course, an intervention cannot be scheduled in case of any constraint violation.

Algorithm 1 Greedy planning algorithm for the *usual* caregiver set

Input: set I of interventions to schedule; set of the *usual* caregivers; caregiver list C_i for each intervention i.

Output: reduced set I of interventions which still have to be scheduled; updated schedule of the involved caregivers.

Initialization

- For each intervention i, determine the caregiver c_i who has the largest priority in the caregiver list C_i.
- Compute the intervention set I_c for which $c_i = c$ (i.e., c is the currently preferred caregiver for all the interventions of I_c).
- Compute the caregiver set C' containing caregivers c such that $I_c \neq \emptyset$ (i.e., the set of caregivers who have the priority for at least one intervention).

While ($C' \neq \emptyset$) **and** ($I \neq \emptyset$), **do**:

1. Randomly select a caregiver c in C'.
2. Use Algorithm A to insert (in the schedule of c) the type-A interventions of I_c at the required time, and remove the scheduled interventions from I_c and I. If not all the type-A interventions have been scheduled, go to step 4.
3. Try to insert (in the schedule of c) all the type-B interventions of I_c using Algorithm B. If not all the type-B interventions can be scheduled, go to step 4 without effectively scheduling the interventions (i.e., without updating I_c and I in this step). Otherwise, schedule all these interventions and remove them from I_c (i.e., $I_c = \emptyset$ at this stage) and I.
4. If $I_c \neq \emptyset$, use Algorithm 3 to try to insert (in schedule of c) as many interventions of I_c as possible. Remove the scheduled interventions from I_c and I.
5. Remove caregiver c from all the list C_i for which $c = c_i$ (as caregiver c is no longer available for such interventions) and update C' accordingly.

The output of Algorithm 1 is the updated schedules of the involved *usual* caregivers, and a set I of interventions which still have to be scheduled. For each intervention i, let R_i be the set of caregivers having the *recommended* skill, and A_i be the set of caregivers having an *authorized* skill (the elements of this set are ranked according to the skill hierarchy, from the lowest to the highest). The pseudocode of Algorithm 2 is self-explanatory. Its goal is to insert additional interventions of I in the schedule of caregivers, while considering the lexicography in the objective-function list. It relies on repeating the following two main steps. First, select the non-already considered intervention i that has the smallest flexibility margin (break ties randomly), which means that all the type-A interventions are considered before the type-B interventions. Second, insert i in the solution according to some priority rules with respect to the caregiver assignment (i.e., favor R_i over A_i) and the start-time determination (i.e., do not use the flexibility margins if it is not necessary, which means that Algorithm A is preferred over Algorithm B). Of course, the intervention is not scheduled at all if a constraint is violated.

Algorithm 2 Greedy insertion algorithm for the remaining interventions

Input: set I of interventions to schedule; schedule of each caregiver (which is initially empty for the caregivers not impacted by Algorithm 1).

Output: updated set I of interventions which are not scheduled at all; updated schedule of the impacted caregivers.

Initialization

- Rank the interventions of I by increasing flexibility margin (i.e., all the type-A interventions are considered first).
- For each intervention i, set $\Omega_i = R_i$ (i.e., start with the caregivers having the *recommended* skills first).
- Set ALGO = Algorithm A (i.e., start with the insertion algorithm that does not take advantage of the flexibility margins).

For each intervention $i \in I$, **do**:

1. For each caregiver $c \in \Omega_i$, do: apply ALGO to try to insert i in the schedule of c; if a feasible solution is returned, evaluate it but do not perform the insertion.
2. Perform the best feasible insertion evaluated in step 1 and remove i from I.
3. If ALGO = Algorithm B, set ALGO = Algorithm A and go to step 5.
4. If (no feasible insertion of i was found) and (i is of type B), do: set ALGO = Algorithm B (to take advantage of the flexibility margins) and go back to step 1.
5. If (intervention i has not been scheduled) and ($\Omega_i = R_i$), set $\Omega_i = A_i$ and go back to step 1 (to try to assign a caregiver with an *authorized* skill because no insertion was found with the *recommended* skill).

For both Algorithms A and B, the input is the intervention i to try to insert in the schedule of the considered caregiver c, whereas the output is a feasibility message if all the constraints are respected (otherwise, a non-feasibility message is returned). In Algorithm A, the possible flexibility margin is not used, which means that the start time of intervention i is exactly set at the required time (if i is of type A) or at the desired time (if i is of type B). In contrast, the use of the flexibility margin is accepted in Algorithm B. Therefore, insertion i can start within an interval defined by the desired time plus or minus its flexibility margin. The best feasible start time (with respect to the objective-function prioritization) is identified in this interval.

The pseudocode of Algorithm 3 is self-explanatory. It tries to insert a set I' of interventions in the schedule of a caregiver c. Two disjoint cases are considered, and the best-encountered solution is returned at the end. If the size of I' does not exceed p (parameter), q_1(parameter) intervention permutations (randomly generated) are investigated. For each considered intervention permutation, the interventions are tried to be inserted one by one in the schedule of c. In contrast, if the size of I' exceeds p, groups of interventions (where possibly overlapping interventions are put in the same group) are first created to reduce the computational effort. Next, q_2 (parameter) intervention permutations (randomly generated) are investigated per group. For each considered permutation, the interventions are tried to be inserted group by group (and one by one in each group) in the schedule of c. After preliminary experiments, and to cope with the allowed time limit imposed by AVASAD (10 min), we set $p = 7$ and $q_1 = q_2 = 30$.

Algorithm 3 Greedy planning algorithm for one selected *usual* caregiver

Input: set I' of type-A and type-B interventions to schedule; current schedule of the *usual* caregiver c to consider.
Output: best-generated solution (i.e., set I' of non-scheduled interventions; updated schedule of caregiver c).
Case 1: if I' contains up to p (parameter) interventions, do:

- Repeat the following process q_1 (parameter) times:
  - Generate a new permutation of the interventions of I' (i.e., interventions of I' are reordered accordingly). If no new permutation can be generated, STOP the "Repeat" process.
  - Try to sequentially insert (in the schedule of c) all the interventions of I' with Algorithm A or B (depending if the considered intervention is of type A or B). Do not perform the insertions but simply evaluate the resulting solution.
  - If all the interventions of I' can be scheduled in the solution, schedule them and STOP the "Repeat" process. Otherwise, update and store the best-encountered solution.

Case 2: if I' contains more than p (parameter) interventions, do:

- Initialization: group together all the interventions that can possibly overlap in time. More precisely, two interventions i and j are put in the same group if they can possibly overlap in time (an intervention k that can possibly overlap with either i or j is also put in the same group, and so on). At the end of this initialization phase, a set $G = (G_1, G_2, \ldots)$ of intervention groups is built.
- **For** each group $G_j \in G$, **do**: Repeat the following process q_2 (parameter) times:
  - Generate a new permutation of the interventions of G_j. If no new permutation can be generated, STOP the "Repeat" process.
  - Try to sequentially insert (in the schedule of c) all the interventions of G_j with Algorithm A or B (depending if the considered intervention is of type A or B). Do not perform the insertions but simply evaluate the resulting solution.
  - If all the interventions of G_j can be scheduled in the solution, schedule them and STOP the "Repeat" process. Otherwise, update and store the best-encountered solution.

---

4 Results

All the algorithms were coded in C++ and run on 3.30 GHz AMD Ryzen 9 5900HX processor with 32 GB of RAM. The computation times are not presented but they all respect the time limit of 10 min imposed by AVASAD. Table 1 presents the results performed on 10 real instances denoted from I1 to I10 (average values are given in the

Table 1 Results

		I1	I2	I3	I4	I5	I6	I7	I8	I9	I10	Average
Charact.	Interventions	322	311	313	311	316	329	301	304	297	300	310
	Type-A interv.	264	247	243	242	248	247	228	227	230	227	240
	Patients	121	117	119	123	120	122	117	120	125	124	121
	Caregivers	31	33	32	34	32	31	34	26	31	29	31
Obj. 1 Top	AVASAD	34	46	38	34	35	47	66	45	55	48	45
	Algo 2	9	9	5	9	7	12	6	7	7	7	8
	Algo 1 & 2	21	22	22	17	18	34	22	33	29	28	25
	Algo 1 & 2 & 3	67	78	57	61	59	97	98	72	89	82	76
Obj. 1 Sum	AVASAD	60	81	68	74	65	93	113	81	101	114	85
	Algo 2	15	27	11	24	25	37	30	20	24	30	24
	Algo 1 & 2	49	50	51	46	52	69	68	63	74	76	60
	Algo 1 & 2 & 3	113	112	107	106	110	147	156	121	142	148	126
Obj. 2	AVASAD	145	136	142	141	148	141	131	111	156	160	141
	Algo 2	218	216	230	225	229	228	217	193	226	221	220
	Algo 1 & 2	213	203	213	214	223	217	203	172	216	209	208
	Algo 1 & 2 & 3	185	186	194	196	202	190	180	157	186	191	187
Obj. 3	AVASAD	230	235	212	206	226	211	218	177	206	226	215
	Algo 2	263	262	268	262	264	267	261	245	257	262	261
	Algo 1 & 2	267	261	268	261	261	268	263	242	261	262	261
	Algo 1 & 2 & 3	256	262	261	255	261	260	262	233	252	258	256
Obj. 3 A	AVASAD	189	186	165	166	179	159	165	136	169	171	169
	Algo 2	217	210	213	203	207	198	200	188	200	198	203
	Algo 1 & 2	220	210	211	203	203	201	200	185	204	199	204
	Algo 1 & 2 & 3	210	209	203	197	201	195	201	175	196	197	198
Obj. 4	AVASAD	6.5	7	6.7	6.8	6.9	6	5.8	5.9	5.9	5.7	6.3
	Algo 2	9.6	10.3	9.5	9.1	9.8	10	9.5	9.2	8.9	8.3	9.4
	Algo 1 & 2	9.7	10.6	10	9.7	10.1	10	9.7	9.3	9.2	8.3	9.7
	Algo 1 & 2 & 3	8.9	10.4	8.9	9.6	9.3	8.9	8.3	8.3	8.5	8.6	9
Obj. 5	AVASAD	4.1	3.4	4.5	3.5	3.8	2.5	4.4	4.7	3.2	3.5	3.8
	Algo 2	5.4	4.2	6	5.7	5.7	3.1	5.8	4.4	6.3	3.5	5
	Algo 1 & 2	5.3	4.6	5.5	5.5	5.2	3.7	5.4	4.9	5.2	3.4	4.9
	Algo 1 & 2 & 3	6.2	3.1	7	6.3	5	3.9	5.1	7.1	6.7	4.1	5.5
Obj. 6	AVASAD	49	53.7	46.8	46	54.2	54.3	43.7	44.7	40.6	42.5	47.6
	Algo 2	14.2	20.2	12.4	13.5	18	18.6	12.9	14.7	11.7	16.2	15.2
	Algo 1 & 2	15.2	19.5	12.5	12.9	19.1	17.2	14.4	17.1	12.8	14.4	15.5
	Algo 1 & 2 & 3	14.5	21.9	16.4	18.4	22.5	22.6	16	15.9	14.4	17.1	18

last column). Their characteristics (number of: interventions, type-A interventions, patients, and caregivers) are first displayed. Next, for each objective function, the results of different combinations of algorithms are compared to the ones of AVASAD. "Algo 1 & 2" allows to measure the benefit of Algorithm 1 (i.e., the advantage of having a dedicated procedure for the *usual* caregivers) when it is combined to Algorithm 2 (which works intervention by intervention instead of caregiver by caregiver).

"Algo 1 & 2 & 3" allows to evaluate the benefit of Algorithm 3 (i.e., the advantage of having a dedicated procedure to insert an intervention set in the schedule of a selected *usual* caregiver) when it is combined to Algorithms 1 and 2. We provide two scores for Obj. 1: "Top" is the number of interventions assigned to the top-ranked caregivers (i.e., the caregivers who are first in the C_i lists); "Sum" is the total number of interventions assigned to *usual* caregivers. We also give two scores for Obj. 3: "A" is the number of scheduled type-A interventions, whereas "3" is simply the number of scheduled interventions (i.e., types A and B).

The following observations can be made. First, given the lexicographic nature of the optimization problem, it can happen that if a method has a better score for a higher-level objective, it can have a worse score for a lower-level objective. Indeed, improvements of higher-level objectives can result in deteriorations of lower-level objectives. Second, both Algorithms 1 and 3 are beneficial when sequentially added to Algorithm 2. Indeed, the two former algorithms are directly linked to Obj. 1 (involving *usual* caregivers). Third, our solutions are much better than the ones of AVASAD for all the objectives, except for Obj. 4 and Obj. 5 which are related to travel and idle times, respectively. It is however normal that more travel time is necessary, on average, if more interventions are scheduled. Last but not least, AVASAD plans to employ the proposed algorithms to schedule its interventions in the near future.

5 Conclusion

In this paper, we propose a solution method combining three insertion algorithms for planning homecare interventions. Different constraint types and six objective functions (ranked lexicographically) are considered and make the problem unique. The experiments are performed for real instances, and the results show the superiority of our solutions when compared to the ones obtained by the involved homecare institute located in Switzerland. Various avenues of research can be envisioned, like the comparison with state-of-the-art approaches for related problems, the consideration of dynamic travel times [14], the possibility to account for overtime, or the design of a way to favor well-balanced workloads for the caregivers.

References

1. Bissoli, D.C., Zufferey, N., Amaral, A.: Lexicographic optimization-based clustering search metaheuristic for the multiobjective flexible job shop scheduling problem. Int. Trans. Oper. Res. **28**(5), 2733–2758 (2021)
2. Caceres-Cruz, J., Arias, P., Guimarans, D., Riera, D., Juan, A.: Rich vehicle routing problem: survey. ACM Comput. Surv. **47**(2), 1–28 (2015)
3. Cappanera, P., Scutella, M.G.: Joint assignment, scheduling, and routing models to home care optimization: a pattern-based approach. Transp. Sci. **49**(4), 830–852 (2015)
4. Euchi, J., Masmoudi, M., Siarry, P.: Home health care routing and scheduling problems: a literature review. 4OR **20**, 351–389 (2022)

5. Fikar, C., Hirsch, P.: Home health care routing and scheduling: a review. Comput. & Oper. Res. **77**, 86–95 (2017)
6. Grenouilleau, F., Lahrichi, N., Rousseau, L.-M.: New decomposition methods for home care scheduling with predefined visits. Comput. & Oper. Res. **115**, 104855 (2020)
7. Hertz, A., Schindl, D., Zufferey, N.: Lower bounding and tabu search procedures for the frequency assignment problem with polarization constraints. 4OR **3**(2), 139–161 (2005)
8. Jiang, H., Gomes, P., Vander Meer, D.: Promoting continuity of care in nurse-patient assignment: a multiple objective heuristic algorithm. Decis. Support. Syst. **167**, 113926 (2023)
9. Labadie, N., Prins, C., Prodhon, C. (eds.): Metaheurisitics for Vehicle Routing Problems. Wiley, New York (2016)
10. Lanzarone, E., Matta, A.: Robust nurse-to-patient assignment in home care services to minimize overtimes under continuity of care. Oper. Res. Health Care **3**(2), 48–58 (2014)
11. Liu, M., Zhao, Y., Xie, X.: Continuity-skill-restricted scheduling and routing problem: formulation, optimization and implications. IISE Trans. **56**(2), 201–220 (2024)
12. Naderi, B., Begen, M.A., Zaric, G.S., Roshanaei, V.: A novel and efficient exact technique for integrated staffing, assignment, routing, and scheduling of home care services under uncertainty. Omega **116**, 102805 (2023)
13. Respen, J., Zufferey, N., Wieser, P.: Three-level inventory deployment for a luxury watch company facing various perturbations. J. Oper. Res. Soc. **68**(10), 1195–1210 (2017)
14. Respen, J., Zufferey, N., Potvin, J.-Y.: Impact of vehicle tracking on a routing problem with dynamic travel times. RAIRO Oper. Res. **53**(2), 401–4140 (2019)
15. Vieira, B., de Armas, J., Ramalhinho, H.: Optimizing an integrated home care problem: a heuristic-based decision-support system. Eng. Appl. Artif. Intell. **114**, 105062 (2022). https://doi.org/10.1016/j.engappai.2022.105062
16. Zufferey, N., Vasquez, M.: A generalized consistent neighborhood search for satellite range scheduling problems. RAIRO Oper. Res. **49**(1), 99–121 (2015)
17. Zufferey, N.: Heuristiques pour les Problèmes de la Coloration des Sommets d'un Graphe et d'Affectation de Fréquences avec Polarités. Ph.D. thesis, École Polytechnique Fédérale de Lausanne (EPFL), Switzerland (2002)
18. Zufferey, N.: Tabu search approaches for two car sequencing problems with smoothing constraints. In: Metaheuristics for Production Systems, pp. 167–190. Springer, Cham, (2016)

Inventory Policy for Perishable Products with Diminishing Freshness

Matan Shnaiderman, Yael Lahav, Elad Harison, and Jacob Kobi Harush

Abstract The main characteristics of perishable products is a predetermined expiration date, after which the economic value of the products diminishes and they cannot be used post expiration. Two types of perishable products exist: First, products that can be consumed prior to the expiration date, but cannot be used after it due to use hazards (e.g., medical products). Second, products whose degree of freshness deteriorates over time until they cannot be used after their shelf-life period (e.g., agricultural products). These two types of products may be perceived as a single type, given the limited shelf-life duration, which may affect their salability. The study presents a model in which a retailer addresses demand for a perishable product that is divisible into multiplicative components according to the freshness level of the product and its selling price. The study also analyzes how quantity optimization vs. profit maximization decisions determine the order quantity. Additionally, the model enables retailers to determine the level of inventory that maximizes their profits.

Keywords Global optimization · Logistics · Supply chain management

1 Introduction and Literature Review

Price plays a major role in influencing the demand rate. Profit-seeking retailers aim to devise pricing strategies that are contingent upon factors such as the demanded quantity and price elasticity in their pursuit of profit maximization. Enterprises can set themselves apart from their competitors through their pricing strategies. Furthermore, pricing strategy plays a significant role in marketing and operations, as presented in [1, 2] and [3]. Due to the wide and diverse selection of products nowadays, there is a distinct difference in consumers' preferences, behaviors, and perceptions

M. Shnaiderman (✉) · Y. Lahav · J. K. Harush
Department of Management, Bar-Ilan University, Ramat-Gan, Israel
e-mail: Matan.Shniderman@biu.ac.il

Y. Lahav · E. Harison
School of Industrial Engineering and Management, Shenkar College of Design and Engineering, Ramat-Gan, Israel

M. Di Francesco et al. (eds.), *Operations Research: Closing the Gap Between Research and Practice*, AIRO Springer Series 15, https://doi.org/10.1007/978-3-031-90095-2_9

regarding products and product pricing. This variation is due to the quality of the product, its shelf life duration, etc. Therefore, the ability of retailers to set prices is a major factor affecting the quantity of sales and decisions regarding dynamic pricing require careful consideration which [4–6] and [7] focused at. Relatedly, [8] examined a dynamic pricing model for organic chemicals based on the credibility theory suggested by [9]. Dynamic pricing is challenging regarding perishable products, such as groceries with a limited shelf life, seasonal apparel, and pharmaceuticals, as consumers prefer fresh products over products that are about to expire. Furthermore, the price differentiation policy may affect consumers buying these products over time. This emphasizes the significance of reference prices, whereby consumers decide on a purchase according to their perception of past prices compared to current prices, and consequently, the demand is sensitive to reference prices. An address to this pricing challenge is provided by [10–16] and [17]. Herbon [18] extended the line of research that presented by [19] by examining a pricing policy problem that combined perishability, dynamic pricing, and retailer uncertainty about consumer sensitivity to product freshness. The developed model included periodic replenishment of inventory, while the decision variable was the timing of price changes and involved several assumptions: a single retailer, deterministic demand for perishable products affected by pricing, and consumer preferences for fresher products over time. Li [20] developed a dynamic pricing and lot-sizing deterministic model related to determining optimal pricing and selling all available inventory while maximizing the total profit over an indeterminate period of time. Thus, they formulated a model based on a multivariate demand function for a perishable product. Our study is based on the representation of linearly decreasing demands [20] and freshness levels [21]. However, the model presented in this paper differs from previous studies by providing additional insights that identify both optimal quantity and maximal profits as interdependent objectives to be considered by economic decision-makers. Our model aims to assist retailers in determining the optimal order quantity for perishable products, where the expiration date is known and expressed as a linearly decreasing level of freshness. The primary objective of this study is to provide retailers with the means to maximize their profits from the sale of perishable products, where the freshness level shapes consumer demand and pricing dynamics. By utilizing this model, retailers can make informed decisions about order quantities, ultimately enhancing their profitability while ensuring efficient management of perishable inventory with known expiration dates.

Section 2 presents the model. Section 3 provides numerical examples. Section 4 discusses the results of the model and provides conclusions and venues for future research.

2 The Model

Following [20], freshness decreases with time t. During a selling season of T periods, a retailer sells one perishable product. The freshness level of the product at each period $1 \leq t \leq T+1$, denoted by f_t, is equal to:

$$f_t = 1 - \frac{t-1}{T}. \tag{1}$$

In (1), it is obtained that the freshness is decreasing in t, and we have $f_1 = 1$ as well as $f_{T+1} = 0$. At the beginning of the season, the retailer orders Q units (the decision variable), and the unit order cost is c. The selling price at period t is p_t, which is affected by the freshness level (as proposed by the model of [21]). It is assumed that the price is proportional to f_t, namely,

$$p_t = P \cdot f_t, \tag{2}$$

The unit order cost is c. Parameter P denotes the initial (ideal) price when the product is completely fresh, and it is assumed that $P > c$. The demand for the product at period t, denoted by D_t, is affected by both the freshness level and the selling price. In particular, the demand is linearly decreasing in p_t [20] and is proportional to f_t [2], that is:

$$D_t = (a - bp_t)f_t, \tag{3}$$

Parameters a and b denote the ideal demand (while the selling price is zero) and the consumers' sensitivity to price growth respectively. It is assumed that $P \leq a/b$, so that the demand is never negative. The demands' orientation is not unambiguous from Eq. (3). As t grows, the freshness, f_t, decreases and, on the one hand, is expected to reduce the demand. However, the price, p_t, is also decreasing. Therefore, on the other hand, the demand may grow. It is possible to show that the sequence of demands is concave. It is increasing while $1 \leq t \leq T\lfloor -\frac{aT-bP}{2bP} \rfloor$ and decreasing while $\lfloor T - \frac{aT-bP}{2bP} \rfloor + 1 \leq t \leq T$.

At the beginning of the season, the retailer orders Q units. Let $0 \leq Q \leq \sum_{t=1}^{T} D_t$. There exists a unique value $1 \leq t^0(Q) \leq T$, such that $\sum_{t=1}^{t^0(Q)-1} D_t < Q \leq \sum_{t=1}^{t^0(Q)} D_t$. The retailer's goal is to maximize its total profit, denoted by $\pi(Q)$, which is equal to

$$\pi(Q) = \sum_{t=1}^{t^0(Q)-1} D_t p_t + \left(Q - \sum_{t=1}^{t^0(Q)-1} D_t \right) p_{t^0(Q)} - cQ,$$

or equivalently,

$$\pi(Q) = \left[p_{t^0(Q)} - c\right]Q + \sum_{t=1}^{t^0(Q)-1} D_t\left(p_t - p_{t^0(Q)}\right). \tag{4}$$

The following proposition is obtained.

Proposition 1 : The profit function (4) is piecewise linear, continuous and concave in $[0,\infty)$.

Proof: For $1 \le t \le T$, denote by I_t the interval $\left(\sum_{\tau=1}^{t-1} D_\tau, \sum_{\tau=1}^{t} D_\tau\right]$. Then $t^0(Q)$ is identically equal to t for every $Q \in I_t$. Also, set $t^0(Q) = T + 1$ for every $Q > \sum_{\tau=1}^{T} D_\tau$. Hence, $t^0(Q)$ is piecewise constant; namely, $\pi(Q)$ is piecewise linear. To demonstrate the continuity of $\pi(Q)$, it is now possible to show that the function is continuous at the point $Q_t = \sum_{\tau=1}^{t} D_\tau$, for every $1 \le t \le T$.

Finally, it is shown that $\pi(Q)$ is concave. The slope of its graph in the interval I_t is $p_t - c$. Since the sequence $(p_t)_{t=1}^{T}$ is decreasing (see (2)), then the sequence of slopes $(p_t - c)_{t=1}^{T}$ decreases as well. □

The model's problem is formulated as follows:

$$\max_{Q \ge 0} \pi(Q) = \left[p_{t^0(Q)} - c\right]Q + \sum_{t=1}^{t^0(Q)-1} D_t\left(p_t - p_{t^0(Q)}\right)$$

subject to (1)–(3).

The objective function of the problem is nondecreasing, while $p_{t^0(Q)} - c \ge 0$, and decreasing otherwise. Consider the set $\{1 \le t \le T | p_t \ge c\}$. According to the model's assumption that $P > c$ (i.e., $p_1 > c$), this set is nonempty and finite. Once $t^* = \max\{1 \le t \le T | p_t \ge c\}$ is defined, then under (1) and (2),

$$t^* = (1 - c/P)T + 1. \tag{5}$$

The optimal order quantity, denoted by Q^*, and the maximal profit $\pi(Q^*)$ can now be calculated according to Corollary 1.

Corollary 1 : The optimal order quantity is

$$Q^* = \sum_{t=1}^{t^*} D_t, \tag{6}$$

which is necessarily positive. If $p_{t^*} > c$, then the optimal solution is unique. If, on the other hand, $p_{t^*} = c$, then every $Q \in \left[\sum_{t=1}^{t^*-1} D_t, \sum_{t=1}^{t^*} D_t\right]$ is optimal. The maximal profit $\pi(Q^*)$, denoted by π^*, is

$$\pi^* = \sum_{t=1}^{t^*} (p_t - c)D_t. \tag{7}$$

According to (5)–(7), as T increases, as P increases up to a/b and as c decreases, t^*, Q^* and π^* increase accordingly.

Proposition 2 examines what happens if the total number of periods, T, is high and tends to infinity.

Proposition 2 : The following properties hold:

$$\lim_{T\to\infty}\frac{t^*}{T} = 1-\frac{c}{P}, \tag{8}$$

$$\lim_{T\to\infty}\frac{1}{T}\sum_{t=1}^{T} D_t = \frac{1}{6}(3a-2bP), \tag{9}$$

$$\lim_{T\to\infty}\frac{Q^*(T)}{T} = \frac{1}{6P^2}\left[(3a-2bc)(P+c)-2bP^2\right](P-c) \tag{10}$$

and

$$\lim_{T\to\infty}\frac{\pi^*}{T} = \frac{1}{12P^2}\left[(2a-bc)(2P+c)-3bP^2\right](P-c)^2. \tag{11}$$

Proof: For every T, there is $\frac{t^*}{T} = \frac{(1-\frac{c}{P})T+1}{T}$ (see (5)), and

$$1-\frac{c}{P} = \frac{((1-\frac{c}{P})T-1)+1}{T} \le \frac{(1-\frac{c}{P})T+1}{T} \le \frac{(1-\frac{c}{P})T+1}{T}$$
$$= \left(1-\frac{c}{P}\right)+\frac{1}{T}.$$

From the sandwich theorem, (8) is now obtained.

Given T, the average demand per period satisfies

$$\frac{1}{T}\sum_{t=1}^{T} D_t = \sum_{t=1}^{T}\frac{1}{T}\left[a-bP\left(1-\frac{t-1}{T}\right)\right]\left(1-\frac{t-1}{T}\right). \tag{12}$$

Consider the function $g(x) = [a-bP(1-x)](1-x)$ in the interval [0,1]. If this interval is participated according to the following partition $\left(0, \frac{1}{T}, \frac{2}{T}, \ldots, \frac{T-1}{T}, 1\right)$, and in each subinterval $\left[\frac{t-1}{T}, \frac{t}{T}\right]$, the left point $(t-1)/T$ is chosen, then the corresponding Riemann sum, $\sum_{t=1}^{T}\frac{1}{T}g\left(\frac{t-1}{T}\right)$, is equal to the RHS of (12). As T tends to infinity, the Riemann sum converges to the integral $\int_0^1 g(x)dx$, which is equal to the RHS of (9).

The proofs of (10) and (11) are analogous. □

From Proposition 2, it is found that if the total number of periods is high, then the number of periods for which it is profitable to order, as well as the total demand, optimal inventory level and maximal profit, can be approximately calculated according to Corollary 2.

Corollary 2 : Assume that the total number of periods, T, is high, then the following approximations hold:

$$t^* \approx t^{\sim} := \left(1 - \frac{c}{P}\right)T,$$

$$\sum_{t=1}^{T} D_t \approx \frac{1}{6}(3a - 2bP)T,$$

$$Q^* \approx Q^{\sim} := \frac{1}{6P^2}\left[(3a - 2bc)(P + c) - 2bP^2\right](P - c)T \tag{13}$$

and

$$\pi^* \approx \pi^{\sim} := \frac{1}{12P^2}\left[(2a - bc)(2P + c) - 3bP^2\right](P - c)^2 T. \tag{14}$$

Next, Propositions 3–4 consider how the approximated optimal order level and the approximated maximal profit are affected by the initial price, P, while the total number of periods is high (the proof of Proposition 4 is omitted).

Proposition 3 : Assume that T is high, then the approximated optimal order quantity, $Q^{\sim}$, is concave in P in the interval $[c, a/b]$. There exists a value $c < P_0 < a/b$ in which that quantity is maximized as follows

$$P_0 = \frac{1}{b}\left[(3a - 2bc)b^2c^2\right]^{1/3}. \tag{15}$$

Proof: Differentiating (13) twice with respect to P, results in

$$\frac{dQ^{\sim}}{dP} = \frac{1}{3P^3}\left[(3a - 2bc)c^2 - bP^3\right]T \tag{16}$$

as well as

$$\frac{d^2Q^{\sim}}{dP^2} = -\frac{1}{P^4}(3a - 2bc)c^2T. \tag{17}$$

As $a \geq bP \geq bc > (2/3)bc$, then (17) is lower than 0, and the concavity of $Q^{\sim}$ is received.

The derivative (16) vanishes at point (15), which necessarily lies between c and a/b while $a \geq bc$. □

Proposition 4 : Assume that T is high, then the approximated maximal profit $\pi^{\sim}$ is quasi-concave in P in the interval $[c, a/b]$. The profit is maximized at point $P_0 < P^*$

$< a/b$, which is implicitly defined by the equation

$$3b(P^*)^3 - (2a - bc)(P^*)^2 - (2a - bc)cP^* - (2a - bc)c^2 = 0. \tag{18}$$

Notably, in (15) and (18), neither P_0 nor P^* depends on the total number of periods, T. Also, it is possible to distinguish a unique dynamic between $Q^{\sim}$ and $\pi^{\sim}$. Despite the maximal quantity results from P_0, the profit obtained will be lower than when a quantity of items is sold that is less than the maximal quantity but the items are sold at a price that is P^*, which is higher than P_0. That is, the profit maximization is reflected when the quantity is not maximal.

This conclusion represents a market dynamic that characterizes a unique strategy for luxury products. Since the market share, i.e., the target audience, for the purchase of this type of product is relatively small at the starting point, the retailer selling items under this category must order a relatively small amount of goods and price it accordingly. This allows the seller to meet the demand, maximize profits according to the existing situation, and reduce the risk of being left with excess inventory.

A clear example of the described case can be attributed to the field of fashion, when clothes produced by luxury brands are considered attractive for the relevant target audience for one season only. Such a scenario is characterized by the rarity effect. Accordingly, due to the rarity of a particular product, expressed in the presented case as a small quantity ordered of a certain item, it is possible to set a high selling price, resulting in higher profits, as shown in Proposition 4.

3 Numerical Examples and Sensitivity Analysis

In what follows, the numerical examples are shown using Maple software to illustrate the exact model of Eqs. (6) and (7) and the approximated model of Eqs. (13) and (14).

Let $T = 7$, $a = 100$, $b = 5$, and $c = 8$. Figure 1 illustrates the concavity of the sequence of demands during the selling season for 3 different values of the initial selling price, P. While $P = 9$, we have $\lfloor T - \frac{aT-bP}{2bP} \rfloor + 1 = 0$, that is, the sequence of demands decreases in t. When $P = 14$, the demand increases and then decreases concavely when the selling price passes that certain threshold value. The maximal demand is obtained in period $t = \lfloor T - \frac{aT-bP}{2bP} \rfloor + 1 = 3$. Similarly, if P grows to 18, the maximal demand is D_4, as $\lfloor T - \frac{aT-bP}{2bP} \rfloor + 1$ becomes 4.

Figure 2 presents the results of Proposition 1, showing the profit as piecewise linear, continuous and concave. Furthermore, Fig. 2 shows the optimal quantity to order in each sales price scenario and the resulting profit. If $P = 9$, then $t^* = \lfloor (1 - \frac{c}{P})T \rfloor + 1 = 1$, that is, the optimal order quantity is $Q^* = D_1 = 55$ (see (6)). When the initial selling price, P, is equal to 14, then $t^* = 4$ and $p_4 = 8 = c$. Therefore, every value in the interval $\left[\sum_{t=1}^{3} D_t, \sum_{t=1}^{4} D_t\right] = [100, 134.3]$ is

Fig. 1 The demand resulting from the number of periods until expiration at different selling price levels

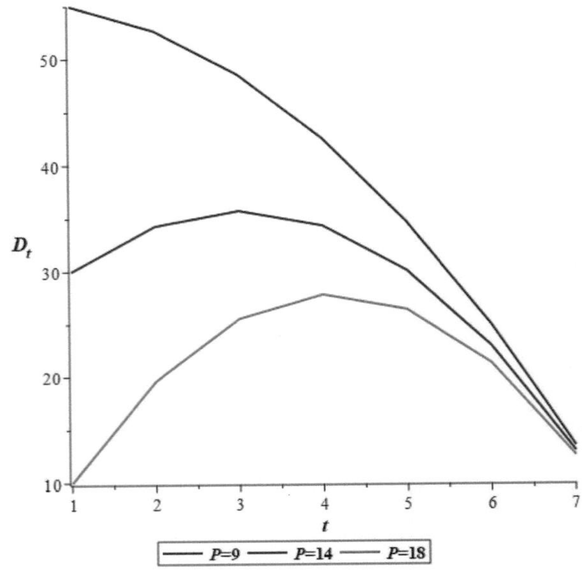

optimal, as Corollary 1 states. If P grows to 18, it is optimal to exactly satisfy the demands of the first 4 periods, namely, $Q^* = \sum_{t=1}^{4} D_t = 82.9$.

Next, the optimal order quantity and maximal profit are considered regarding how they are affected by the order cost c, in different selling price scenarios:

Fig. 2 The profit resulting from the quantity ordered at different selling price levels

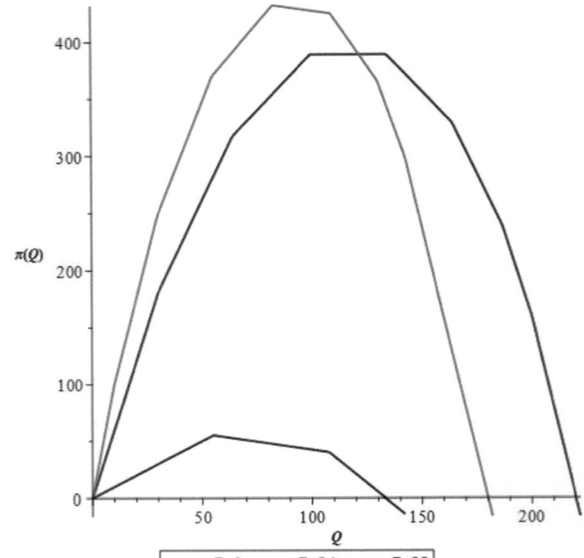

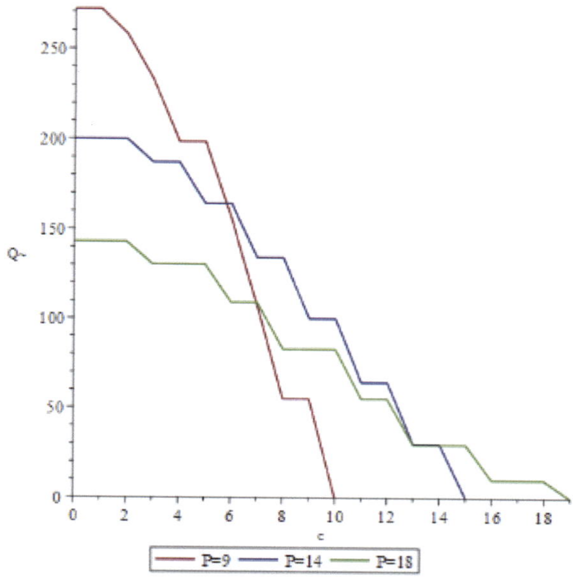

Fig. 3 The optimal quantity resulting from different order costs in different values of selling price

From Figs. 3 and 4, it can be concluded that, as the cost of order increases, thereby increasing the ratio between the cost of order and the selling price (c/p), both the optimal quantity and the maximal profit decrease. In these two analyses, an interesting situation can be seen where the graphs representing different levels of selling prices cross each other. The meaning of this crossing refers to an indifference between these price levels regarding the specific order cost. According to this occurrence, at the point of a specific order cost, a sales price can be determined that is reflected in this scenario, since in both cases, the optimal quantity (according to Fig. 3) and the maximal profit (according to Fig. 4) will be the same.

Finally, the approximated optimal quantity and profit changes resulting from the sale prices changing over a range between c (i.e. 8) and a/b (i.e. 20) for different high values of T are examined. Figures 5 and 6 verify Propositions 3 and 4 respectively, meaning that $Q^{\sim}$ is concave and $\pi^{\sim}$ is quasi-concave in P in the interval [14, 19].

The maximal optimal order quantities and the maximal profits are obtained while P is equal to $P_0 = 14.1$ and to $P^* = 17.7$ respectively, for every value of T. When $T = 20$, the maximal optimal order quantity $Q^{\sim}$ is equal to 293.9. The profit $\pi^{\sim}$ in this case is 879.3. If P increases to P^*, then the maximal profit grows to 1123.6, while the optimal order quantity is lower and equal to 260.6. When the total number of periods (T) increases to 50, the maximal optimal order quantity is 734.8, and the profit in this case is 2198.2. Once P grows to P^*, the order quantity is reduced to 651.4, but the profit increases to 2808.9. Similarly, if $T = 100$, then under P_0 we have $Q^{\sim} = 1469.7$ and $\pi^{\sim} = 4396.5$, while under P^* we obtain $Q^{\sim} = 1302.8$ as well as $\pi^{\sim} = 5617.8$. To summarize, as P increases from P_0 to P^*, the optimal order quantities are reduced in 11.3%, but the maximal profits grow in 27.8%.

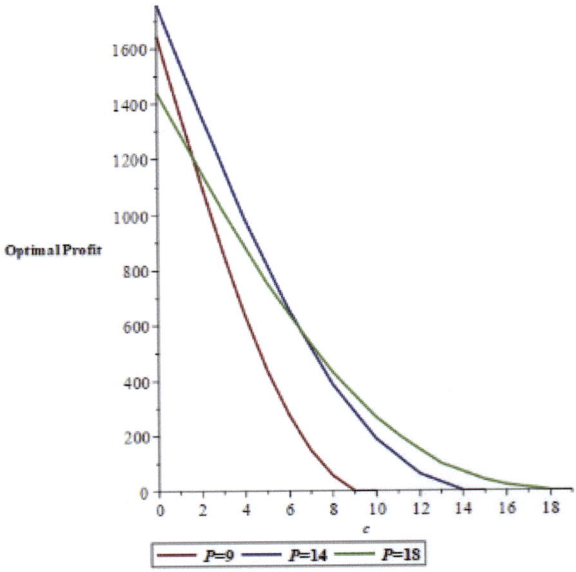

Fig. 4 The maximal profit resulting from different order costs in different values of selling price

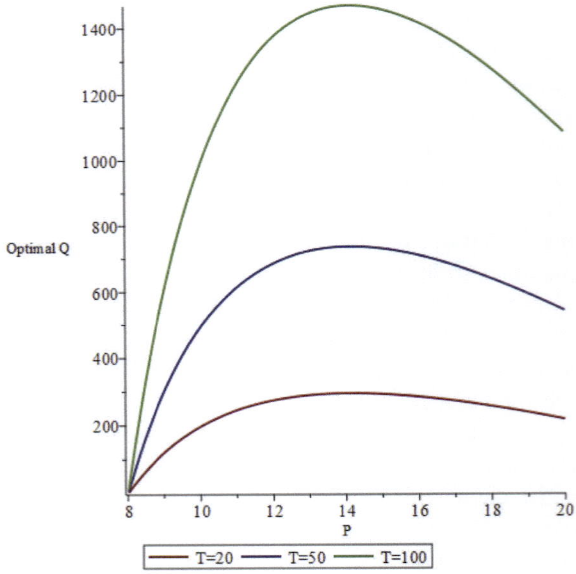

Fig. 5 The optimal quantity resulting from different selling prices in different values of T

Fig. 6 The maximal profit resulting from different selling prices in different values of T

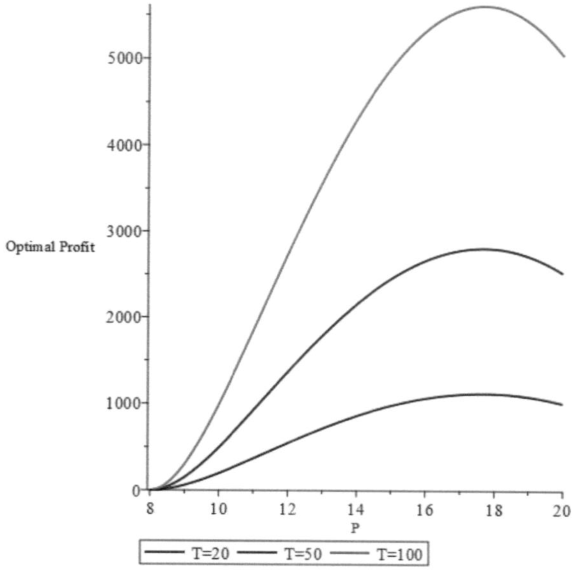

4 Discussion and Concluding Remarks

By using analytical means, this study proves that demand is concave for perishable products over time and depends on product freshness (see Proposition 1) and the selling price. A major finding from the proposed model suggests that the retailer must determine a decision either by profit maximization or sales maximization.

This study opens new venues for research that explores the rarity effect on luxury goods: For example, examining how a retailer can maximize profits while considering freshness reduction with a known expiration date, which may affect demand and selling price. This phenomenon characterizes the market of perishable products. Furthermore, the profit equation is piecewise linear, continuous and concave. Following these findings, the optimal quantity leads to maximal profits. Advanced versions of the model can be developed to include other aspects such as storage costs and safety stocks, among others.

In the dynamic analysis, the selling quantity must be smaller than the demand at the optimal price to maximize profits. This finding is in accordance with the fact that the maximal profit, π^*, is quasi-concave in P. If the optimal quantity is equal to the demand level, when the optimal order quantity Q^* is concave in P, the price and profits are lower.

Based on the findings made within this study, a key managerial insight formulated to help retailers. According to this insight, the retailer should adopt the appropriate selling strategy for the perishable products he sells. From the research findings, two strategies are proposed: (1) selling an optimal quantity or (2) maximizing profits.

For example—retailers in the food market would prefer to sell a larger quantity while compromising on the selling price, in order to avoid a scenario in which they will be left with expired food inventory due to a short period of time for expiration according to the nature of this industry.

On the other hand, in the fashion market, especially luxury fashion, the drop in freshness is seasonal and therefore the retailer can adopt a strategy designed to maximize profits and includes keeping a limited inventory of the products and selling them at a more expensive selling price.

Future research should examine the effect of the exponentially decreasing freshness compared to its linear decrease, as presented in this study, as well as other forms of expression of the freshness dynamics. In addition, a memory factor, such as a reference price, could be used when examining demand behavior to analyze the impact of price changes on demand, sales and profits.

Glossary

$\boldsymbol{T}$ The total number of periods during a selling season
$\boldsymbol{T}$ The relative time unit from the total number of periods T
$\boldsymbol{D_t}$ The demand for the product at period t
$\boldsymbol{f_t}$ The freshness level of the product at period t
$\boldsymbol{p_t}$ The selling price at period t
$\boldsymbol{P}$ The initial selling price when the product is completely fresh
$\boldsymbol{P_0}$ The initial selling price that maximizes the optimal order quantity
$\boldsymbol{P^*}$ The optimal initial selling price that maximizes the retailer's profits
$\boldsymbol{C}$ The unit order cost
$\boldsymbol{t^0(Q)}$ The period during which the inventory has run out
$\boldsymbol{t^*}$ The maximal period during which the selling price is greater than or equal to the order cost
$\boldsymbol{Q}$ The number of units ordered at the beginning of the season
$\boldsymbol{Q^*}$ The exact optimal number of units ordered at the beginning of the season
$\boldsymbol{\pi(Q)}$ The total profit resulting from the total quantity Q ordered
$\boldsymbol{\pi^*}$ The exact total maximal profit resulting from the total quantity Q ordered
$\boldsymbol{Q^{\sim}}$ The approximated optimal number of units ordered at the beginning of the season
$\boldsymbol{\pi^{\sim}}$ The approximated total maximal profit resulting from the total quantity Q ordered

References

1. Chenavaz, R.Y.: Modeling the impact of product quality on dynamic pricing and advertising policies. Eur. J. Oper. Res. **284**(3), 990–1001 (2020)
2. Subrahmanyam, S.: Premium pricing strategy and customer retention–an analysis. Int. J. Early Child. **14**(03) (2022)

3. Lu, Z.J.: Strategic pricing of new pharmaceuticals. Rev. Econ. Stat. **80**(1), 108–118 (1998)
4. Luo, C.: Dynamic pricing and energy management strategy for EV charging stations under uncertainties (2018). arXiv:1801.02783
5. Priester, A.: A special price just for you: effects of personalized dynamic pricing on consumer fairness perceptions. J. Revenue Pricing Manag. **19**(2), 99–112 (2020)
6. Victor, V.: Factors influencing consumer behavior and prospective purchase decisions in a dynamic pricing environment—an exploratory factor analysis approach. Soc. Sci. **7**(9), 153 (2018)
7. Hyginus, O.O.: Pricing strategy as a factor for sales performance of consumable goods: evidence from consumable goods dealers in Wukari local government area, Taraba State, Nigeria. Noble Int. J. Bus. Manag. Res. **3**(3), 48–61 (2019)
8. Xiong, Y.: Dynamic pricing model and algorithm for perishable products with fuzzy demand. Appl. Stoch. Model. Bus. Ind. **26**(6), 758–774 (2010)
9. Liu, B.: Uncertainty theory: an introduction to its axiomatic foundation. Physica-Verlag, Heidelberg (2004)
10. Abrate, G.: The impact of dynamic price variability on revenue maximization. Tour. Manag. **74**, 224–233 (2019)
11. Avinadav, T.: Optimal inventory policy for a perishable item with demand function sensitive to price and time. Int. J. Prod. Econ. **144**(2), 497–506 (2013)
12. Chen, N.: Perishable good dynamic pricing under competition: an empirical study in the airline markets (2018). SSRN 3228392
13. Dye, C.Y.: Optimal joint dynamic pricing, advertising and inventory control model for perishable items with psychic stock effect. Eur. J. Oper. Res. **283**(2), 576–587 (2020)
14. Hsieh, T.P.: Optimal dynamic pricing for deteriorating items with reference price effects when inventories stimulate demand. Eur. J. Oper. Res. **262**(1), 136–150 (2017)
15. Xue, M.: Optimal dynamic pricing for deteriorating items with reference-price effects. Int. J. Syst. Sci. **47**(9), 2022–2031 (2016)
16. Zhang, J.: Durable goods pricing with reference price effects. Omega **91**, 102018 (2020)
17. Chernonog, T.: Pricing and advertising in a supply chain of perishable products under asymmetric information. Int. J. Prod. Econ. **209**, 249–264 (2019)
18. Herbon, A.: Dynamic pricing vs. acquiring information on consumers' heterogeneous sensitivity to product freshness. Int. J. Prod. Res. **52**(3), 918–933 (2014)
19. You, P.S.: Ordering and pricing of service products in an advance sales system with price-dependent demand. Eur. J. Oper. Res. **170**(1), 57–71 (2006)
20. Li, R.: Pricing and lot-sizing decisions for perishable goods when demand depends on selling price, reference price, product freshness, and displayed stocks. Eur. J. Oper. Res. **270**(3), 1099–1108 (2018)
21. Sebatjane, M.: A three-echelon supply chain for economic growing quantity model with price-and freshness-dependent demand: pricing, ordering and shipment decisions. Oper. Res. Perspect. **7**, 100153 (2020)

Unassigned Distance Geometry and the Buckminsterfullerene

Leo Liberti

Abstract The Buckminsterfullerene is an inorganic molecule consisting of 60 carbon atoms in the shape of a soccer ball. It was used in [15] to showcase algorithms that find the correct shape of a protein from limited data (length of inter-atomic distances) without any further chemical experiment: in that case, by means of a complicated constructive heuristic based on genetic algorithms. In this paper we show that we can reconstruct the Buckminsterfullerene structure by means of mathematical programming, standard solver software, and little else.

Keywords Mathematical programming · Molecule · Diagonal dominance · Matrix cones

1 Introduction

The authors of [15] state that "*ab initio* structure solution of nanostructured materials [that cannot be solved using crystallographic methods] is feasible using diffraction data in combination with distance geometry methods". They then proceed to find the structure of Buckminsterfullerene (a highly symmetric molecule consisting of 60 carbon atoms [25]) using a heuristic method, called LIGA, that employs distance values in order to construct small rigid polytopes which are then combined into larger structures. The combination is achieved by means of a genetic algorithm approach that involve tournaments for selection of the fittest. Some backtracking also occurs.

The input to LIGA (number of atoms and list of all inter-atomic distance values) is obtained from powder diffraction data: we note that the graph underlying the distance values is not a part of the input. The experiment in [15, Fig. 2c–d] suggests that the

This paper was partly supported by the ANR PRCI project "MultiBioStruct".

L. Liberti (✉)
LIX CNRS Ecole Polytechnique, Institut Polytechnique de Paris, Palaiseau, France
e-mail: liberti@lix.polytechnique.fr

M. Di Francesco et al. (eds.), *Operations Research: Closing the Gap Between Research and Practice*, AIRO Springer Series 15, https://doi.org/10.1007/978-3-031-90095-2_10

algorithmic reconstruction process yields some errors; but that it suffices to increase the multiplicities of the distance values in the list to obtain the correct shape.

In this paper we show, using Mathematical Programming (MP) methods, that a correct structure of the Buckminsterfullerene can be obtained from the original distance list, without having to add multiple copies of each distance value, and without having to create a new algorithm. Most of the work is carried out on one of two MP formulations involved in the reconstruction problem, as explained below. The only algorithmic contribution is given by a "vanilla" MultiStart (MS) approach.

Our interest in the Buckminsterfullerene only extends so far as it constitutes a well-known and interesting instance of the Unassigned Distance Geometry Problem (UDGP), defined below. Many people have managed to reconstruct the shape of the Buckminsterfullerene: most of them have used heuristic optimization methods to find "good" minima of an energy formula derived from the partial differential equations that define the atomic force field [6, 10, 22]. We follow a different path, namely that of [15], which takes as input the values of the inter-atomic distances of the molecule.

1.1 Two Problems in Distance Geometry

We now introduce the Distance Geometry Problem (DGP) and its "unassigned" variant, the UDGP. The input of the DGP is an integer K and a simple edge-weighted undirected graph $G = (V, E, d)$ where $d : E \to \mathbb{R}_+$. The DGP asks whether there exists a *realization* $x : V \to \mathbb{R}^K$ of G in $\mathbb{R}^K$ satisfying

$$\forall \{i, j\} \in E \quad \|x_i - x_j\|_2^2 = d_{ij}^2. \tag{1}$$

In other words, the DGP asks us to draw the graph G in K dimensions such that edges are drawn as segments having length equal to the weight [19]. The DGP is **NP**-hard [24], it is in **NP** if $K = 1$ but it is unknown whether it is in **NP** for $K > 1$ [3]. It has applications in designing time synchronization protocols in wireless networks, in the localization of mobile sensors, in the determination of protein structure from Nuclear Magnetic Resonance (NMR) experiments, and in many other fields. See [9, 20] for more information about the DGP.

The UDGP removes the adjacency information from the graph edges: its input is a pair of integers K, n and a list δ of m positive scalars. It asks whether there exists an assignment α mapping $[m] = \{1, \ldots, m\}$ to the set $\bar{E} = \{\{i, j\} \mid i, j \in [n]\}$, and a realization $x : [n] \to \mathbb{R}^K$ such that

$$\forall \ell \le m \text{ with } \alpha(\ell) = \{i, j\} \quad \|x_i - x_j\|_2^2 = \delta_\ell^2. \tag{2}$$

Although the UDGP is not as well known as the DGP, in many applications it is the problem that is actually closest to the experiments leading to the use of the DGP. For example, the determination of protein structures from distances involves the solution of a DGP, but the actual raw input from the NMR experiments is closer to a list of values δ than to a graph G. In fact, G is inferred from δ because the assignment α is computed first, together with a very imprecise realization [26], see Sect. 3.1 below.

The UDGP has applications in the study of DNA [8, 17], where it is known as the *turnpike* or *beltway* or *partial digest* problem. It is the natural modelling framework for the study of nanostructures [4] such as the Buckminsterfullerene. It also arises in the context of protein structure [26], but in this setting $|\delta|$ (the length of the sequence δ) is much smaller than the number of atom pairs $N = n(n-1)/2$, which yields very imprecise assignment functions α. Conversely, it is often the case that all (or almost all) pairwise distances can be measured by NMR interaction in molecules that are much smaller than proteins (e.g. the Buckminsterfullerene) [15]. In turn, this yields much more precise assignments. The UDGP is **NP**-hard even if $m = N$ [17].

Both DGP and UDGP can be defined for any K, and there exist many applications for $K \in \{1, 2, 3\}$, as well as some for larger values of K. In this paper we fix $K = 3$, since the Buckminsterfullerene is realized in 3D space.

1.2 Goal of This Paper

In this paper we solve the UDGP instance known as "Buckminsterfullerene" [16] by means of MP, a formal language for specifying and solving optimization problems. We recall an existing Mixed-Integer Nonlinear Programming (MINLP) formulation of the UDGP, and propose a new Mixed-Integer Quadratically Constrained Programming (MIQCP) formulation. These can only be solved for tiny inputs much smaller than the Buckminsterfullerene. We therefore decompose the decision on the assignment α and the realization x: we construct an approximate Mixed-Integer Linear Programming (MILP) formulation of the UDGP which can be solved to larger sizes. We obtain an approximate assignment α' and an approximate realization, which we discard. We then use α' to define an edge-weighted graph $G_{\alpha'}$, and obtain a more precise realization by solving the DGP on $G_{\alpha'}$. Finally, we show the perfectly reconstructed shape of the Buckminsterfullerene from a distance list δ of size N.

2 MINLP Formulations for the UDGP

The UDGP aims at finding an assignment $\alpha : [m] \to \bar{E}$ and a realization x consistent with each other. We note that realizations are defined as functions from V to $\mathbb{R}^K$ that satisfy the distance Eq. (1). Therefore x can be represented by an $n \times K$ matrix, where $n = |V|$: every row x_i of x is a vector in $\mathbb{R}^K$ for $i \in V$. We can encode the assignment α by means of binary decision variables $y_{ij\ell}$ such that $y_{ij\ell} = 1$ iff $\alpha(\ell) = \{i, j\}$. The assignment α satisfies the following constraints: each distance value δ_ℓ for $\ell \leq m$ is assigned to exactly one edge $\{i, j\}$, and each pair $\{i, j\}$ is assigned to at most one ℓ (some pairs may remain unassigned).

2.1 Exact MINLP Formulations

These considerations yield the following "natural" MP formulation for the UDGP, proposed in [11]:

$$\left.\begin{array}{lr} \min\limits_{x,y} & \sum\limits_{i<j\le n}\sum\limits_{\ell\le m} y_{ij\ell}(\|x_i - x_j\|_2^2 - \delta_\ell^2)^2 \\ \forall \ell\le m & \sum\limits_{i<j\le n} y_{ij\ell} = 1 \\ \forall i<j\le n & \sum\limits_{\ell\le m} y_{ij\ell} \le 1 \\ & x\in\mathbb{R}^{nK} \quad\wedge\quad y\in\{0,1\}^{Nm} \end{array}\right\} \tag{3}$$

Since Eq. (3) has nonlinear terms and integer variables, it is a MINLP formulation with a quintic polynomial as objective.

We propose a new MIQCP formulation. It decreases the polynomial degree of Eq. (3) from five to two.

$$\left.\begin{array}{ll} \min\limits_{x,y,z} & \sum\limits_{i<j\le n}\sum\limits_{\ell\le m} (z^+_{ij\ell} + z^-_{ij\ell}) \\ \forall i<j\le n, \ell\le m & -z^-_{ij\ell} - M(1-y_{ij\ell}) \le \|x_i - x_j\|_2^2 - \delta_\ell^2 \le z^+_{ij\ell} + M(1-y_{ij\ell}) \\ \forall \ell\le m & \sum\limits_{i<j\le n} y_{ij\ell} = 1 \\ \forall i<j\le n & \sum\limits_{\ell\le m} y_{ij\ell} \le 1 \\ & x\in\mathbb{R}^{nK} \wedge y\in\{0,1\}^{Nm} \wedge z\ge 0, \end{array}\right\} \tag{4}$$

where $M = (\sum_{\ell\le m}\delta_\ell)^2$ (see [5, Prop. 2.2]). "Sandwiching" the distance constraints allows constraint activity whenever $\alpha(\ell) = \{i, j\}$, and trivialize the constraint otherwise. The slack variables z^-, z^+ allow the solver to find an approximately feasible solution with imprecise distances.

2.2 An Approximate MILP Formulation

The only nonlinear part of Eq. (4) is the Euclidean distance term $\|x_i - x_j\|_2$ in the middle member of the distance constraints. We have:

$$\|x_i - x_j\|_2 = \|x_i\|_2 + \|x_j\|_2 - 2\langle x_i, x_j\rangle = \langle x_i, x_i\rangle + \langle x_j, x_j\rangle - \langle x_i, x_j\rangle.$$

Now we linearize the nonlinear (inner product) terms: we create an $n\times n$ variable matrix X, and we replace each term $\langle x_i, x_j\rangle$ by X_{ij} (which, by commutativity, yields a symmetric matrix X). For all $i < j$ and ℓ, we obtain the linearized distance constraint

$$-z^-_{ij\ell} - M(1 - y_{ij\ell}) \le X_{ii} + X_{jj} - 2X_{ij} - \delta_\ell^2 \le z^+_{ij\ell} + M(1 - y_{ij\ell}).$$

So far, this is the usual first step in the construction of the Semidefinite Programming (SDP) relaxation for MP formulations with quadratic terms. However, instead of imposing the positive semidefinite (PSD) constraint $X \succeq 0$, we restrict our attention to a polyhedral inner approximation thereof, that of *Diagonally Dominant* (DD) matrices. An $n \times n$ matrix is DD if it satisfies:

$$\forall i \leq n \quad X_{ii} \geq \sum_{j \neq i} |X_{ij}|. \tag{5}$$

By Gershgorin's Circle Theorem [13], every eigenvalue λ_i of a matrix A is contained in an interval $[A_{ii} - \sum_{j \neq i} |A_{ij}|\,,\ A_{ii} + \sum_{j \neq i} |A_{ij}|]$. By the definition of DD matrices (Eq. (5)), if X is DD, then the lower interval bound $X_{ii} - \sum_{j \neq i} |X_{ij}|$ is non-negative, which implies that every eigenvalue of X is non-negative, which means that X is PSD. We note that the converse need not hold: many PSD matrices are not DD.

The DD cone is polyhedral: its extreme rays are given by the rank-one matrices $e_i e_i^\top$ and $(e_i \pm e_j)(e_i \pm e_j)^\top$ [2]. We therefore proceed to linearize the absolute value term in Eq. (5) by means of an additional $n \times n$ matrix variable T, obtaining the linear reformulation [1, Thm. 3.9]:

$$\forall i \leq n \quad X_{ii} \geq \sum_{j \neq i} T_{ij} \qquad \wedge \qquad -T \leq X \leq T.$$

We remark that linear programming over the DD cone is also known as DD Programming (DDP).

We now put everything back together to yield the following MIDDP formulation:

$$\left.\begin{array}{lrl}
\min\limits_{X,T,y,z} & \sum\limits_{i<j\leq n} \sum\limits_{\ell \leq m} (z^+_{ij\ell} + z^-_{ij\ell}) & \\
\forall i < j \leq n, \ell \leq m & -z^-_{ij\ell} - M(1-y_{ij\ell}) \leq X_{ii} + X_{jj} - 2X_{ij} - d^2_\ell \leq z^+_{ij\ell} + M(1-y_{ij\ell}) & \\
\forall \ell \leq m & \sum\limits_{i<j\leq n} y_{ij\ell} = 1 & \\
\forall i < j \leq n & \sum\limits_{\ell \leq m} y_{ij\ell} \leq 1 & \\
\forall i \leq n & X_{ii} \geq \sum\limits_{j \leq n : j \neq i} T_{ij} & \\
 & -T \leq X \leq T & \\
 & X, T \in \mathbb{R}^{n \times n} \ \wedge\ y \in \{0,1\}^{Nm} \ \wedge\ z \geq 0. &
\end{array}\right\} \tag{6}$$

We note that Eq. (6) is a MILP, for which there exist technically viable solvers even for relatively large sizes [14]. We also remark that, although this reformulation may not immediately appear simple, it is based on a set of well-known building blocks.

Since not every PSD matrix is DD, a DDP is an inner approximation of the corresponding SDP, which means that, potentially, the DDP might be infeasible even if the SDP is not. This is not an issue for the MIDDP formulation in Eq. (6) since it is always feasible: it suffices to satisfy the DD constraints first, and then choose large enough slack variables z^+, z^- to accommodate the error.

In practice, solving Eq. (6) yields an $n \times n$ matrix X. There are two ways to obtain an $n \times K$ (approximate) realization of δ.

1. *Principal Component Analysis* (PCA). Let $X = P \Lambda P^\top$ be a spectral decomposition of X. The diagonal matrix Λ is non-negative because X is DD and therefore PSD, so that $\sqrt{\Lambda}$ is a real matrix. Let Λ^K be the matrix obtained by Λ by zeroing all of the diagonal entries aside from the K largest. Then $\hat{x} = P\sqrt{\Lambda^K}$ has the property that $\bar{X} = \hat{x}\hat{x}^\top = P\Lambda^K P^\top$ is a best rank-K approximation of X.
2. *Barvinok's "naive algorithm"* generalized to K dimensions [21]. Proceeding as for PCA, consider the factor $F = P\sqrt{\Lambda}$ of X, then sample (componentwise) an $n \times K$ matrix $Z \sim \mathsf{Normal}(0, 1/\sqrt{K})$. The $n \times K$ matrix $\check{x} = FZ$ has good proximity properties with the constraints $\|x_i - x_j\|_2 = \bar{y}_{ij\ell}\delta_\ell$ with arbitrarily high probability (here $\bar{y}$ are the assignment variables for α having value 1 after solving the MIDDP: they simply select the correct ℓ to match to the $\{i, j\}$).

Either of these two ways would allow us to retrieve an approximate realization x from the DD matrix X, but experiments in this yielded very wrong realizations.

3 A Matheuristic for the UDGP

Matheuristics are heuristic algorithms based on MP formulations [23]. Our matheuristic decomposes the decision on the assignment α and the realization x.

3.1 The Way of Proteins and Their Scale

This decomposition idea is not new: the DGP application to protein conformation derives the distance values from NMR experiments, which in fact yield an ambiguous assignment. This assignment is disambiguated in two phases. The first rests on chemical considerations. The second phase is based on a metaheuristic (usually simulated annealing) which returns an assignment together with a poor quality realization, which is discarded. This assignment, however, is used to define a weighted graph that becomes the input of a DGP. The solution of this DGP provides the realization of interest, and the shape of the protein.

There are two technical differences with the Buckminsterfullerene: powder diffraction data does not yield any assignment (not even an ambiguous one), and the first disambiguation phase does not apply because of the atomic uniformity of the Buckminsterfullerene: it consists entirely of carbon atoms.

The main difference w.r.t. methodologies for computing the shape of proteins and of inorganic molecules such as the Buckminsterfullerene, however, is one of scale. Proteins are much larger than the range of distances that can be measured using NMR, which is around 5.5Å. As a consequence, only very few distances can be obtained. The protein graphs that provide the DGP input by means of which we compute realizations are very sparse.

The Buckminsterfullerene—and similar molecules—are much smaller than proteins. As a consequence, we can often obtain lists δ of distance values that have length comparable with the total number N of index pairs. Measurements are never free from errors, so having "complete" lists of distances does not mean that the list is correct. The authors of [15] imply that a source of error is that δ may contain *more* distances than needed: for each given distance value, its number of occurrences in δ may be overestimated. We do not simulate this type of error here, but we do look at how the structure reconstruction changes as $|\delta|$ grows to reach N.

3.2 *The Algorithm*

For the above reasons, we propose a different methodology, based on MP, to implement the decomposition between α and x.

The matheuristic we propose is as follows:

1. compute an assignment α and an approximate realization x' using the MIDDP (MILP) formulation Eq. (6);
2. discard the realization x'
3. construct the weighted graph G from α
4. define a DGP instance on G and solve it, to obtain a realization x^*
5. return x^*.

3.2.1 Solving the MIDDP

Even if MILP solvers are technically very advanced, the MIDDP formulation in Eq. (6) is very challenging to solve at the instance scale yielded by the Buckminsterfullerene. We therefore solve it "locally", which, for a MILP solver, means setting a resource limit (either number of Branch-and-Bound nodes, or time). The solution returned by the solver is an approximate assignment.

3.2.2 Solving the DGP

The DGP consists in finding a realization x satisfying Eq. (1). Many different formulations of the DGP have been discussed in the literature [18, 20]. In this case we use the quartic formulation:

$$\min_{x \in \mathbb{R}^{nK}} \sum_{\{i,j\} \in E} \left(\|x_i - x_j\|_2^2 - d_{ij}^2\right)^2. \tag{7}$$

It is unconstrained and nonconvex in the x variables. As for the MIDDP, solving Eq. (7) exactly is out of the technological reach of global optimization solvers, at least at the scale of the Buckminsterfullerene. We therefore use a simple MS heuristic: at each iteration we sample a random starting point, then call the local optimization solver IPOPT [7] to identify a close local minimum, choosing the best over the allowed iterations.

4 Finding the Shape of the Buckminsterfullerene

We implemented our formulation using AMPL [12]. We used a smaller M value in solving the MIDDP than the one mentioned in [5, Prop. 2.2], since it is far from being tight. We set $M = \sum_\ell \delta_\ell^2$ in our computational experiments.

We solved the MIDDP formulation using CPLEX 22.1 [14]. The local solver deployed on Eq. (7) is IPOPT 3.4.12. All experiments were carried out on an Apple M1 Max with 64GB RAM running MacOS Ventura 13.6.4. All software was natively compiled for the ARM64 architecture.

CPLEX was given a time limit of 360s (six minutes—extending to one hour yielded no appreciable difference). The MS algorithm used to solve the DGP in Eq. (7) was allowed to run for 10 iterations. IPOPT was given no time limit.

The Buckminsterfullerene has $n = 60$ vertices and $N = 1770$ distance values, many of them with multiplicities, given its molecular symmetry. We computed the exact realization according to [25] and extracted a complete list δ of its inter-atomic distances. From this instance we also generated instances with smaller $|\delta|$ through a probabilistic Erdős-Rényi edge removal choice. We examined instances with $|\delta| \in [0.1\,N, 1.0\,N]$, with $1.0\,N$ corresponding to the complete list. The original realization of the Buckminsterfullerene is given in Fig. 1 (left). We also show a realization for $|\delta| = 0.5\,N$ (Fig. 1 center), which is rather typical of the whole range of probabilistically generated instances. Lastly, the perfectly reconstructed molecule corresponding to $|\delta| = 1770$ is shown in Fig. 1 (right). It may not appear identical because it is a rotation of the original: but the error returned by the solver was zero.

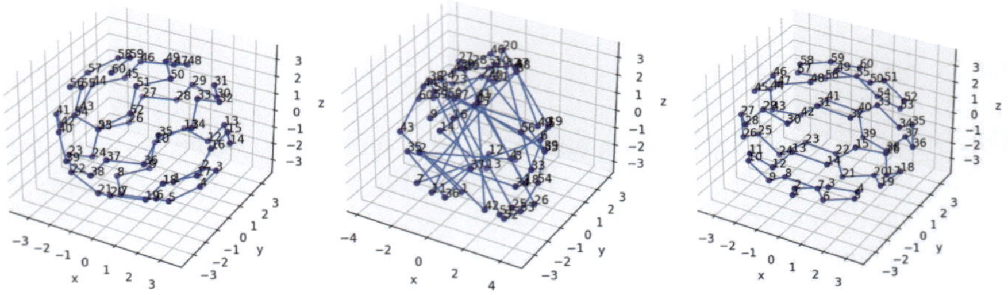

Fig. 1 Left: the Buckminsterfullerene molecule with a spiral order on the vertices. Center: wrong realizations obtained for $|\delta| = 0.5\,N$. Right: the realization for $|\delta| = N$ yields a perfect reconstruction (rotated)

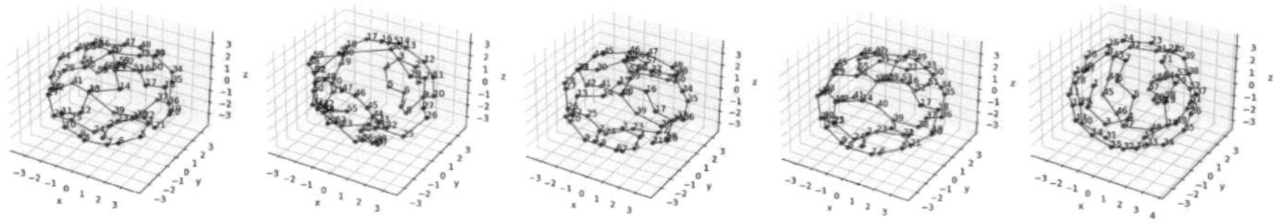

Fig. 2 Slightly wrong realizations for $|\delta| \in [0.995N, 0.999N]$

By trial-and-error we narrowed down an interval of interest for $|\delta|$, i.e. when the resulting shape is not completely wrong but not yet correct, to $|\delta| \in [0.995\,N, 0.999\,N]$. For this range we present numerical experiments based on the Mean Distance Error (MDE), i.e. $(1/|E|) \sum_{\{i,j\}\in E} |\ \|x_i - x_j\|_2 - d_{ij}\ |$ computed on the reconstructed weighted graph.

Density	0.995	0.996	0.997	0.998	0.999
MDE	0.722	1.057	0.778	0.552	0.603

The realizations corresponding to above densities are shown in Fig. 2. As one can see, the realizations have the correct overall shape with some local errors.

We report the mean errors $(1/N)\|\mathsf{Adj}(G') - \mathsf{Adj}(G^*)\|_1$ of the reconstructed graphs G' w.r.t. the correct Buckminsterfullerene graph G^*, where $\mathsf{Adj}(G)$ is the upper triangular part of weighted adjacency matrix of G.

Density	0.995	0.996	0.997	0.998	0.999
Mean error	1.043	1.291	1.129	0.851	0.720

Finally, the CPU time taken for each instance to be solved was between 1000s and 1200s.

5 Conclusion

The proposed methodology, simpler than that of [15], allowed us to reconstruct the Buckminsterfullerene perfectly from the list of all its distances. It also allowed us to reconstruct locally wrong realizations with very few missing distances. This is in constrast to [15], which required more than N distances (same distance values but with higher multiplicities) in order to reconstruct the molecule perfectly.

References

1. Ahmadi, A., Majumdar, A.: DSOS and SDSOS optimization: more tractable alternatives to sum of squares and semidefinite optimization. SIAM J. Appl. Algebr. Geom. **3**(2), 193–230 (2019)
2. Barker, G., Carlson, D.: Cones of diagonally dominant matrices. Pac. J. Math. **57**(1), 15–32 (1975)
3. Beeker, N., Gaubert, S., Glusa, C., Liberti, L.: Is the distance geometry problem in NP? In: Mucherino, A., Lavor, C., Liberti, L., Maculan, N. (eds.) Distance Geometry: Theory. Methods, and Applications, pp. 85–94. Springer, New York (2013)
4. Billinge, S., Duxbury, P., Gonçalves, D., Lavor, C., Mucherino, A.: Assigned and unassigned distance geometry: applications to biological molecules and nanostructures. 4OR **14**, 337–376 (2016)
5. Bruglieri, M., Cordone, R., Liberti, L.: Maximum feasible subsystems of distance geometry constraints. J. Glob. Optim. **83**, 29–47 (2022)
6. Cheng, Y.-H., Liao, J.-H., Zhao, Y.-J., Ni, J., Yang, X.-B.: Theoretical investigations on stable structures of $C_{60-n}N_n$ (n = 2-12): symmetry, model interaction, and global optimization. Carbon **154**, 140–149 (2019)
7. COIN-OR. Introduction to IPOPT: a tutorial for downloading, installing, and using IPOPT (2006)
8. Dakić, T.: On the turnpike problem. Ph.D. thesis, Simon Fraser University (2000)
9. Dokmanić, I., Parhizkar, R., Ranieri, J., Vetterli, M.: Euclidean distance matrices: essential theory, algorithms and applications. IEEE Signal Process. Mag. **1053–5888**, 12–30 (2015)
10. Doye, J., Leary, R., Locatelli, M., Schoen, F.: Global optimization of Morse clusters by potential energy transformations. INFORMS J. Comput. **16**(4), 371–379 (2004)
11. Duxbury, P., Lavor, C., Liberti, L., de Salles-Neto, L.: Unassigned distance geometry and molecular conformation problems. J. Glob. Optim. **83**, 73–82 (2022)
12. Fourer, R., Gay, D.: The AMPL Book. Duxbury Press, Pacific Grove (2002)
13. Gershgorin, S.: über die Abgrenzung der Eigenwerte einer Matrix. Zvesti Akademii Nauk SSSR. Otdelenie Fizicheskikh i Matematicheskikh Nauk **6**, 749–754 (1931)
14. IBM. ILOG CPLEX 22.1 User's Manual. IBM (2022)
15. Juhás, P., Cherba, D., Duxbury, P., Punch, W., Billinge, S.: Ab initio determination of solid-state nanostructure. Nature **440**(30), 655–658 (2006)
16. Kroto, H., Heath, J., O'Brien, S., Curl, R., Smalley, R.: C_60: buckminsterfullerene. Nature **318**, 162–163 (1985)
17. Lemke, P., Skiena, S., Smith, W.: Reconstructing sets from interpoint distances. In: Aronov, B., et al. (eds.) Discrete and Computational Geometry. Algorithms and Combinatorics, vol. 25, pp. 597–631. Springer, Berlin (2003)
18. Liberti, L.: Distance geometry and data science. TOP **28**, 271–339 (2020)
19. Liberti, L., Lavor, C.: Euclidean Distance Geometry: An Introduction. Springer, New York (2017)
20. Liberti, L., Lavor, C., Maculan, N., Mucherino, A.: Euclidean distance geometry and applications. SIAM Rev. **56**(1), 3–69 (2014)
21. Liberti, L., Vu, K.: Barvinok's naive algorithm in distance geometry. Oper. Res. Lett. **46**, 476–481 (2018)
22. Locatelli, M., Schoen, F.: Efficient algorithms for large scale global optimization: Lennard-Jones clusters. Comput. Optim. Appl. **26**, 173–190 (2003)
23. Maniezzo, V., Stützle, T., Voß, S. (eds.): Matheuristics: Hybridizing Metaheuristics and Mathematical Programming. Volume 10 of Annals of Information Systems. Springer, New YorkD (2009)

24. Saxe, J.: Embeddability of weighted graphs in k-space is strongly NP-hard. In: Proceedings of 17th Allerton Conference in Communications, Control and Computing, pp. 480–489 (1979)
25. Senn, P.: Computation of the cartesian coordinates of Buckminsterfullerene. J. Chem. Educ. **72**(4), 302–303 (1995)
26. Wüthrich, K.: NMR studies of structure and function of biological macromolecules (Nobel lecture). Angewandte Chemie **42**, 3340–3363 (2003)

Stop-Skipping to Reduce Delays in Train Rescheduling

Sonia Di Cola, Bianca Pascariu, Paola Pellegrini, and Valentina Cacchiani

Abstract In railway systems, unexpected disturbances often make the planned train schedule infeasible. The real-time Railway Traffic Management Problem (rtRTMP) consists in applying retiming, reordering and rerouting of trains with the aim of minimizing delay propagation. In this work, we additionally consider the option of skipping stops to further reduce delays. We extend the RECIFE-MILP model, a Mixed Integer Linear Programming based heuristic for the rtRTMP with microscopic network description, by including stop-skipping. We evaluate the benefit and impact of this rescheduling measure on train and passenger delays in a suburban railway network.

Keywords Stop-skipping · Train rescheduling · MILP model

1 Introduction

Unexpected delays and disturbances occur on a daily basis and cause conflicts between trains running on the railway infrastructure, and, as a consequence, inconvenience to passengers. Train rescheduling [3] consists in retiming, reordering and rerouting trains to determine a new conflict-free schedule with the goal of

S. Di Cola (✉)
Department of Multi-Actor Systems, Delft University of Technology, Delft, The Netherlands
e-mail: s.dicola@tudelft.nl

B. Pascariu · P. Pellegrini
Université Gustave Eiffel, Lille, France
e-mail: bianca.pascariu@univ-eiffel.fr

P. Pellegrini
e-mail: paola.pellegrini@univ-eiffel.fr

V. Cacchiani
DEI, University of Bologna, Bologna, Italy
e-mail: valentina.cacchiani@unibo.it

M. Di Francesco et al. (eds.), *Operations Research: Closing the Gap Between Research and Practice*, AIRO Springer Series 15, https://doi.org/10.1007/978-3-031-90095-2_11

minimizing delay propagation and inconvenience to the passengers. Recently, there has been a noticeable increase in enhancing passengers' experience and focusing on passenger-oriented rescheduling approaches [7].

An additional rescheduling measure that has recently been investigated is that of skipping stops, which is particularly relevant for suburban railway or metro systems with high train frequency. In this case, Train Operators can decide to skip some stops so that train delays are reduced and passengers can reach their destination faster, thus decreasing their delays as well. Clearly, skipping stops may also negatively affect some passengers who needed to get on/off the train at a skipped stop: these passengers have to get on the following train or have to get off at the previous stop and wait for the following train. Therefore, an appropriate trade-off has to be obtained between delay minimization and stop-skipping.

The main contribution of this work is that we extend the RECIFE-MILP model [6], a Mixed Integer Linear Programming (MILP) based heuristic for the rtRTMP, by including stop-skipping. This model uses a microscopic level of detail to describe the railway network. In particular, the infrastructure is modeled considering *track-circuits*, i.e., track segments on which the presence of a train is automatically detected [6]. Sequences of track-circuits are grouped into block sections, which start and end with a light signal. Several block sections can share track-circuits, for example in presence of a switch: two block sections may have the same entry signal, share the first track-circuits, split through the switch and have different exit signals. A sequence of block sections composes a route. We consider a suburban railway network and compare the new model with the classical RECIFE-MILP to evaluate how train delays can be reduced by stop-skipping. In addition, we embed the new model in a framework to evaluate the impact of stop-skipping on the passengers' delays at destination by considering the assignment of passenger demand to trains computed as in [9].

In Sect. 2 we review the related literature. In Sect. 3 we present the new model and the framework for passengers' evaluation. In Sect. 4 we report computational results on real-life instances. Finally, we draw some conclusions and present directions for future research in Sect. 5.

2 Literature Review

The literature on train rescheduling is extremely broad. In this section, we focus only on works that apply stop-skipping to train rescheduling during operations. The real-time stop-skipping technique was first used in [10] who developed a simulation study to evaluate two stop-skipping policies by considering passenger distribution and disruption duration. Several years later, in [1], an ILP model was proposed with stop-skipping and including rolling stock constraints with the goal of minimizing the passengers' waiting time and the number of skipped stops in a dense railway system. A macroscopic level of detail of the network was employed. Recently, in [2], the model by [1] was extended by considering decisions on canceling or short-turning trains and adding stops, and by including multiple objectives. A simulation-optimization approach was proposed to derive multiple non-dominated solutions. In [5], stop-skipping is adopted to effectively reduce the number of passengers stranded on the platform to avoid overcrowded situations during the recovery period after a disruption in a bi-directional metro line. An optimization model that includes passenger flows is proposed with a macroscopic description of the network. An iterative heuristic algorithm, based on a model decomposition, is developed to solve the problem.

Recently, to address the rescheduling problem during complete track blockages, in [11], a MILP model that employs flexible stopping (i.e., skipping stops and adding stops) and flexible short-turning (i.e., a rolling stock unit turns at a station to be used for another train service in the opposite direction), combined with retiming, reordering, and canceling, has been proposed. The model considers networks with double-track and single-track lines described at a macroscopic level, and passenger demand is handled in a static way. The objective is to minimize passenger delays. To further extend previous works, dynamic passenger flows have been included in [4, 12]. In [12], a passenger-oriented timetable rescheduling model is proposed to cope with railway disruptions. This model accounts for flexible stopping (adding and skipping stops) and for passenger re-routing with passenger transfers. The railway infrastructure is described at a macroscopic level. An adapted fix-and-optimize algorithm is developed to solve the problem. In [4], the stop-skipping technique is considered while taking into account rolling stock constraints and dynamic passenger flows. A model predictive control algorithm is applied that solves the problem in a rolling horizon setting. The method is tested on a metro line described at a macroscopic level.

From the literature analysis, we can see that the railway infrastructure is described at a macroscopic level of detail. In this work, we extend the RECIFE-MILP model by adding the possibility of skipping stops while considering a very detailed description of the railway network.

3 Stop-Skipping and Passenger Delays Evaluation

This section is devoted to presenting the RECIFE-MILP model extended by stop-skipping options (Sect. 3.1) and the framework used to evaluate the effects on passengers' delays (Sect. 3.2).

3.1 Stop-Skipping in RECIFE-MILP

In this section, we report the RECIFE-MILP model enhanced with the new possibility of skipping stops. We first introduce all sets, parameters and variables used in the classical version of RECIFE-MILP [6]. Then, we present the new sets, parameters and variables and show the complete MILP model.

As already mentioned, RECIFE-MILP considers a detailed description of the railway infrastructure. In particular, the infrastructure is modeled considering track-circuits. The model takes as input the actual traffic state and computes new train schedules by allowing for reordering, retiming and rerouting trains with the goal of minimizing train delays [6]. RECIFE-MILP uses the following sets and parameters:

- T, R_t: the set of trains and of routes available to train $t \in T$;
- TC^r, TC_t: the set of track-circuits belonging to route $r \in R$ and of those which can be used by train $t \in T$;
- S_t, $TCS_{t,s}$: the set of stations where $t \in T$ has a scheduled stop and set of track-circuits that can be used by t for stopping at $s \in S_t$;
- S: the set of stations;
- $sched_t$: scheduled arrival time of train $t \in T$ at destination;
- $init_t$, $exit_t$: earliest time at which train $t \in T$ can be operated and earliest time at which it can reach its destination given $init_t$, the route assigned in the timetable and the intermediate stops;
- $rt_{r,t,tc}$, $ct_{r,t,tc}$: running and clearing time of $tc \in TC^r$ along $r \in R$ for train $t \in T$;
- $ref_{r,tc}$: reference track-circuit for the reservation of $tc \in TC^r$ along $r \in R$, depending on block section structure and interlocking system;
- for_{bs}, rel_{bs}: formation and release time for block section bs;
- $dw_{t,s}$, $a_{t,s}$, $d_{t,s}$: minimum dwell time, scheduled arrival and scheduled departure times for train $t \in T$ at station $s \in S_t$;
- $p_{r,tc}$, $s_{r,tc}$: track-circuits preceding and following $tc \in TC^r$ along $r \in R$;
- w_t: weight of train t delay;
- M: a large constant.

RECIFE-MILP employs the following decision variables:

- $sU_{t,tc}$, $eU_{t,tc}$: continuous positive variables representing the time at which $tc \in TC_t$ starts and ends being utilized by $t \in T$;
- $x_{t,r}$: binary variable equal to 1 if train $t \in T$ uses route $r \in R_t$, 0 otherwise;
- $y_{t,t',tc}$: binary variable equal to 1 if train $t \in T$ utilizes track-circuit tc before train t', such that index t is smaller than index t' ($t \prec t'$), with $tc \in TC_t \cap TC_{t'}$, and 0 otherwise.
- $o_{t,r,tc}$: time in which $t \in T$ starts the occupation of $tc \in TC^r$ along $r \in R_t$;
- $l_{t,r,tc}$: longer stay of the head of $t \in T$ on $tc \in TC^r$ along $r \in R_t$, due to dwell time and scheduling decisions (delay);
- D_t: delay suffered by train t when exiting the infrastructure, if the exit does not correspond to a station.

To include the possibility of skipping stops, we introduce the following sets and parameters:

- S_{skip}: set of stations that can be skipped;
- $\Delta dec_{t,r,tc}$: deceleration time of train t on tc along route r;
- $\Delta acc_{t,r,tc}$: acceleration time of train t on tc along route r;
- w^s: penalty for skipping a stop at station s.

In addition, we introduce the following decision variables:

- $k_{t,s}$: binary variable equal to 1 if train $t \in T$ skips its stop at station $s \in S_{skip} \cap S_t$, 0 otherwise. Defined for all the trains on the stations that can be skipped;
- $art_{t,r,tc} \in \mathbb{R}_+$: actual running time of train t on tc along route r.

Differently from the RECIFE-MILP model from the literature, when including the option of skipping stops, we have to account for the shorter running times that a train can achieve since it does not have to decelerate and accelerate after a stop. In addition, if a stop is skipped it is not necessary to respect the scheduled departure and dwell times at that stop.

The stop-skipping RECIFE-MILP objective function is to minimize the weighted sum of train delays and penalties for skipping stops:

$$\min \sum_{t \in T} \left(w_t D_t + \sum_{s \in S_t \cap S_{skip}} w^s k_{t,s} \right). \tag{1}$$

Obviously, skipping a stop causes inconvenience to passengers who should get on or off the train at that stop. Therefore, we penalize skipping stops. On the other hand, skipping a stop can reduce delays for other passengers. The weights of the two terms in the objective function are used to find a trade-off between the two conflicting objectives. The values of w^s ($s \in S_{skip}$) can be selected based on the relevance of stop s according to the experience of the practitioners (e.g., if it is a stop where several passengers transfer to other lines the corresponding penalty should be high) but can also be chosen based on the current train delays and predicted passenger flows. In Sect. 4, we consider several values for w^s ($s \in S_{skip}$) to analyze their impact on train and passenger delays.

Constraints (2) set the start time for the occupation of track circuit tc by train t along route r to be at least the earliest time for operating train t, while constraints (3) set this time to 0 if the train does not use route r:

$$o_{t,r,tc} \geq init_t\, x_{t,r} \qquad \forall t \in T, r \in R_t, tc \in TC^r, \tag{2}$$

$$o_{t,r,tc} \leq M x_{t,r} \qquad \forall t \in T, r \in R_t, tc \in TC^r, \tag{3}$$

Constraints (4)–(8) define the actual running time of train t on track circuit tc along route r by considering the effect of stop-skipping. In particular, the actual running time corresponds to the scheduled running time if the stop cannot be skipped. On the contrary, if a stop is skipped, acceleration and deceleration times do not have to be included, thus the actual running time is smaller than the scheduled one:

$$art_{t,r,tc} = rt_{r,t,tc}x_{t,r}$$
$$\forall t \in T, r \in R_t, tc \in TC^r : \nexists s \in S_{skip}, tc \in TCS_{t,s} \vee p_{r,tc} \in TCS_{t,s} \quad (4)$$
$$art_{t,r,tc} \geq rt_{r,t,tc} - \Delta dec_{t,r,tc}k_{t,s} - M(1 - x_{t,r})$$
$$\forall t \in T, r \in R_t, s \in S_{skip}, tc \in TC^r \cap TCS_{t,s}, \quad (5)$$
$$art_{t,r,tc} \leq rt_{r,t,tc} - \Delta dec_{t,r,tc}k_{t,s} + M(1 - x_{t,r})$$
$$\forall t \in T, r \in R_t, s \in S_{skip}, tc \in TC^r \cap TCS_{t,s}, \quad (6)$$
$$art_{t,r,tc} \geq rt_{r,t,tc} - \Delta acc_{t,r,tc}k_{t,s} - M(1 - x_{t,r})$$
$$\forall t \in T, r \in R_t, s \in S_{skip}, tc \in TC^r, p_{r,tc} \in TCS_{t,s}, \quad (7)$$
$$art_{t,r,tc} \leq rt_{r,t,tc} - \Delta acc_{t,r,tc}k_{t,s} + M(1 - x_{t,r})$$
$$\forall t \in T, r \in R_t, s \in S_{skip}, tc \in TC^r, p_{r,tc} \in TCS_{t,s}, \quad (8)$$

Constraints (9) define the start time of occupation of track circuit tc by train t along route r based on the potential longer stay $l_{t,r,tc}$ on the track circuit and on the actual running time $art_{r,t,tc}$. Constraints (10)–(11) impose the start time of the occupation of the track circuit following a stop to be at least the scheduled departure time, unless the stop is skipped:

$$o_{t,r,s_{r,tc}} = o_{t,r,tc} + l_{t,r,tc} + art_{r,t,tc} \qquad \forall t \in T, r \in R_t, tc \in TC^r \quad (9)$$
$$o_{t,r,s_{r,tc}} \geq d_{t,s}x_{t,r} \qquad \forall t \in T, r \in R_t, s \in S_t \setminus S_{skip}, tc \in TCS_{t,s} \cap TC^r \quad (10)$$
$$o_{t,r,s_{r,tc}} \geq d_{t,s}(x_{t,r} - k_{t,s}) \qquad \forall t \in T, r \in R_t, s \in S_t \cap S_{skip}, tc \in TCS_{t,s} \cap TC^r \quad (11)$$

The longer stay of train t on track circuit tc along route r is required to be at least the minimum dwell time (constraints (12)) unless the stop is skipped (constraints (13)):

$$l_{t,r,tc} \geq dw_{t,s}x_{t,r} \qquad \forall t \in T, r \in R_t, s \in S_t \setminus S_{skip}, tc \in TCS_{t,s} \cap TC^r \quad (12)$$
$$l_{t,r,tc} \geq dw_{t,s}(x_{t,r} - k_{t,s}) \qquad \forall t \in T, r \in R_t, s \in S_t \cap S_{skip}, tc \in TCS_{t,s} \cap TC^r \quad (13)$$

Constraints (14) define the delay of train t at destination (to be minimized in the objective function), while constraints (15) require to select one route for each train $t \in T$:

$$D_t \geq \sum_{r \in R_t} o_{t,r,tc_\infty} - sched_t \qquad \forall t \in T, \quad (14)$$

$$\sum_{r \in R_t} x_{t,r} = 1 \qquad \forall t \in T \quad (15)$$

Constraints are used to define the start and end utilization times of track circuit tc by train t:

$$sU_{t,tc} = \sum_{r \in R_t : tc \in TC^r} \left(o_{t,r,ref_{r,tc}} - for_{bs_{r,tc}} x_{t,r}\right) \qquad \forall t \in T,\ tc \in TC_t \quad (16)$$

$$eU_{t,tc} = \sum_{\substack{r \in R_t: \\ tc \in TC^r}} o_{t,r,tc} + \sum_{\substack{tc' \in TC^r: \\ tc \in OTC_{t,r,tc}}} l_{t,r,tc'} + art_{r,t,tc} + (ct_{r,t,tc} + rel_{bs_{r,tc}})x_{t,r}$$

$$\forall t \in T, tc \in TC_t, \quad (17)$$

$$eU_{t,tc} - M(1 - y_{t,t',tc}) \leq sU_{t',tc} \qquad \forall t, t' \in T, t \prec t',\ tc \in TC_t \cap TC_{t'}, \quad (18)$$

$$eU_{t',tc} - My_{t,t',tc} \leq sU_{t,tc} \qquad \forall t, t' \in T, t \prec t',\ tc \in TC_t \cap TC_{t'}. \quad (19)$$

3.2 *Passenger Delays Evaluation*

The enhanced RECIFE-MILP model is embedded in the framework illustrated in Fig. 1. The goal of this framework is to evaluate the impact of train rescheduling strategies on the passenger delays. Indeed, objective function (1) minimizes the weighted sum of train delays at destination and penalties for stop-skipping. However, depending on which and how many stops are skipped, some passengers may incur high delays or even may not be able to reach their destination within the considered rescheduling horizon. In particular, passengers may not be able to reach their destination if no train stops at their origin or destination station after their arrival in the system: we say that they are *stranded passengers* who should reach their destination outside the considered rescheduling period. Therefore, we want to investigate how passenger trips are affected by skipping stops, and evaluate whether passengers can have alternative trains to reach their destination. Moreover, we want to assess if

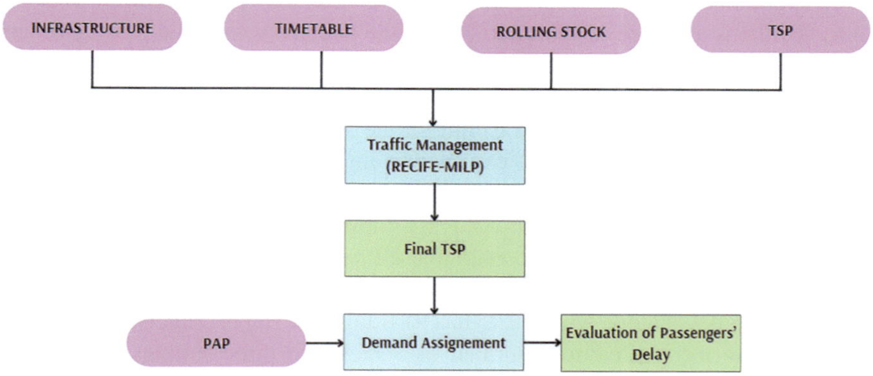

Fig. 1 Proposed framework's illustration

skipping stops may bring an overall gain, in terms of delays, both for passengers that needed to get on/off at the skipped stops and for passengers who were not affected by skipping stops.

The input data, highlighted in pink, consists of the planned timetable, the infrastructure and rolling stock description, the Traffic State Prediction (TSP) and the Passenger Assignment Plan (PAP). The TSP provides the information regarding the traffic flow which needs to be optimized, including the observed perturbations and delays. The PAP includes the relevant data regarding the passengers, such as their number, the origin-destination (OD) combination for each passenger group, and all the train-based paths, i.e., the subset of trains that can be employed for an OD pair. The specific format of these data is detailed in [8]. In particular, for each passenger group, the PAP indicates the desired time to arrive at the destination under unperturbed conditions and the actual arrival time considering the current traffic status.

Starting from the timetable, infrastructure, rolling stock and TSP, the traffic flow is optimized by RECIFE-MILP, with a focus on train delays. This leads to the generation of a new TSP that reflects the optimized traffic. Based on the final TSP and PAP, the demand assignment is conducted [9]: each passenger group is assigned to trains, and the difference between desired and actual arrival time is computed.

The described procedure is first executed with the classical RECIFE-MILP, i.e., stop-skipping is not allowed. Subsequently, the stop-skipping constraints are incorporated: several penalty values w^s ($s \in S_{skip}$) are considered, in order to determine a good trade-off penalty to improve both train and passenger delays.

4 Computational Results

We conduct computational experiments on a real-life case study related to the part of the Copenhagen suburban railway network shown in Fig. 2. To generate perturbed traffic situations, delays between 5 and 15 min are randomly applied to 20% of the trains. Traffic is assumed to evolve following the planned routes and passing orders until the optimization occurs. For the TSP, we consider two instances (*Inst1* and *Inst2*) representing 40 min of traffic starting at 8 a.m. and 9 a.m., respectively: in each instance, all trains that are traveling in the infrastructure at the starting time are considered, as well as all trains expected to enter within the next 40 min. For the PAP, we consider arrival of passengers within an interval of 20 min, according to the historical demand on September 9, 2017. By considering 40 min of traffic and only 20 min of passenger arrivals, we ensure that all passengers entering have available trains to complete their journey, at least when no stops are skipped. *Inst1* has 89 trains and 251 passengers, while *Inst2* has 91 trains and 76 passengers.

We execute the procedure described in Sect. 3.2 in three settings. Firstly, we impose that *None* of the stops can be skipped. Then, for the stop-skipping case, we evaluate two alternative scenarios. In the first one, *No HL-LY*, we require that HL (Hellerup) and LY (Lyngby) stations cannot be skipped: in fact, HL is used by many

Fig. 2 Part of Copenhagen suburban railway used as a case study

passengers to transfer to other lines and LY is the destination of many passengers. Remark that the a priori selection of which stops can be skipped or not is often done in the practice. It is typically based on experience and historical data on passenger demand. In the second scenario, *All*, we allow for skipping any stop. We execute the procedure for both scenarios by considering in turn a single value for w^s ($s \in S_{skip}$) in set $\{10, 25, 50, 75, 100\}$. We impose a time limit of 180 s for each execution: in the first at most 30 s, a solution with fixed timetable routes is obtained; then this solution is used as a warm-start for the routing and scheduling problem solution in the remaining time. The results are reported in Table 1. For each instance, we report the solution objective value, the sum of train delays and the sum and average of passenger delays (in seconds), the percentage *Strd%* of stranded passengers who could not reach their destination due to stop-skipping, the number of skipped stops and the optimality gap when the time limit is reached. The total number of stops that can be skipped (skippable stations) for all train services is reported in brackets for each case. We show in gray train and passenger delays when they are smaller than in the no-skip case.

By looking at the results in Table 1, we can observe that stop-skipping allows a significant reduction of train delays in both instances and with almost all penalty values, thanks to the reduction in train running and dwell times. On the other hand, when many stops are skipped (i.e., when the penalty is equal to 10 or 25), the percentage of stranded passengers is not negligible, reaching 24%.

If we impose that stops HL and LY cannot be skipped, the percentage of stranded passengers is reduced with respect to the case in which all stops can be skipped, but no benefit on passenger delays is obtained for *Inst1*. On the contrary, if all stops can

Table 1 Experimental results: grey cells show delay reduction obtained when skipping stops

Inst1 (89 trains, 251 pax)							
Skippable stations	Pen	Obj	Tr. Del.	Pax Del. (avg.)	Strd%	N. Sk.	Gap
None (0)	-	13700	13700	39128 (156)	0.00	0	0.13
No HL-LY (153)	10	12061	11551	43897 (192)	8.74	51	0.16
	25	12855	11930	39852 (166)	4.37	37	0.17
	50	13349	12699	39470 (162)	2.90	13	0.17
	75	13532	13157	40976 (166)	1.63	5	0.15
	100	13680	13280	40507 (164)	1.62	4	0.15
All (234)	10	11427	10577	59933 (281)	15.00	85	0.16
	25	12494	11219	55111 (246)	10.80	51	0.20
	50	13325	12675	37121 (157)	6.00	13	0.19
	75	13508	13133	40440 (169)	4.60	5	0.17
	100	13616	13216	40812 (171)	5.00	4	0.17

Inst2 (91 trains, 76 pax)							
Skippable stations	Pen	Obj	Tr. Del.	Pax Del. (avg.)	Strd%	N. Sk.	Gap
None (0)	-	18527	18527	15254 (201)	0.00	0	0.11
No HL-LY (155)	10	16624	16014	15843 (404)	9.00	61	0.11
	25	17428	16328	16196 (218)	2.30	44	0.13
	50	18024	17274	15095 (201)	0.88	15	0.13
	75	18309	18159	14876 (186)	0.04	2	0.13
	100	18340	18240	14831 (196)	0.00	1	0.12
All (234)	10	15763	14853	23369 (229)	23.90	91	0.16
	25	17269	15744	16196 (218)	2.30	61	0.20
	50	18173	17623	14930 (200)	2.04	11	0.19
	75	18062	17237	13818 (196)	2.00	11	0.17
	100	18394	18294	14869 (195)	0.04	1	0.17

be skipped, passenger delays can be reduced for both instances when the penalty is 50, thus suggesting that an intermediate value is the most appropriate to reach a good trade-off between stranded passengers and delay reduction.

For all settings, the performance of RECIFE-MILP in terms of optimality gap is similar, hence stop-skipping can be handled in the model even when all stops can be skipped. By looking at the solution objective value, we can see that it almost always improves when allowing all stops to be canceled, with a few exceptions due the heuristic nature of the solution approach.

We underline that the choice of which stops to skip (and not only the number of stops to skip) is very relevant: indeed, in *Inst1* with penalty 50, the number of skipped stops is 13 both when HL and LY cannot be skipped and when all stops can be skipped, but only in the latter case an improvement is obtained on the passenger delays, both in total and in average. This suggests that it is hard to decide a priori the set of stops that can be skipped to reduce passenger delays, while the decision should be made by directly embedding passenger choices inside the model. Similarly, it would be useful to add to the objective function a penalization accounting for the stranded passengers. Obviously, this would imply considering the passenger flows within the model and thus solving a model with a larger number of variables and constraints, which may be harder especially in a real-time setting.

5 Conclusions

In this paper, we presented a variant of a state-of-the-art algorithm for railway traffic management. In this variant, we allow the algorithm to skip planned stops to reduce delay propagation. Stop-skipping is a quite frequently applied measure in urban railway, but few algorithms exist including it in the optimization. To the best of our knowledge, this is the first paper in which this measure is optimized when considering a microscopic representation of the railway infrastructure.

In the experimental analysis, we assessed the system performance in various contexts: we considered the possibility of skipping any stop, or we impose that stops at some critical stations must be performed. We assessed the performance in terms of both train and passenger delays, and we show that in some cases it is possible to reduce both train and passenger delays by wisely selecting the stops to skip. However, this wise selection is not trivial and a priori decisions are not always beneficial.

As future work, one could consider including information on static passenger demand to partially evaluate the impact of skipping stops or including passenger delay evaluation with a feedback mechanism to dynamically adjust the penalty values used in the objective function. Embedding passenger routing or passenger delay evaluation directly within the model is also an interesting research direction to be exploited. Furthermore, it could be useful to develop machine learning techniques to compute the penalties associated with skipping stops based on observed and historical data of passenger demand in the network.

References

1. Altazin, E., Dauzère-Pérès, S., Ramond, F., Tréfond, S.: Rescheduling through stop-skipping in dense railway system. Transp. Res. Part C: Emerg. Technol. **79** (2017). https://doi.org/10.1016/j.trc.2017.03.012
2. Altazin, E., Dauzère-Pérès, S., Ramond, F., Tréfond, S.: A multi-objective optimization-simulation approach for real time rescheduling in dense railway systems. Eur. J. Oper. Res. **286**(2), 662–672 (2020). https://doi.org/10.1016/j.ejor.2020.03.034
3. Cacchiani, V., Huisman, D., Kidd, M., Kroon, L., Toth, P., Veelenturf, L., Wagenaar, J.: An overview of recovery models and algorithms for real-time railway rescheduling. Transp. Res. Part B: Methodol. **63**, 15–37 (2014). https://doi.org/10.1016/j.trb.2014.01.009
4. Chen, Z., Li, S., D'Ariano, A., Yang, L.: Real-time optimization for train regulation and stop-skipping adjustment strategy of urban rail transit lines. Omega **110**, 102631 (2022). https://doi.org/10.1016/j.omega.2022.102631

5. Gao, Y., Kroon, L., Schmidt, M., Yang, L.: Rescheduling a metro line in an over-crowded situation after disruptions. Transp. Res. Part B: Methodol. **93**, 425–449 (2016). https://doi.org/10.1016/j.trb.2016.08.011
6. Pellegrini, P., Marlière, G., Pesenti, R., Rodriguez, J.: RECIFE-MILP: An effective MILP-based heuristic for the real-time railway traffic management problem. IEEE Trans. Intell. Transp. Syst. **16**(5), 2609–2619 (2015). https://doi.org/10.1109/TITS.2015.2414294
7. Sharma, B., Pellegrini, P., Rodriguez, J., Chaudhary, N.: A review of passenger-oriented railway rescheduling approaches. Eur. Transp. Res. Rev. **15**(1), 14 (2023). https://doi.org/10.1186/s12544-023-00587-0
8. SORTEDMOBILITY: Technical Deliverable Report 4.1 Data Exchange Format and Software Interfaces (2023). https://www.sortedmobility.eu/download
9. SORTEDMOBILITY: Technical Deliverable Report 2.2 Data-Driven Operational Prediction Model (2024). https://www.sortedmobility.eu/download
10. Sun, A., Hickman, M.: The real–time stop–skipping problem. J. Intell. Transp. Syst. **9**(2), 91–109 (2005). https://doi.org/10.1080/15472450590934642
11. Zhu, Y., Goverde, R.: Railway timetable rescheduling with flexible stopping and flexible short-turning during disruptions. Transp. Res. Part B: Methodol. **123** (2019). https://doi.org/10.1016/j.trb.2019.02.015
12. Zhu, Y., Goverde, R.M.: Integrated timetable rescheduling and passenger reassignment during railway disruptions. Transp. Res. Part B: Methodol. **140**, 282–314 (2020). https://doi.org/10.1016/j.trb.2020.09.001

On the Distribution of the Random Risk Under Alternative Loss Functions

Fulvio De Santis, Stefania Gubbiotti, and Francesco Mariani

Abstract In standard statistical decision theory the average performance of a decision is quantified by the risk function, that depends on the parameter of the model. The risk function may be summarized by the Bayes risk, its average with respect to a prior distribution assigned to the parameter. However, the expected value might not be a good representative of risk's randomness. We here propose to study the whole distribution of the risk function and to report additional summaries such as the median and the mode. For the specific set up of point estimation of a normal variance, we provide closed-form expressions for expected value, cumulative distribution and density functions of the risk under quadratic loss. A simulation-based approach is proposed when closed-form expressions of the distribution and its summaries are not available, for instance when alternative loss functions are considered. Finally, the problem of sample size determination is considered.

Keywords Bayes risk · Point estimation · Prior distributions · Sample size determination · Statistical decision theory

1 Introduction

Statistical decision functions—such as point and set estimates or test statistics—are typically compared in terms of loss functions. In frequentist decision theory, the risk function is the expected value of the loss with respect to the sampling distribution of the data and it depends on the parameter of the model. The risk function is usually summarized by optimality criteria, i.e. by applying a suitable real-valued functional

F. De Santis · S. Gubbiotti
Department of Statistical Sciences, Sapienza University of Rome, Rome, Italy
e-mail: fulvio.desantis@uniroma1.it

S. Gubbiotti
e-mail: stefania.gubbiotti@uniroma1.it

F. Mariani (✉)
Department of Statistics, University of Bologna, Bologna, Italy
e-mail: francesco.mariani20@unibo.it

M. Di Francesco et al. (eds.), *Operations Research: Closing the Gap Between Research and Practice*, AIRO Springer Series 15, https://doi.org/10.1007/978-3-031-90095-2_12

(see for instance [1]). We here focus on Bayesian criteria that require a prior probability distribution for the parameter of the model, that induces a distribution on the risk function as well. Among several alternatives, the most popular Bayesian criterion is the Bayes risk, the expected value of the risk function with respect to the prior. However, relying solely on averaging might be a limitation, since the expected value is not always a good summary of the entire distribution of a random variable. In the context of clinical trials, this point has been raised by several authors for the power function of a test, a quantity that is closely related to the risk. For instance [3–6] argue that examining the whole distribution of the random power guarantees a better insight on the probability of success of an experiment. In this article we borrow this idea and we propose to study the probability distribution of the risk function induced by a prior in the context of point estimation. Inspection of the shape of the random risk distribution allows one to assess the impact of prior assumptions and of sample size on the quality of a candidate point estimate. This analysis has also potential uses for the selection of the sample size and, in general, experimental design. Specifically, the proposed approach is developed for point estimation of the variance of the normal model. We provide the explicit expressions for the expected value, the cumulative distribution function (cdf) and the probability density function (pdf) of the risk under the quadratic loss function. For comparison, we also consider an asymmetric loss function, whose risk function is not available in closed form. Hence, we propose a simulation scheme that can be used in this and similar cases.

The outline of the paper is as follows. In Sect. 2 we introduce notations and formalize the problem. In Sect. 3 we derive explicit results for the normal variance case under quadratic loss. In Sect. 3.1 we sketch the algorithm that can be used for simulation when alternative loss functions are considered. Section 4 addresses the problem of sample size determination based on suitable summaries of the random risk. All the previous ideas are illustrated through an example concerning an experiment on the variability of the dissolving time of a drug in gastric juice. Finally, Sect. 5 contains a discussion and hints at future developments of this work.

2 Methodology

Let $\mathbf{X}_n = (X_1, \ldots, X_n)$ be a random sample, $f_n(\mathbf{x}_n|\theta)$ its distribution depending on an unknown parameter $\theta \in \Omega \subseteq \mathbb{R}$, where Ω is the parameter space. Let $d(\mathbf{X}_n)$ be a point estimator of θ, $d(\mathbf{x}_n)$ the corresponding estimate and $\mathcal{D}$ the set of all estimators. Adopting the frequentist approach to decision theory, the performance of $d(\cdot)$ is evaluated by its *risk function*

$$R(\theta, d) = \mathbb{E}_\theta[\mathbb{L}(\theta, d(\mathbf{X}_n))],$$

where $\mathbb{L}(\theta, d(\mathbf{X}_n))$ is the *loss function* of d and $\mathbb{E}_\theta[\cdot]$ denotes the expected value with respect to the sampling distribution $f_n(\cdot|\theta)$. For simplicity, we here assume that $\mathbb{L}(\theta, d) \geq 0$, for any pair (θ, d). For any estimator d, $R(\theta, d)$ is a function of θ. Following a hybrid frequentist-Bayesian approach, we assign a prior distribution $\pi(\theta)$ to the parameter that is now treated as a random variable Θ. The Bayes risk of d is defined as

$$r_\pi(d) = \mathbb{E}_\pi[R(\Theta, d)],$$

i.e. the expected value of $R(\Theta, d)$ with respect to $\pi(\theta)$. We assume that Θ is an absolute continuous random variable and that $\pi(\theta)$ is a density function. Then,

$$r_\pi(d) = \int_\Omega R(\theta, d)\pi(\theta)d\theta.$$

The Bayes risk can be used to evaluate a given estimator, to compare alternative estimators and to identify optimal estimators $d^\star$ such that $r_\pi(d^\star) = \inf_{d\in\mathcal{D}} r_\pi(d)$. Let now

$$Y = R(\Theta, d)$$

be the random risk function and $\mathcal{Y}$ its sample space. In the following section we derive the cdf $G(\cdot)$ and pdf $g(\cdot)$ of Y under quadratic loss function in a relevant model.

3 Normal Variance Case

Suppose that $X_i|\theta \sim \mathrm{N}(\mu_0, \theta)$, i.i.d, $i = 1, \ldots, n$, with μ_0 known and $\theta > 0$. Let us assume a conjugate prior for Θ, i.e. an inverse gamma distribution of parameters (α, β), with cdf $\mathbb{F}\pi(\cdot)$ and pdf

$$\pi(\theta) = \frac{\beta^\alpha}{\Gamma(\alpha)} \frac{1}{\theta^{\alpha+1}} \exp\left\{-\frac{\beta}{\theta}\right\}, \quad \theta > 0.$$

Recall that $\mathbb{E}_\pi(\Theta) = \beta/(\alpha - 1)$, $\mathbb{E}_\pi(\Theta^2) = \beta^2/[(\alpha - 1)(\alpha - 2)]$, $\mathbb{V}_\pi(\Theta) = \beta^2/[(\alpha - 1)^2(\alpha - 2)]$, where $\mathbb{V}_\pi(\Theta)$ denotes the variance of Θ with respect to $\pi(\theta)$. In the application we elicit the prior distribution in terms of its mean and variance, as follows: we set values of α and β such that $\mathbb{E}_\pi(\Theta) = m$ and $\mathbb{V}_\pi(\Theta) = v$ that yield $\alpha = \frac{m^2}{v} + 2$ and $\beta = m\left(\frac{m^2}{v} + 1\right)$.

Consider $d(\mathbf{X}_n) = S_0^2 = \sum_{i=1}^n (X_i - \mu_0)^2/n$, the maximum likelihood estimator (`mle`) of θ, and assume the quadratic loss $\mathbb{L}(\theta, d) = (\theta - d)^2$. In this case it is well known that S_0^2 is the uniformly minimum variance unbiased estimator (`umvue`) of θ and that $R(\theta, d) = 2\theta^2/n$. It is easy to check that

$$\mathbb{E}_\pi[Y] = r_\pi(d) = \frac{2}{n}\mathbb{E}_\pi(\Theta^2);$$
$$G(y) = \mathbb{F}_\pi\left(\sqrt{\frac{n}{2}y}\right), \qquad y \geq 0;$$
$$g(y) = \frac{1}{2}\sqrt{\frac{n}{2y}}\,\pi\left(\sqrt{\frac{n}{2}y}\right), \quad y \geq 0. \tag{1}$$

Example: gastric dissolving time of a drug. Let us consider an experiment on the variability of the dissolving time (in seconds) of a drug in gastric juice, as described in Zar [7]. Let the dissolving time be distributed as $\mathrm{N}(\mu_0, \theta)$ with known mean $\mu_0 = 45$ seconds and unknown variance θ. The purpose of the experiment is to estimate θ. For instance, this can be useful as a preliminary estimate of the variance to be used in a subsequent trial. As an example we specify an inverse gamma prior for Θ based on the historical data reported by Zar [7]. Specifically, we set $m = 2.69$ that is the observed value of the variance in the dataset. Then, in order to obtain a prior density that assigns 0.95 probability to the interval centered on m, $m \pm 2\sqrt{v} = (0.69, 4.69)$, we set $v = 1$. This implies that $\alpha = 9.24$ and $\beta = 22.15$. The inverse gamma density is plotted in Fig. 1 panel (a). Assuming a quadratic loss function, we obtain the density of the risk function Y according to (1) and we plot the curves of $g(y)$ for several increasing values of the sample size in panels (b)-(d). We also report in Table 1 the values of three alternative summaries of $g(y)$, e.g. expectation $\mathbb{E}(Y) = r_\pi(d)$, median $\mathbb{M}\mathrm{e}(Y)$ and mode $\mathbb{M}\mathrm{o}(Y)$, for the different sample sizes. For moderate values of n (for instance $n = 25$ in panel (b) and $n = 50$ in panel (c)), since the distribution is highly skewed, it does not seem to be well summarized by $r_\pi(d)$ which is quite different from median and mode; when the sample size increases, the distribution of Y tends to concentrate on lower and lower values and the three summaries take values closer and closer to 0. In summary, results show that the Bayes risk tends to overestimate the random risk, whereas the median and the mode are more representative of its distribution, in particular for small values of n.

3.1 *Alternative Loss Functions*

Although the quadratic loss is the most widely used, sometimes we may be willing to assign asymmetric penalizations to values of θ. This can be achieved, for instance, by considering the loss function

$$\mathbb{L}_b(\theta, d) = \frac{(\theta - d)^2}{b\theta + 1}, \qquad b \geq 0,$$

that has been previously discussed in Casella and Berger [2]. As shown in Fig. 1 (left panel), $\mathbb{L}_b$ is asymmetric for $b > 0$: it penalizes more values of θ less than $d(\mathbf{x}_n)$

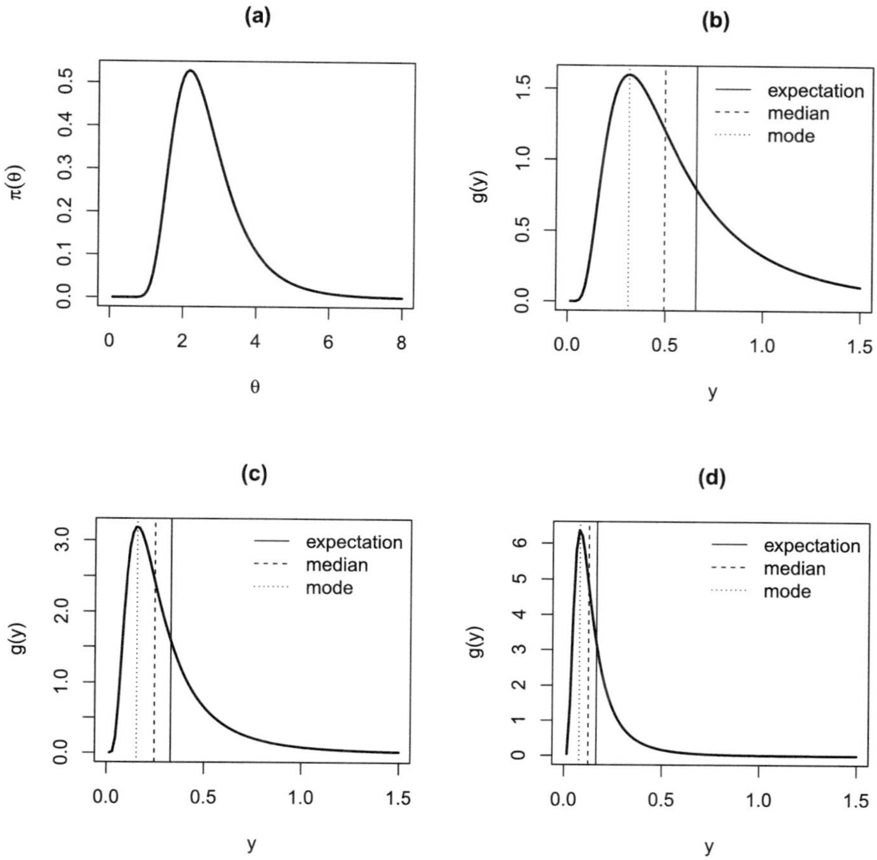

Fig. 1 Example (gastric dissolving time of a drug): **a** inverse gamma prior for Θ with $\alpha = 9.24$ and $\beta = 22.15$ and corresponding plots of the distribution of the risk function under quadratic loss, $g(y)$, for **b** $n = 25$, **c** $n = 50$, **d** $n = 100$

Table 1 Summaries of $g(y)$ under the quadratic loss for several values of n

n	$\mathbb{E}(Y)$	$\mathbb{M}\text{e}(Y)$	$\mathbb{M}\text{o}(Y)$
25	0.659	0.493	0.311
50	0.329	0.247	0.156
100	0.165	0.124	0.078

than those larger. Note that, for $b = 0$, the standard quadratic loss is retrieved. For each $b > 0$, $\mathbb{L}_0(\theta, d) \geq \mathbb{L}_b(\theta, d)$, $\forall \theta > 0$. Furthermore, the larger the value of b, the stronger the penalization.

With this loss function, the explicit expression of $R(\Theta, d)$ cannot be obtained analytically but we can resort to Monte Carlo simulation to approximate its density and the corresponding summaries, as sketched below.

Table 2 Summaries of $g(y)$ under the asymmetric loss for several values of b

b	$\mathbb{E}(Y)$	$\mathbb{M}\mathrm{e}(Y)$	$\mathbb{M}\mathrm{o}(Y)$
0.0	0.659	0.493	0.311
0.5	0.251	0.220	0.187
1.0	0.158	0.142	0.125

1. Draw M values $\theta^{(1)}, \ldots, \theta^{(j)}, \ldots, \theta^{(M)}$ from $\pi(\cdot)$;
2. for each $\theta^{(j)}$, draw N values $d^{(1)}, \ldots, d^{(h)}, \ldots, d^{(N)}$ from $f_d(\cdot|\theta^{(j)})$, the sampling distribution of d with $\theta = \theta^{(j)}$;
3. approximate $R(\theta^{(j)}, d)$ with $\sum_1^N \mathbb{L}_b(\theta^{(j)}, d^{(h)})/N$;
4. obtain the empirical distribution of $Y = R(\Theta, d)$ from the M values of $R(\theta^{(j)}, d)$, $j = 1, \ldots, M$;
5. obtain $\mathbb{E}(Y)$ as $\sum_1^M R(\theta^{(j)}, d)/M$ [and, similarly, other summaries of $g(y)$].

The above algorithm can be generalized to any alternative choice of the loss function by simply adapting step 3.

Note that for our specific case where $d(\mathbf{X}_n) = S_0^2$, $f_d(\cdot|\theta)$ is the cdf of a gamma distribution of parameters $\left(\frac{n}{2}, \frac{n}{2\theta^2}\right)$.

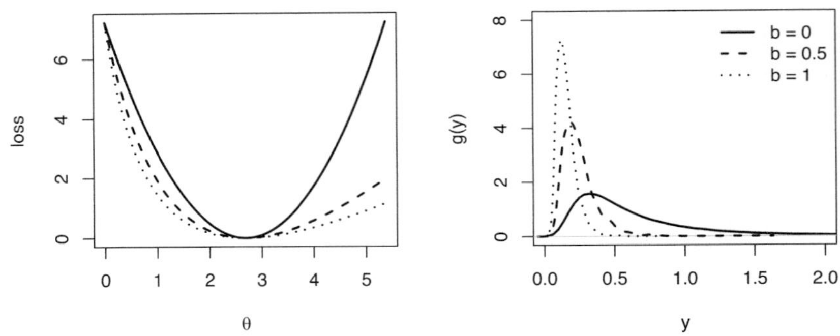

Example: gastric dissolving time of a drug (continued). In the same setup of the example of Sect. 3, we now consider the asymmetric loss function and show the plots of $g(y)$ to assess the impact of different choices of b, when $m = 2.69$, $v = 1$, $n = 25$ (Fig. 1, right panel). The three densities are right-skewed. As expected, the larger b, the more concentrated the corresponding density $g(y)$ on small values of y: it is then the quadratic loss to yield larger risks. The discrepancy between the values of $\mathbb{E}(Y)$, $\mathbb{M}\mathrm{e}(Y)$, $\mathbb{M}\mathrm{o}(Y)$ decreases as b increases (see Table 2); for instance, for $b = 1$, these summaries are very close even for a relatively small sample size ($n = 25$).

4 Sample Size Determination

The prior distribution of Y can be used to define criteria for sample size determination (SSD). In order to highlight the dependence of the risk function on the sample size, we use now the notation $R_n(\Theta, d) = Y_n$. We consider decisions that are consistent with respect to risk functions, that is such that $\lim_{n\to\infty} R_n(\theta, d) = 0, \forall\theta \in \Omega$. This implies that, as $n \to \infty$, $R_n(\Theta, d) \to 0$ pointwise.

Let us suppose that we use the asymmetric loss $\mathbb{L}_b$, with $b \geq 0$ and that we summarize the distribution of the random Y_n using the expectation $\mathbb{E}(Y_n)$, the median $\mathbb{M}\mathrm{e}(Y_n)$ or the mode $\mathbb{M}\mathrm{o}(Y_n)$. Note that consistency of Y_n implies that these three quantities tend to 0 as n tends to infinity. Thus we define the following optimality criteria for SSD:

$$n^\star_{ex} = \min\left(n \in \mathbb{N} : \ \mathbb{E}(Y_n) \leq \lambda\right),$$
$$n^\star_{me} = \min\left(n \in \mathbb{N} : \ \mathbb{M}\mathrm{e}(Y_n) \leq \lambda\right),$$
$$n^\star_{mo} = \min\left(n \in \mathbb{N} : \ \mathbb{M}\mathrm{o}(Y_n) \leq \lambda\right),$$

where the threshold λ represents the maximum tolerable risk.

Example: gastric dissolving time of a drug (continued). Let us now suppose that we want to design a new experiment on the dissolving time of a drug in gastric juice. As an example, we focus on criterion based on the median. Under the assumptions described in the example of Sect. 3.1, Fig. 2 shows the curves of $\mathbb{M}\mathrm{e}(Y_n)$ with respect to n for $b = 0, 0.5, 1$. Consistently with the behavior of $g(y)$ described in Fig. 1 (right panel), as b increases the density of the random risk tends to be concentrated on lower values and its median tends to decrease. Therefore, the optimal sample sizes $n^\star_{me}$ are smaller and smaller, for a given threshold. For illustrative purposes, we set $\lambda = 0.2$ and report in Table 3 the values obtained using the three SSD criteria. As expected, for each SSD criterion the larger b the smaller the optimal sample size; for each value of b, $n^\star_{ex} > n^\star_{me} > n^\star_{mo}$, due to the right-skewness of $g(y)$. In summary, the expectation criterion yields the most conservative choice of the sample size.

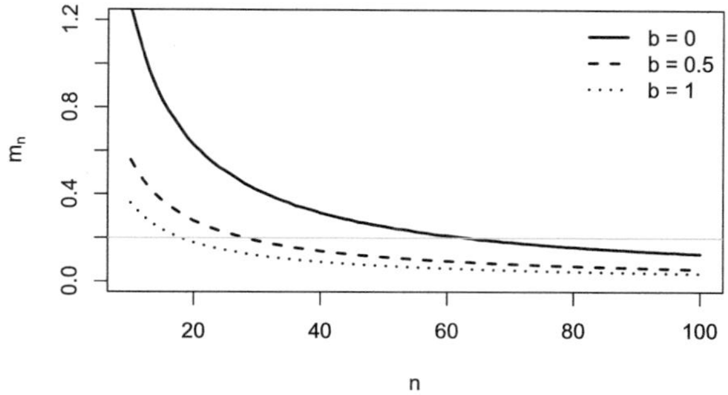

Fig. 2 Plot of the median of the random risk as a function of n for several values of b, $\lambda = 0.2$

Table 3 Optimal sample sizes obtained using different summaries of $g(y)$ under the asymmetric loss for several values of b

b	$n^{\star}_{ex}$	$n^{\star}_{me}$	$n^{\star}_{mo}$
0.0	82	62	42
0.5	32	28	23
1.0	20	18	16

5 Concluding Remarks

In this article, we investigate the entire distribution of the random risk of a point estimator induced by a design prior assigned to the unknown parameter of interest, thus under a Bayesian-frequentist approach. Our point is that the Bayes risk, which is typically used to summarize the random risk, may not be a well-representative summary and a deeper inspection of the whole distribution of the risk is needed. Specifically, we focus on the normal variance conjugate model and provide closed-form expressions for the cdf, pdf and expected value of the random risk under quadratic loss; for those cases when explicit expressions are not available, a simulation algorithm to approximate the density and the summaries of the random risk is provided. We illustrate our ideas through an application to gastric dissolving time of a drug. When the risk function is induced by a quadratic loss, results show that the Bayes risk tends to overestimate the random risk, especially for low values of the sample size, while median and mode seem to be better summaries. As the sample size increases, the distribution of the random risk tends to concentrate on values closer to zero and mean, median and mode tend to coincide. When we account for a modification of the quadratic loss function, we observe that the discrepancy between the three summaries decreases as the asymmetry in the loss function increases, and mean, median and mode tend to assume similar values even when the sample size is low. Finally, also in order to highlight the drawbacks concerning the bad choice of the summary of the random risk, we try to address the sample size determination problem: by overestimating the random risk, the Bayes risk yields an extremely conservative criterion for sample size determination. In conclusion, note that we here focused on the problem related to point estimation. As a matter of future research, a similar analysis can be conducted for hypotheses testing by considering a suitable loss function (for instance, the standard 0–1 loss). In this case the original idea of [5] of studying the whole distribution of the random power could be embedded in a decision-theoretic framework, where the main focus is on the random risk instead of the power.

References

1. Lehman, E.L., Casella, G.: Theory of Point Estimation, 2nd edn. Springer, New York (1998)
2. Casella, G., Berger, R.L.: Statistical Inference, 2nd edn. Duxbury Press, Pacific Grove (2002)
3. De Santis, F., Gubbiotti, S.: On the distribution of the power function for the scale parameter of exponential families. Stat. Med. **43**(10), 1973–1992 (2024). https://doi.org/10.1002/sim.10043
4. Liu, C.C., Yu, R.X.: Epistemic uncertainty in Bayesian predictive probabilities. J. Biopharm. Stat. (2023). https://doi.org/10.1080/10543406.2023.2204943
5. Rufibach, K., Burger, H. U., Abt., M.: Bayesian predictive power: choice of prior and some recommendations for its use as probability of success in drug development. Pharm. Stat. **15**(5), 438–446 (2016). https://doi.org/10.1002/pst.1764
6. Spiegelhalter, D.J., Freedman, L.S.: A predictive approach to selecting the size of a clinical trial, based on subjective clinical opinion. Stat. Med. **5**(1), 1–13 (1986). https://doi.org/10.1002/sim.4780050103
7. Zar, J.H.: Biostatistical Analysis. Pearson (2018)

A Mixed-Integer Linear Program to Create the Shifts in a Supermarket

Nicolò Gusmeroli and Andrea Bettinelli

Abstract The shift design and the personnel scheduling problem is known to be a difficult problem. It is a real-world problem which has lots of applications in the organization of companies. Solutions are usually found by dividing the problem in two steps: first the shifts are created, then the employees are assigned to them by respecting a bunch of constraints. The assignment of different tasks increases the complexity, since we have to consider the skills of the single employee necessary to perform any activity. In this paper we present a mixed-integer linear programming formulation which models together the shift creation and the construction of rosters for employees, with the objective of minimizing the amount of uncovered demand. Finally we provide the results for three real-world instances, confirming that this approach is promising.

Keywords Scheduling · Combinatorial optimization · Mixed-integer liner programming

1 Introduction

Building employee rosters while respecting legal and organizational constraints to satisfy personnel requirements is an NP-complete problem, see [1]. In the past, companies mainly adopted two techniques to simplify the problem: one was to fix the shift hours beforehand and then the employees were assigned in these shifts, in the other they repeated the schedules in a cyclic way over the weeks by shifting the employees, so the planning was done only once. This second approach left the possibility that some worker was not available, which was solved manually at any

N. Gusmeroli
Dipartimento di Elettronica, Informazione e Bioingegneria, Politecnico di Milano, Milan, Italy
e-mail: nicolo.gusmeroli@optit.net

A. Bettinelli (✉)
OPTIT srl, Bologna, Italy
e-mail: andrea.bettinelli@optit.net

M. Di Francesco et al. (eds.), *Operations Research: Closing the Gap Between Research and Practice*, AIRO Springer Series 15, https://doi.org/10.1007/978-3-031-90095-2_13

occurrence. Nowadays the situation has changed since the employees want to have more freedom in the assignment of the shifts, hence the companies started to let people decide the rest day in the week and to give preferences on the working hours. This flexibility made the mathematical modelling of this problem even more difficult and, most likely, less studied.

The problem studied consists in designing individual schedules for a group of heterogeneous workers, which amounts to fixing days-off, creating shifts, and assigning available fixed tasks within these shifts, by respecting both the governor and the company rules, and it is called the Shift-Design Personnel Task Scheduling Problem (SDPTSP). Due to the difficulty of the problem the related literature work is scarce: to the best of our knowledge only few approaches have been proposed and no paper ever presented an explicit mathematical formulation. We aim to mind this gap by formulating a mixed-integer linear programming formulation (MILP) for the SDPTSP. The motivation for this work comes from a real-world application, namely the creation and the assignment of the shifts to the employees of a large Italian retail company, hence we consider additional constraints for our specific use case. The aim of this collaboration is threefold: automatize the rostering process, reduce the work of the managers, and improve the current schedules. Together with our customer we developed a model which was solved with standard algorithms, i.e., not specific for this problem, so there is big room for improvements. We were able to achieve the current demand satisfaction in few minutes, while, today, managers need some hours. Moreover, by setting a time limit of one hour, we obtained solutions that outperformed by far the ones of our customer, both in the missing demand and in the constraints violation (currently managers can violate some of them to have better schedules).

The paper is organized as follows. In the rest of this section we mention the related work. Section 2 presents the mixed-integer linear programming formulation for the SDPTSP. In Sect. 3 we show some experimental results. Finally, Sect. 4 concludes the paper and introduces the future research directions.

1.1 Literature Review

The literature work related to personnel, staff, roster, or crew scheduling, is huge: there are thousands of papers studying several different variants of this problem. Due to the vast literature, we focus only on the few works with big importance for our study case, by mentioning the related problems and suggesting some references.

The problem of creating shifts is called Minimum Shift Design problem (MSD) and various solution methods have been proposed, for more details see [2]. The assignment of employees to fixed shifts, called personnel scheduling, has been introduced in the 1950s by Dantzig [3]. Solution methods have been classified into several categories, we refer to Alfares [4] for a comprehensive survey. The problems concerned with assigning a set of tasks with fixed times to a heterogeneous workforce having predetermined working times fall under the name of Personnel Task

Scheduling problems (PTSP), see [5]. Considering also the design of the shifts makes the situation harder and defines the Shift-Design Personnel Task Scheduling Problem (SDPTSP). As far as we know, only two papers dealt with this problem, they used as objective function an Equity criterion, for which they proposed two different approaches: one based on constraint programming [6] and a two-step metaheuristic [7].

The mentioned problems have lots of real-world applications, Ernst et al. [8] presents some of the possible applications areas. For a very broad and detailed survey on personnel scheduling and rostering problems we refer to [9], another interesting review combining managerial insights and technical knowledge is [10].

2 Definition of the MILP for the SDPTSP

We consider the Shift-Design Personnel Scheduling Task Problem: this means that given the two sets of the employees, with the individual skills and the time availability, and of the demands, with the associated task and times, we want to design the shifts, by fixing the times and the tasks, and to assign them to the different employees. The objective is to cover as much of the demand as possible while respecting legal obligations (max daily hours, max daily span, ...) and company constraints (min time on same activity, min work before break, ...).

We denote by R, A, D, and T the sets of employees, activities, days, and time slots, respectively.

The set of days is defined as $D = \{1, \ldots, |D|\}$, where $|D|$ is the planning horizon. The set of time slots, denoted by T, contains numbers representing the starting minute. For example, if the interval length is 30 min, so the time step is $ts = 30$, we represent the time slot 8–8:30 by the number $16 = (8 \cdot 60)/ts$, the time slot 8:30–9 by $17 = (8 \cdot 60 + 30)/ts$, ... , hence the set of time slots is $T = \{16, 17, \ldots\}$.

To simplify the formulation, we denote by D_k the set of the first k days, i.e., $D_k = \{1, \ldots, k\}$, and its complementary, so the set of all days but the first k, by $D_{\overline{k}} = D \setminus D_k$. In the same way D^k is the set of the last k days, and the set of all the days but the last k is denoted by $D^{\overline{k}}$. Similarly we define T_k (resp. T^k) as the set of the first (resp. last) k time slots and $T_{\overline{k}}$ (resp. $T^{\overline{k}}$) as the set of time slots but the first (resp. last) k.

Finally, we denote by tsD the number of time slots in a day, so $tsD = 1440/ts$.

2.1 Variables

The main variable that we use is the assignment of employee r to activity a at time slot t of day d, so $x(r, a, t, d) = 1$ if r does a at t of d and 0 otherwise. Another important information is the *change of work status* of employee r at time slot t of day d on activity a, i.e., $y(r, a, t, d) = 1$ (resp. -1) if r starts (stops) a at t of d, while $y(r, a, t, d) = 0$ means that r on day d either works both at t and t-1 on a or that he does not perform a in any of these two intervals. We also need a variable indicating whether employee r works on day d, we have that $z(r, d) = 1$ if r works on at least one time slot of d and $z(r, d) = 0$ otherwise. Moreover for each employee r and for each day d we define the begin (resp. end) of work, which is represented by the first (resp. last) time slot with some activity assigned, which is defined $b(r, d)$ (resp. $e(r, d)$).

The demands are defined by exploiting the activity a, the day d, and the start and end time, namely t_1 and t_2 (such that $t_1 < t_2$). It follows that $dem(a, d, t_1, t_2) = k$ means that on day d between time slots t_1 and t_2 employees should work k minutes on activity a. In order to keep the problem feasible, we add a nonnegative slack on each demand, so $\alpha(a, d, t_1, t_2)$ are the minutes of demand $dem(a, d, t_1, t_2)$ which is not satisfied.

2.2 Constraints

In this section we present all the constraints of the model, by exploiting their mathematical formulation. They can be divided into three different groups: problem defining, i.e., giving the relationships between variables, legal, i.e., fixed by the law, and company, i.e., defined by our customer to have better schedules. In order to guarantee feasibility the satisfaction of the demand is modeled as a soft constraint. The compatibility between employees and activities depends on the individual skills of each worker and are defined by the compatibility matrix, which is created in a preprocessing step by matching the skills of the employees and the requirement for each activity. in this way we avoid to consider the skills in the model.

All the variables are derived from the assignment variable $x(r, a, t, d)$, the constraints defining these relationships are

($D1$) definition of y

$$y(r, a, t, d) = x(r, a, t, d) \qquad \forall\, r \in R, a \in A,\ t \in T_1,\ d \in D$$
$$y(r, a, t, d) = x(r, a, t, d) - x(r, a, t\text{-}1, d) \qquad \forall\, r \in R, a \in A,\ t \in T_{\overline{1}},\ d \in D$$

($D2$) definition of z

$$\sum_{a \in A} \sum_{t \in T} x(r, a, t, d) \leq z(r, d) \cdot M \qquad \forall\, r \in R,\ d \in D$$
$$\sum_{a \in A} \sum_{t \in T} x(r, a, t, d) \geq z(r, d) \qquad \forall\, r \in R,\ d \in D$$

where M is a sufficiently big number, e.g., $M = |A| \cdot |T|$,

($D3$) definition of b and e

$$b(r, d) \leq t + tsD \cdot \big(1 - \sum_{a \in A} x(r, a, t, d)\big) \qquad \forall\, r \in R,\ t \in T,\ d \in D$$

$$e(r, d) \geq (t + 1) \cdot \sum_{a \in A} x(r, a, t, d) \qquad \forall\, r \in R,\ t \in T,\ d \in D$$

($D4$) each employee can perform simultaneously at most one activity

$$\sum_{a \in A} x(r, a, t, d) \leq 1 \qquad \forall\, r \in R,\ t \in T,\ d \in D$$

($D5$) correctness of start and end times, so

$$b(r, d) \leq e(r, d) \qquad \forall\, r \in R,\ d \in D$$

The legal constraints depend on the work laws of the country, for us they are

($L1$) maximum daily hours (k_{L1} time slots)

$$\sum_{a \in A} \sum_{t \in T} x(r, a, t, d) \leq k_{L1} \qquad \forall\, r \in R,\ d \in D$$

($L2$) maximum working hours in the planning horizon (k_{L2} time slots)

$$\sum_{a \in A} \sum_{t \in T} \sum_{d \in D} x(r, a, t, d) \leq k_{L2} \qquad \forall\, r \in R$$

($L3$) maximum k_{L3} consecutive working days

$$\sum_{d' \in [d, d+k_{L3}]} z(r, d') \leq k_{L3} \qquad \forall\, r \in R,\ d \in D^{\overline{k_{L3}}}$$

($L4$) maximum consecutive working time, namely every $k_{L4} + j_{L4}$ time slots each employee must be off for at least j_{L4} slots, which is formulated

$$\sum_{t' \in [t, t+k_{L4}+j_{L4})} \sum_{a \in A} x(r, a, t', d) \leq k_{L4} \qquad \forall\, r \in R,\ t \in T^{\overline{k_{L4}+j_{L4}-1}}$$

($L5$) maximum daily span (k_{L5} time slots)

$$e(r, d) - b(r, d) \leq k_{L5} \qquad \forall\, r \in R,\ d \in D$$

($L6$) minimum rest between two working days (k_{L6} time slots)

$$tsD + b(r, d) - e(r, d\text{-}1) \geq k_{L6} \qquad \forall\, r \in R,\ d \in D_{\overline{1}}$$

During several meetings with our customer we defined and implemented their internal constraints, i.e., the company constraints, which are

($G1$) minimum working time after a break (k_{G1} time slots)

$$\sum_{t' \in [t, t+k_{G1})} \sum_{a \in A} x(r, a, t', d) \geq \sum_{a \in A} y(r, a, t, d) \cdot k_{G1} \quad \forall r \in R,\ t \in T^{\overline{k_{G1}\text{-}1}},\ d \in D$$

($G2$) demand satisfaction, formulated as

$$\sum_{t \in [t1, t2)} \sum_{r \in R} x(r, a, t, d) \cdot ts + \alpha(d, a, t_1, t_2) \geq dem(d, a, t_1, t_2) \quad \forall a \in A,\ d \in D,\ t_1, t_2 \in T$$

($G3$) minimum consecutive working time on the same activity ($k_{G3,a}$ time slots)

$$\sum_{t' \in [t, t+k_{G3,a})} x(r, a, t, d) \geq y(r, a, t, d) \cdot k_{G3,a} \quad \forall r \in R,\ a \in A,\ t \in T^{\overline{k_{G3,a}-1}},\ d \in D$$

($G4$) compatibility between employee and activity

$$x(r, a, t, d) \leq cRA(r, a) \quad \forall r \in R,\ a \in A,\ t \in T,\ d \in D$$

where cRA is the compatibility matrix matching employees and activities such that $cRA(r, a) = 1$ if r can do a and 0 otherwise

($G5$) compatibility between employee and time slot

$$x(r, a, t, d) \leq cRTD(r, t, d) \quad \forall r \in R,\ a \in A,\ t \in T,\ d \in D$$

where $cRTD$ is the availability matrix such that $cRTD(r, t, d) = 1$ if r is available to work at t of d and 0 otherwise

($G6$) the checkout management comprises two different activities, i.e., normal working *opCAS* and closure *clCAS*, and it has a specific daily rule: "each employee who works at the checkout has to do only one closure rigth after its last opening": so, for each day, if an employee works on the checkout (do at least one slot *opCAS*), then he must do one slot *clCAS* right after its last *opCAS* slot; this is modelled by the following constraints

- an employee can do *clCAS* only if, the same day, he works also on *opCAS*

$$\sum_{t \in T} x(r, opCAS, t, d) \leq \sum_{t \in T} x(r, clCAS, t, d) \cdot M \quad \forall r \in R,\ d \in D$$

for M sufficiently big, e.g., $M = |T|$

- each employee does at most one *clCAS* per day

$$\sum_{t \in T} x(r, clCAS, t, d) \leq 1 \qquad \forall\, r \in R,\ d \in D$$

- after doing *clCAS*, the employee cannot do *opCAS* until the next day

$$\sum_{t' \geq t} x(r, opCAS, t', d) \leq M(1 - x(r, clCAS, t, d)) \qquad \forall\, r \in R,\ t \in T,\ d \in D$$

for M sufficiently big

- right before *clCAS* the employee must do a *opCAS* time slot

$$x(r, clCAS, t, d) \leq x(r, opCAS, t\text{-}1, d) \qquad \forall\, r \in R,\ t \in T_{\bar{1}},\ d \in D$$

It is important to note that some constraints, e.g., $(L6)$, have to consider the previous schedules; in the results of the next section we also add the historical information even though we did not report the exact formulation of these constraints throughout this section.

2.3 Objective Function

In the optimization process several objective functions might be used, in general the aim is to minimize the costs (declined either as total working days or as uncovered demand).

The company we are working with asked us, as objective function, to cover the most demand possible in order to avoid the employees doing overwork. They also specified that the activities have different priorities, so we gave a penalty to the slack for each activity a, denoted $p(a)$. Moreover, their assignment of employees to tasks depends on the individual work experience, so an employee should be assigned to most suited activity even if he can perform several of them. In order to catch this, we added a multiplicative factor which depends on the matching between employee r and activity a, denoted $c(r, a)$. Thus, the objective function used in the model is

$$\min \sum_{a \in A} \sum_{d \in D} \sum_{t_1, t_2 \in T} p(a) \cdot \alpha(a, d, t_1, t_2) + \sum_{r \in R} \sum_{a \in A} \sum_{t \in T} \sum_{d \in D} x(r, a, t, d) \cdot c(r, a)$$

3 Instances and Computational Aspects

The main scope of this paper is to introduce the mixed-integer linear programming formulation for the Shift-Design Personnel Task Scheduling Problem, with the specific constraints for our case study. In order to validate this formulation and to provide relevance to this problem, we solve three benchmark instances corresponding to three different planning weeks of our customer with a granularity of 15 min (so $|T| = 48$, and $|D| = 7$), and we compare the solution with the actual schedules. The instances have small-medium size, because $|R| = 66$, and $|A| = 72$.

The following results are obtained by using the commercial solver Xpress [11], with version 8.11 on a standard laptop Intel Core i7-8550 at 1.8 GHz with 16 GB RAM running Windows 10.

3.1 Resolution Steps

During the analysis phase and the study of the first results we were able to find some interesting considerations that led us to reduce the size of the problem. From a computational point of view, the most useful was the *identity* of some activities, i.e., there are some tasks which can be performed by the same set of employees, with nonoverlapping demand, and for which all the parameters are equal. Hence, we merged the identical tasks into macro-activities to reduce the size of the problem. By doing this as a pre-processing step we were able to simplify the problem and to obtain better results: we decreased the number of activities, on average, by 29%, and we improved the best solution by 37%; the detailed results can be seen in Table 1. After the optimization, the assignment to these macro-activities is redistributed on the original tasks by a simple backward assignment procedure.

As it is easy to imagine, the main difficulty of a commercial solver on these kind of problems is to give big improvements on the best solution: since there are many feasible solutions with similar values, the incumbent continuously decreases but very slowly. We solved this problem by implementing a base heuristic algorithm,

Table 1 Computational results of Optit for the instances given by our customer with a time limit of one hour

Instance	Merge	Heur	Best sol	Best bound	Gap (%)
W20	N	N	701 384	23 073	97
W20	Y	N	369 069	23 041	94
W20	N	Y	87 691	23 073	74
W20	Y	Y	95 517	23 041	76
W21	N	N	556 049	4 780	99
W21	Y	N	459 375	4 778	99
W21	N	Y	103 632	4780	95
W21	Y	Y	77 633	4 778	94
W22	N	N	693 821	31 009	95
W22	Y	N	297 904	30 976	89
W22	N	Y	169 001	31 009	82
W22	Y	Y	159 152	30 976	80

it produced a feasible assignment in few seconds which was then passed to the solver as a warm-start. The heuristic we developed uses a greedy approach: it creates the shifts for all the employees satisfying the compatibilities of tasks and hours, then it checks for all the conflicts on the constraints and it fixes them by changing the tasks, in this phase the priority is given to activities with higher demand.

3.2 *Experimental Results*

In this section we present the results obtained on a set of three instances provided by the company we are working for, which correspond to three week of plannings of a medium-big size retail store.

Since the results we obtained were already rewarding, we agreed with our customer to set a time limit of one hour, which is far less the actual time needed (around 6 hours) by the store managers.

In Table 1 we validated the computational approaches presented in Sect. 3.1 by exploiting the results for the different combinations of the pre-processing techniques explained. The second column indicates whether *identical* activities have been merged, while the third if the greedy heuristic has been used as a warm start. From these results it follows that using the heuristic decreases the best solution on average by 78%, while merging the activities gives an improvement of 37%. The total processing time is around 1 minute, so it is negligible compared to the time limit.

We claim that these results are far from being optimal, and it can be seen from the gaps in Table 1, this is mainly because our approach is not problem-dependent: we used a commercial solver on a general formulation without tuning the internal parameters and we did not implement specific heuristics. Anyway, these first results are very promising for us and rewarding for our customer.

Since these instances are provided by the company we are working with, there is no benchmark result to check the quality of the solution. The only comparison we can make is with the current plannings of our customer.

In Table 2 we compare our results (case O) with the real schedules of our customer (case C). In the third column we provide the total missing hours to cover the demand, i.e., the sum of the slacks without any multiplicative parameter, while the fourth

Table 2 Comparison of Optit results (case O) with actual planning (case C)

Instance	Case	$\sum \alpha$ [h]	Violations	Dept dem %
W20	C	374	579	86%
W20	O	59	0	96%
W21	C	366	544	86%
W21	O	62	0	95%
W22	C	356	744	86%
W22	O	110	0	95%

indicates the number of violated constraints; since we use an exact model, we do not violate any constraint, but in the real life managers do it quite often. In the fifth column we give a percentage of saturation of the whole demand by considering the departments and not the single tasks, this is done because sometimes the managers assign the employees to only one task, but then they can freely shift activity within the same department if needed.

It is easy to see that the missing demand in our solution is much smaller than the actual planning. If we consider the single activities we have a huge decrease of around 70%. With respect to the associated departments, we can see that on average our demand satisfaction is 10% more, which still gives a big improvement. Another important indicator of the quality of our solutions are the constraints violated in the current weekly schedules (around 600), which for us are 0 since we use an exact model.

It follows from the results that, in comparison with the actual planning, we are outperforming the current schedules both with respect to the covered demand and to the violation of the constraints.

4 Conclusion

The Shift-Design Personnel Task Scheduling Problem has been studied only by few authors in the literature and, to the best of our knowledge, no exact methods were proposed to solve this problem. The main work of this paper has been to give the first MILP formulation for this problem, defining the generic constraints given by the law. Our study case is a big size Italian retail company, so we also implemented some constraints for this specific case, to provide them with better results. In order to validate our model we studied three different weeks of planning on a store, and we provided the results, by using the commercial solver Xpress. Moreover, we explained how to decrease the magnitude of the problem, in order to tackle more difficult, i.e., bigger, instances. Finally, we compared our results with the actual planning. We were able to decrease the uncovered demand by 70%, which corresponds to a huge saving. At the same time our solutions respected all the constraints, while, as of today, the schedules of our customer violate around 600 constraints every week. Considering the department and not the single tasks, the improvement is lower, but still very significant since we cover 10% more demand. It is also important to say that as of today managers need six hours every week for the planning, while we set a time limit of 60 min. Hence we obtained better results in a less amount of time.

As it can be imagined the company we are working with is very satisfied with these first results, but there is still ongoing work in different directions. As a first step, we are trying to catch some other needs, in order to add more specific constraints and to provide even better results. Another direction of future work is the development of specific techniques to speed up the computational times and, hence, to solve bigger instances. We saw that, by passing a feasible schedule to the solver as a warm-start, the quality of the solution improved drastically in a short amount of time, even with our simple greedy heuristic: we were able to reach the same results in few minutes instead of an hour. Thus, implementing a smarter heuristic is one of the tasks with higher priority to further improve the current solutions.

References

1. Smet, P., Wauters, T., Mihaylov, M., Berghe, G.V.: The shift minimisation personnel task scheduling problem: a new hybrid approach and computational insights. Omega **46**, 64–73 (2014)
2. Di Gaspero, L., Gärtner, J., Kortsarz, G., Musliu, N., Schaerf, A., Slany, W.: The minimum shift design problem. Ann. Oper. Res. **155**, 79–105 (2007)
3. Dantzig, G.B.: A comment on Edie's "Traffic delays at toll booths." J. Oper. Res. Soc. Am. **2**, 339–341 (1954)
4. Alfares, H.K.: Survey, categorization, and comparison of recent tour scheduling literature. Ann. Oper. Res. **127**, 145–175 (2004)
5. Krishnamoorthy, M., Ernst, A.T.: The personnel task scheduling problem. Optimization methods and applications, pp. 343–368 (2001)
6. Lapègue, T., Bellenguez-Morineau, O., Prot, D.: A constraint-based approach for the shift design personnel task scheduling problem with equity. Comput. & Oper. Res. **40**, 2450–2465 (2013)
7. Prot, D., Lapègue, T., Bellenguez-Morineau, O.: A two-phase method for the shift design and personnel task scheduling problem with equity objective. Int. J. Prod. Res. **53**, 7286–7298 (2015)
8. Ernst, A.T., Jiang, H., Krishnamoorthy, M., Sier, D.: Staff scheduling and rostering: a review of applications, methods and models. Eur. J. Oper. Res. **153**, 3–27 (2004)
9. Van den Bergh, J., Beliën, J., De Bruecker, P., Demeulemeester, E., De Boeck, L.: Personnel scheduling: a literature review. Eur. J. Oper. Res. **226**, 367–385 (2013)
10. De Bruecker, P., Van den Bergh, J., Beliën, J., Demeulemeester, E.: Workforce planning incorporating skills: state of the art. Eur. J. Oper. Res. **243**, 1–16 (2015)
11. FICO Xpress Optimization Suite, v. 8.11

A Greedy Heuristic for the Equitable Allocation of Edible Surplus Food

Fuzail Dawood and Jan H. van Vuuren

Abstract The strategic allocation of surplus food by *foodbanks* to vulnerable populations presents an opportunity to aid in the daily decision making of foodbank managers by assisting them in making three critical daily decisions in support of achieving operational efficiency. These decisions are which *beneficiary organisations* to serve, which food items to allocate to each beneficiary organisation, and the quantity of each of these items to supply to each beneficiary organisation on any particular day. In this paper, we present a greedy heuristic for the equitable food allocation among beneficiary organisations, constrained by the unpredictable nature of donations entering the foodbank warehouse. The strategic prioritisation of beneficiary organisations is considered in the heuristic, as are nutritional requirements, inventory management, and diverse nutritional distribution. The heuristic is aimed at addressing the challenge of utilising limited resources and supporting the critical decisions that foodbanks face on a daily basis towards meeting the demands of vulnerable populations. By applying the heuristic in a real-world case study, it is demonstrated that the allocation of food waste may increase the average quality of monthly deliveries across all beneficiary organisations by up to 13.40%.

Keywords Multiple-resource allocation · Greedy heuristic · Foodbank · Decision support

1 Introduction

In 2022, it was estimated that approximately US $1 trillion worth of edible food is discarded by retailers and consumers annually. Moreover, the *United Nations Food and Agriculture Organisation* (FAO) has estimated that reversing this trend has the

---

F. Dawood (✉) · J. H. van Vuuren
Stellenbosch Unit for Operations Research in Engineering, Stellenbosch University, Stellenbosch, South Africa
e-mail: 21582068@sun.ac.za

J. H. van Vuuren
e-mail: vuuren@sun.ac.za

M. Di Francesco et al. (eds.), *Operations Research: Closing the Gap Between Research and Practice*, AIRO Springer Series 15, https://doi.org/10.1007/978-3-031-90095-2_14

potential to feed two billion people—more than twice the global number of undernourished people [16]. To this end, non-profit organisations known as *foodbanks*, receive food or monetary donations that are redirected towards vulnerable populations. These foodbanks play a pivotal role in the acquisition and allocation of surplus food [6]. As such, the efficiency of foodbank operations and their logistics is critical in the fight against hunger.

The typical foodbank model involves three key participants, namely *donors* (such as farmers, manufacturers, and retailers) who supply damaged or surplus food to the foodbank, the foodbank itself which receives and manages these donations, and various vetted *beneficiary organisations* (BOs) (such as orphanages, old age homes, and homeless shelters) to which the food is distributed. Foodbank models operate within highly unpredictable supply chains in which resources are typically very limited [2]. A major source of this unpredictability results from the nature of food supply, which depends solely on donations. More specifically, the uncertainty associated with food donations derives from variabilities in the frequency, type, quality, and quantity of food donated. This results in a situation where foodbanks are often unable to employ equitable strategies in respect of how much of each product each BO should receive daily. In addition to the aforementioned uncertainty, the operations of foodbanks typically rely on volunteers, leading to inefficiencies resulting from an often inexperienced and unaccountable workforce with a high turnover rate. As a result, managing the operations of a foodbank can be a complex task and often results in sub-optimal allocation to some BOs [12].

In recent years, there has been a burgeoning of programmes globally aimed at distributing wasted and damaged food items to people in need. In particular, *Foodbank Australia* provides 65 million meals per year, *OzHarvest* distributes more than 5 500 tons of food per year, *Foodbank Central and Eastern North Carolina* distributes 33 000 tons of food per year, *Feeding America* feeds 46 million people per year, *Second Harvest Canada* provides 11 million meals per year, and *FoodForward South Africa* (FFSA) distributes more than 22 000 tons of food per year [5, 13]. Despite these impressive contributions, it is an unfortunate reality that the nutritional demands of BOs often exceed the available supply. This imbalance is often not due to the mismanagement of foodbanks, but rather speaks volumes about the degree to which many populations experience chronic hunger, and that the mere existence of foodbanks means that millions of people globally are able to receive some relief which they otherwise would not. As the scale of foodbank operations grows, so too does the demand for accompanying, sophisticated quantitative decision support frameworks. Logistically, the typical foodbank model comprises two main components, namely food allocation and food distribution. The former, which is the focus of this paper, is concerned with the fair allocation of food among BOs, while the latter is concerned with determining the most cost-effective transportation schedules [13].

In this paper, we propose a *greedy heuristic* for multiple-resource allocation. Greedy heuristics are commonly employed in the field of optimisation in order to obtain a feasible solution in a computationally efficient manner [15]. The principle governing such a heuristic is the pursuit of that decision aspect yielding a locally optimal partial solution at each decision point, in the hope of reaching a global optimum [10]. The drawback of this type of heuristic, however, is that it may often not lead to the global optimum since the optimisation approach is short-sighted and does not consider the full extent of the solution space [7]. The solution presented here employs data provided by FFSA, the largest foodbank in South Africa, and serves as decision support for the foodbank manager in support of which BOs to serve, which food items to allocate to these BOs, and the quantity of each item to supply each day. While previous research has been focussed on equitable multiple-resource allocation, no study has, to the best of our knowledge, employed a BO-specific demand based on nutritional goals, nor has an effective monitoring metric been introduced to assess the relative urgency or deprivation experienced by each BO.

The remainder of the paper is organised as follows. A brief review of the literature on fair resource allocation in the context of food waste is provided in Sect. 2. Thereafter, the methodology adopted in our proposed greedy heuristic is discussed in Sect. 3. Numerical results obtained when applying the heuristic in a realistic case study are presented in Sect. 4, after which the paper closes in Sect. 5 with a brief conclusion and suggestions for future work.

2 Related Work

The majority of the literature on foodbank decision support is focused on the distribution component of foodbank logistics. These problems are largely concerned with the optimal placement of depots and the cost-efficient transportation of food to BOs, and most often neglect to address the problem of equitable food allocation [13].

Various modelling approaches have been adopted in the literature in order to address the problem of equitable allocation of scarce resources. The objective in these models vary in nature, taking the form of minimising the range between maximum and minimum values, minimising variance, minimising the coefficient of variation, minimising the sum of absolute deviations, minimising the largest deviation, or minimising the mean absolute deviation. Pursuing these objectives may lead to enhanced equitability, but the resulting solutions often vary significantly [14]. Fianu and Davis [4] derived a Markov decision process model capable of allocating an uncertain supply of food equitably by employing the *pounds distributed per person in poverty* (PPIP) indicator as a measure of equitability. In particular, their model is able to (i) identify an optimal food allocation that maximises equitability in respect of meeting the PPIP indicator, (ii) estimate the unmet demand in each region of a geographic area, and (iii) determine the likelihood of each region's demand remaining unmet.

Lien et al. [8] modelled the *sequential resource allocation* problem in the context of a dynamic programming framework. In this context, the demand of each BO is observed sequentially, and the aim is to maximise equitable allocation and effective service. The authors demonstrated that employing a model in which a waste-minimisation objective is embedded, in fact, leads to inequitable allocation due to the fact that the model tends to pivot towards an extreme allocation policy as opposed to perfectly satisfying each BO's *fill rate* (the ratio of the food amount allocated to the food demand). Instead, they proposed that an objective function be adopted in which the aim is to maximise the expected minimum fill rate across all BOs, since this objective is expected to yield the best overall results with respect to both equitability maximisation and waste minimisation. Much like the subsequent approach of Orgut et al. [9], the aforementioned approach of Lien *et al.* is limited to the allocation of a single resource, making it difficult to adopt in a foodbank context comprising multiple heterogeneous food items.

Alkaabneh et al. [1] proposed a framework that directly addresses the effectiveness and efficiency of the resource allocation challenges faced by foodbanks while also indirectly taking into account a measure of equitability in performance. This was incorporated into a dynamic programming model aimed at determining an optimal quantity of each food item to be allocated to each BO. The effectiveness of an allocation was determined by the nutritional value provided, while the measure of efficiency was based on the benefit derived by the BOs. More specifically, equitability was measured by how evenly food items had been allocated to each BO in proportion to their demand.

In a recent survey paper, covering more than 300 research papers related to the operations of foodbanks and the optimisation of their supply chains, Rivera et al. [14] claimed that the focus of recent models had been on the nutritional quality of food allocated as opposed to the quantity of allocated food. Moreover, the authors found that food deprivation[1] had not been addressed in any single review paper, despite being a critical metric often advocated in the field of humanitarian logistics with respect to the degree of urgency at which vulnerable populations have to be aided.

3 Methodology

The greedy heuristic that we propose functions by sequentially allocating available units of food items to a ranked list of BOs. This ranked list may be initialised by employing a data-driven approach towards spatially quantifying food security, as exemplified by Dawood and Van Vuuren [3]. Moreover, the demand estimation of each BO is based on the nutritional requirements of the beneficiaries served by the BO, and each allocation made by the foodbank has an associated *benefit* for the BO which, in turn, has an effect on the perceived *urgency* of the BO, signalling to the foodbank which BOs are most vulnerable at any given point in time. In particular, a proportion ϕ_f of each food item f available in the warehouse is allocated proportionally to the demand of each BO. The aim of the foodbank is assumed to be to provide at least one delivery per month to each BO which satisfies approximately 30–50% of the BO's total demand. As such, an approach towards pursuing equitability is to serve BOs in a manner that increases the mean benefit enjoyed by all BOs over a

[1] Food consumption that does not satisfy the minimum required dietary goal [11].

period of 30 days. The variables employed are demand, benefit, and urgency, which are defined as described in the remainder of this section.

3.1 Demand

One of the challenges faced by foodbanks is the absence of a formal system for quantifying the specific nutritional requirements of each BO. In the case where the foodbank has access to data pertaining to the number of beneficiaries belonging to each gender/age cohort in each BO, it is possible to obtain an estimate of the minimum number of calories and kilograms of food items required by each BO, by employing an appropriate lookup table. An example of this approach may be found in [17], where nutritional guidelines are specified for each of two genders indexed by i, each of seven age cohorts indexed by j, and each of four macro-nutrients indexed by k (proteins, carbohydrates, fats, and fibres). This may serve as a theoretical static demand estimate that the foodbank aims to satisfy at least once a month. The total monthly (static) demand for each BO may then be defined as

$$D_{\beta}^{Static} = n_{i,j}^{\beta} \sum_{k=1}^{4} G_{i,j,k}^{\beta}, \quad i \in \{0, 1\}, \quad j \in [1, 7], \quad k \in [1, 4], \tag{1}$$

where $n_{i,j}^{\beta}$ denotes the number of individuals of gender i and age cohort j represented by BO β, and where $G_{i,j,k}^{\beta}$ denotes the corresponding nutritional goal, measured in kilograms, for macro-nutrient k.

3.2 Benefit

The perceived benefit enjoyed by a BO in respect of each delivery is defined as the ratio $B(t) \in [0, 1]$ of the food quantity delivered to the food demand for the current period, and represents the degree to which the demand is fulfilled. This quantity is sometimes referred to as the fill rate [8] and may be expressed as

$$B(t) = \frac{x_{t,\beta}}{D_{t,\beta}}, \tag{2}$$

where $x_{t,\beta}$ denotes the amount of food delivered to BO β during time interval t and $D_{t,\beta}$ denotes the demand of BO β during time interval t. Thus, a ratio $B(t) = 0$ corresponds to no delivery, while a ratio satisfying $0 < B(t) < 1$ corresponds to a partial delivery. Finally, a ratio $B(t) = 1$ corresponds to a perfect delivery, while $B(t) > 1$ corresponds to an excess delivery. In the case where the foodbank delivers in excess of the demand of a BO (that is, if $x_{t,\beta} > D_{t,\beta}$), then the benefit function value is capped at the value $B(t) = 1$ so as to ensure that delivering excess food to a BO is not mistakenly interpreted as providing additional benefit to the BO while, in fact, the action takes benefit away from another BO. This becomes increasingly important in the context of a resource allocation problem in which the model should not be incentivised to over-deliver to any of the BOs, for this will likely be to the detriment of some other BO. A benefit-levelling perspective should therefore rather be adopted in which the aim is to deliver perfectly ($x_{t,\beta} = D_{t,\beta}$) to as many BOs as possible.

3.3 Urgency

The urgency $U_\beta(t) \in [0, 1]$ of a BO β during time interval t serves as a relative measure of its vulnerability or deprivation at that time. This metric may be utilised by a foodbank as a means to monitor and assess the vulnerability levels of each BO over time. Under the assumption that the foodbank aims to provide one delivery per month to each BO, the urgency function, modelled as an exponential growth function, is configured such that a BO which receives a perfect delivery is associated with an urgency of $U_\beta(t) = 1$ after 30 days have elapsed. Accordingly, the rate at which the urgency of a BO increases over time is proportional to the benefit derived from a delivery. For example, if a benefit value $B(t) = 0.8$ is derived from a delivery, then the time it takes until $U_\beta(t) = 1$ is 24 days. The urgency $U_\beta(t)$ of each BO β at time t may be evaluated based on one of the three delivery scenarios:

No delivery. For a BO which does not receive a delivery during time period t (i.e. $x_{t,\beta} = 0$), the corresponding benefit is equal to zero. As a result, the urgency experienced after a period of length t following the last delivery at time $t_{delivery}$ may be expressed as

$$U_\beta(t) = U_\beta(t_{delivery}) \times (1 - B(t_{delivery}))^{t/c}. \tag{3}$$

Solving for the time scaling factor in (3) yields

$$c = t \times \frac{\ln(1 - B(t_{delivery}))}{\ln\left(\frac{U_\beta(t)}{U_\beta(t-1)}\right)}. \tag{4}$$

In accordance with the requirement that the time it takes until a BO β is associated with an urgency of $U_\beta(t) = 1$ exactly 30 days after having received a delivery is proportional to the benefit derived from a perfect delivery, the expression in (4) has to be configured so that $t = B(t_{delivery}) \times 30$ days in order to determine the number of days before the full benefit derived from the last delivery has been consumed and will result in the value $U_\beta(t) = 1$. Accordingly, the expression in (4) becomes

$$c = B(t_{delivery}) \times 30 \times \frac{\ln(1 - B(t_{delivery}))}{\ln\left(\frac{1}{U_\beta(t-1)}\right)}. \tag{5}$$

The term $1 - B(t_{delivery})$ is present in the expression above to ensure that deliveries exhibiting greater values of $B(t)$ result in a slower growth rate, while smaller values of $B(t)$ result in a faster growth rate. Moreover, since $U_\beta(t)$ is interpreted as a value in the range $[0, 1]$, a value of $U_\beta(t) > 1$ is capped at a value of 1 to ensure that the upper limit of the range is not violated.

Partial delivery. A BO that receives a partial delivery (i.e. $0 < x_{t,\beta} < D_{t,\beta}$) corresponds to a benefit function value $0 < B(t) < 1$, indicating that the foodbank has under-delivered, since only part of the demand for time period t has been satisfied. As such,

$$U_\beta(t) = \left[U_\beta(t-1) - B(t)\right] \times (1 - B(t))^{t/c}, \tag{6}$$

where the scaling factor for time is once again determined by the expression in (5). Moreover, the first factor in (6) is calculated as the difference between the urgency registered for the previous time period and the benefit derived from the delivery during the current period. This difference represents the amount by which the urgency is reduced as a result of the delivery.

Perfect or excess delivery. A BO that receives a perfect delivery or an excess delivery (i.e. $x_{t,\beta} \geq D_{t,\beta}$) corresponds to a benefit value $B(t) = 1$, indicating that the foodbank has satisfied a BO's demand exactly, or has delivered in excess of the BO's demand. In the case of the latter delivery, a benefit $B(t) > 1$ is not assigned, as mentioned, since the benefit function is constrained to the range $[0, 1]$ for proper interpretation purposes as well as to avoid incentivising future assignment strategies which may seek to maximise the benefit of a few BOs at the cost of other poorly served BOs. In addition to potentially disadvantaging some BOs, incentivising a delivery in excess of a BO's demand has the potential to disrupt the delivery schedule in an already highly uncertain supply environment. In the case of a perfect delivery or excess delivery, therefore,

$$U_\beta(t) = 0.001 \times (1 - 0.999)^{t/-30},$$

where the values 0.001 and 0.999 are conveniently used as proxies for the values 0 and 1, respectively. Furthermore, the value of the denominator in the exponent, -30, is based on the expression in (5).

3.4 *The Greedy Heuristic*

For each time period, Algorithm 1 is invoked so as to determine the greedily assigned quantities of food items to each BO as well as the revised ranked list in which BOs should be prioritised for service during the upcoming period.

For each food item f available at the foodbank during time period t, the assignment strategy involves allocating a proportion ϕ_f (defined in line 2) to the highest-ranked BO based on its static demand $D_{\alpha,\beta}^{Static}$ which is a proportion α of the total BO demand that the foodbank seeks to satisfy. Allocating the available food items according to the ratio ϕ_f ensures diversity among food allocations, so that no single food item is allocated disproportionately relative to the other food items. From an equitable allocation standpoint for BOs, this approach further ensures that if, for example, there is an abundance of a desirable food item f in the warehouse, then most BOs served on a particular day will receive a portion of it until its quantity is depleted. In the case that ϕ_f is not adopted, however, the system is at risk of fully satisfying a BOs demand with a desirable food item, potentially leaving very little to nothing of it to be allocated to other BOs. This process is repeated for each BO in order of decreasing priority in the ranked list of BOs $\mathcal{B}^t$ until the demand of all BOs has been met or until all food items in the set $\mathcal{W}^t$ have been allocated. The inventory levels of the foodbank are monitored for every allocation and partially served BOs are served first during the succeeding period, after which unserved BOs are served, and finally, the fully served BOs (line 20). This approach ensures that no BO is served twice or more before any other BO has been served once. Furthermore, the proposed greedy approach is aimed at satisfying the minimum weight of food required to meet the nutritional needs of each BO by fulfilling its full demand, and so it allocates food items in a manner that ensures as diverse an allocation as possible, it monitors the inventory level, and it accounts for the strategic prioritisation of BOs in a fair manner. It is important to reiterate that allocations are made based on the weight required to satisfy the minimum nutritional goals of a BO, as opposed to satisfying the nutritional goals themselves. This comes as a result of data limitations. More importantly, however, in an environment characterised by time-sensitive decisions and high supply uncertainty, it is impractical to wait for or anticipate the donation of food items belonging to a particular food group in order to satisfy the nutritional demand of a BO exactly, since during this time, it is inevitable that other BOs will be starved and perishable items are at greater risk of being rendered inedible, thereby decreasing the foodbank's inventory.

4 Numerical Results

In order to evaluate the efficacy of our greedy heuristic, it was necessary to obtain a baseline scenario. As such, the allocations of the Western Cape region's branch of FFSA were assessed over a two-year period. It was found that 62 donors provided 1 651 distinct food items to 124 BOs during this period. The same period was also considered during a retrospective hypothetical application of our greedy allocation approach, described in the previous section, in order to gauge whether or not this simple allocation strategy would be able to improve upon the *status quo* in respect of the average monthly urgency and benefit of each BO. The means and standard deviations of the 30-day urgency and 30-day benefit indicators are tabulated in Table 1 for both scenarios.

Algorithm 1 Greedy food allocation heuristic

Input: A set $\mathcal{W}^t$ of food items available in the foodbank's warehouse at time t, and a ranked list $\mathcal{B}^t$ of BOs, each with an associated demand $D_{\alpha,\beta}^{Static}$, as defined in (1).

Output: A greedy allocation of food items $f \in \mathcal{W}^t$ to a maximum number of BOs during time t, and a revised ranked list $\mathcal{B}^{t+1}$ for time period $t + 1$.

$TotalStockRemaining \leftarrow$ The sum of all food items' weight in $\mathcal{W}^t$ at the start of the day (i.e. at time t)
$\boldsymbol{\phi} \leftarrow$ The ratios between the weights of all $f \in \mathcal{W}^t$
for each BO in $\mathcal{B}^t$ **do**
  $TotalStockAssigned_f \leftarrow 0$
  while $TotalStockRemaining > 0$ AND $TotalStockAssigned_f < D_{\alpha,\beta}^{Static}$ **do**
    for each food item $f \in \mathcal{W}$ **do**
      if Remaining weight of $f > 0$ **then**
        $StockAssigned_f \leftarrow \phi_f \times D_{\alpha,\beta}^{Static}$
        $TotalStockAssigned_f \leftarrow TotalStockAssigned_f + StockAssigned_f$
        $TotalStockRemaining \leftarrow TotalStockRemaining - StockAssigned_f$
      end if
    end for
  end while
  Append BO to *FullyServedBOs* list
  if $TotalStockRemaining = 0$ AND $TotalStockAssigned_f < D_\beta$ **then**
    Append BO to *PartiallyServedBOs* list
  end if
end for
Append BOs $\notin$ {*FullyServedBOs, PartiallyServedBOs*} to *UnservedBOs* list
$\mathcal{B}^{t+1} \leftarrow$ {*PartiallyServedBOs, UnservedBOs, FullyServedBOs*}

Table 1 The means and standard deviations of the 30-day urgency and benefit indicators in the baseline and greedy allocation scenarios

Performance indicator	Statistic	Baseline	Greedy	$\lvert\Delta\rvert$
30-day urgency	Mean	0.754	0.708	0.046
	Standard deviation	0.221	0.060	0.161
30-day benefit	Mean	0.372	0.506	0.134
	Standard deviation	0.241	0.067	0.147

When comparing the baseline scenario with the greedy scenario, a 4.6% decrease in the mean urgency is observed, indicating a diminished sense of immediacy for food among BOs, accompanied by a 16.1% reduction in urgency variability, suggesting more uniform experiences of need across BOs. Furthermore, a 13.4% increase in mean benefit is indicative of an improvement in the quality of each delivery, on average, while a 14.7% reduction in benefit variability is indicative of a more equitable allocation.

A scatter-plot of the 30-day mean urgency and benefit experienced by each BO over the study period is provided in Fig. 1 for both scenarios. In this plot, each point represents a BO, whilst the two scenarios are distinguished by the colour shading of the points. It is evident from the plot that the greedy approach exhibits superior performance with respect to the two axes independently. This is because of the

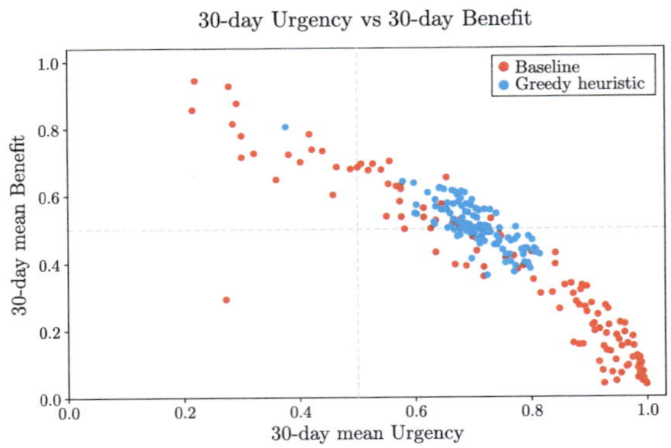

Fig. 1 A scatter-plot of the urgency and benefit experienced by all BOs in both the baseline and greedy allocation scenarios

Table 2 A comparison between the quadrants of the scatter plot in Fig. 1

Quadrant	Baseline	Greedy	|Δ|
1 (Top left)	14.52% (18)	0.81% (1)	13.71% (17)
2 (Top right)	17.74% (22)	54.03% (67)	36.29% (45)
3 (Bottom right)	66.94% (83)	45.16% (56)	21.77% (27)
4 (Bottom left)	0.81% (1)	0% (0)	0.81% (1)

homogeneity of the blue-shaded cluster of points, which suggests that the perceived experience of urgency and benefit by each BO is significantly more similar in the greedy allocation scenario than in the baseline scenario.

A summary of the proportion (and count) of BOs in each quadrant of the plot in Fig. 1 is provided in Table 2 for both scenarios.

While the ultimate objective would be to determine an allocation that is able to position as many BOs as possible in the first (top left) quadrant, and as few in the third (bottom right) quadrant as possible, the limitations of adopting a greedy approach towards food allocation only allows for partial improvement at best. A total of 13.71% of all BOs were displaced from the first quadrant, while an additional 36.29% of all BOs were included in the second quadrant. Moreover, 21.77% of all BOs were displaced from the third quadrant, while only one BO was displaced from the fourth quadrant, which represents a significant improvement in the quality of food allocation.

5 Conclusion

In this paper, we introduced a greedy heuristic for distributing surplus food to BOs within a foodbanking context. Employing two-year's worth of FFSA delivery data, this approach was shown to reduce the monthly urgency and increase the monthly benefit markedly across all BOs when compared with the existing system's baseline performance. More specifically, we observed a 4.6% reduction in monthly urgency, reflecting a decrease in immediate need, alongside a 13.4% increase in monthly benefit, reflecting an improvement in the quality of food deliveries with respect to meeting the demand of BOs. Moreover, accompanying decreases in the variability of the monthly urgency and benefit values underscore a move towards more equitable distribution among BOs. While the model presented here is simple in its approach, it is significantly limited by the uncertainty of supply by donors, thereby raising a question as to whether the model performs better, worse, or the same under a different set of donation conditions. As such, a recommendation for future work is to evaluate the robustness of the model under varying degrees of supply, perhaps by Monte Carlo simulation. Regardless of this limitation, the proposed heuristic still represents a simple, fair, and effective approach towards strategically allocating food surplus to those in need. The quality of the allocation achieved by the heuristic suggests a potential for further enhancements, warranting further exploration of more sophisticated optimisation approaches which may yield even better improvements in the equitable allocation of wasted food.

References

1. Alkaabneh, F., Diabat, A., Gao, H.: A unified framework for efficient, effective, and fair resource allocation by food banks using an approximate dynamic programming approach. Omega **100**(1), 1–10 (2021)
2. Ataseven, C., Nair, A., Ferguson, M.: An examination of the relationship between intellectual capital and supply chain integration in humanitarian aid organisations: a survey-based investigation of food banks. Decis. Sci. **49**(5), 827–862 (2018)
3. Dawood, F., van Vuuren, J.H.: A multi-dimensional spatial index for the quantification of food insecurity. J. Agric. Food Res. **14**(1), 1–12 (2023)
4. Fianu, S., Davis, L.B.: A Markov decision process model for equitable distribution of supplies under uncertainty. Eur. J. Oper. Res. **264**(3), 1101–1115 (2018)
5. FoodForward SA 2022 Impact: Leading change in the cost of living crisis. Electronic press release (2022). https://foodforwardsa.org/2022/12/12/foodforward-sa-2022-impact-leading-change-in-the-cost-of-living-crisis/
6. González-Torre, P.L., Coque, J.: How is a food bank managed? Different profiles in Spain. Agric. Human Values **33**(1), 89–100 (2016)
7. Hao, J.K., Wu, Q.: A review on algorithms for maximum clique problems. Eur. J. Oper. Res. **242**(3), 693–709 (2015)
8. Lien, R.W., Iravani, S.M., Smilowitz, K.R.: Sequential resource allocation for nonprofit operations. Oper. Res. **62**(2), 301–317 (2014)
9. Orgut, I.S., Ivy, J.S., Uzsoy, R., Hale, C.: Robust optimisation approaches for the equitable and effective distribution of donated food. Eur. J. Oper. Res. **269**(2), 516–531 (2018)
10. Peng, S., Zhou, Y., Cao, L., Yu, S., Niu, J., Jia, W.: Influence analysis in social networks: a survey. J. Netw. Comput. Appl. **106**(1), 17–32 (2018)
11. Pérez-Sales, P., Petersen, H.D.: Forensic Assessment of Psychological Torture. In: Houck, M.M. (ed.) Encyclopedia of Forensic Sciences, 3rd edn., pp. 493–502. Elsevier, Amsterdam (2023)
12. Reusken, M., Cruijssen, F., Fleuren, H.: A food bank supply chain model: Optimising investments to maximise food assistance. Int. J. Prod. Econ. **261**(1), 1–7 (2023)

13. Rey, D., Almi'ania, K., Nair, D.J.: Exact and heuristic algorithms for finding envy-free allocations in food rescue pickup and delivery logistics. Transp. Res. Part E: Log. Transp. Rev. **112**(1), 19–46 (2018)
14. Rivera, A.F., Smith, N.R., Ruiz, A.: A systematic literature review of food banks' supply chain operations with a focus on optimization models. J. Humanit. Log. Supply Chain Manag. **13**(1), 10–25 (2023)
15. Roy, V., Desai, A.: A heuristic approach of web users decision-making using deep learning models. In: Cognitive Big Data Intelligence with a Metaheuristic Approach, pp. 117–130. Academic, Massachusetts (MA) (2001). https://doi.org/10.1016/B978-0-323-85117-6.00015-7
16. United Nations World Food Program USA: Eight facts to know about food waste and hunger. Electronic press release (2022). https://www.wfpusa.org/articles/8-facts-to-know-about-food-waste-and-hunger/
17. United States Department of Health and Human Services: Nutritional goals for each age/sex group used in assessing adequacy of USDA food patterns at various calorie levels. Electronic press release (2015). https://health.gov/sites/default/files/2019-09/Appendix-E3-1-Table-A4.pdf

Probabilistic Branch-and-Bound for Clusterwise Linear Regression

Andrea Fois, Luca Insolia, Luca Consolini, Fabrizio Laurini, Marco Locatelli, and Marco Riani

Abstract Clusterwise linear regression, a supervised learning technique that aims at finding latent groups with distinct linear relationships, has numerous applications across diverse scientific and applied domains. However, it often leads to optimization challenges due to the combinatorial nature of the problem. This paper introduces `pclustreg`, a novel approach based on probabilistic branch-and-bound optimization that maximizes the log-likelihood of a Gaussian mixture model. We show that, under suitable conditions, `pclustreg` guarantees (with a user-defined probability $s < 1$) a solution with log-likelihood at least as good as the infeasible oracle estimator, which knows the true cluster assignments. Additionally, `pclustreg` can be used as a computationally lean heuristic, as showcased on simulated and real-world datasets.

Keywords Clusterwise regression · Mixed-integer · Probabilistic branch-and-bound · Maximum likelihood estimation · Gaussian mixture

---

A. Fois (✉) · L. Consolini · M. Locatelli
Department of Engineering and Architecture, University of Parma, Parma, Italy
e-mail: andrea.fois@unipr.it

L. Consolini
e-mail: luca.consolini@unipr.it

M. Locatelli
e-mail: marco.locatelli@unipr.it

L. Insolia
Geneva School of Economics and Management, University of Geneva, Geneva, Switzerland
e-mail: luca.insolia@sns.it

F. Laurini · M. Riani
Department of Economics and Management, University of Parma, Parma, Italy
e-mail: fabrizio.laurini@unipr.it

M. Riani
e-mail: marco.riani@unipr.it

M. Di Francesco et al. (eds.), *Operations Research: Closing the Gap Between Research and Practice*, AIRO Springer Series 15, https://doi.org/10.1007/978-3-031-90095-2_15

1 Introduction

Mixture models offer a powerful probabilistic framework for capturing the inherent heterogeneity in a population, where data may arise from unobserved subpopulations with distinct characteristics. In the realm of supervised learning, clusterwise regression models (CRM) provide a valuable tool for uncovering relationships within such heterogeneous populations [1]. They model the relationships between a response and a set of explanatory variables coming from several latent components, and their diverse applicability extends across various domains, including engineering [2], economics [3], environmental sciences [4], and sociology [5]. Unlike traditional regression models, which assume a single underlying relationship for all data points, CRM techniques aim to group the data into distinct clusters, each governed by its own regression model.

This paper focuses on clusterwise linear regression [6], a specific class of CRM well-suited for modeling linear relationships within clustered data arising from a Gaussian mixture. Estimating the parameters of such models presents a significant challenge due to the complexity of the parameter space and the potential presence of multiple local optima. Existing optimization methods primarily rely on the Expectation-Maximization (EM) algorithm [7], which, due to its local search nature, can be susceptible to convergence to local optima and does not allow the assessment of the solution quality. To address these issues, several mathematical optimization approaches have been put forward to estimate the parameters of a wide spectrum of models [8].

The focus of this work is on the maximization of the likelihood of Gaussian mixture models. Existing approaches try to find either a local optimum without any quality guarantees or the global optimum. On the other hand, our approach finds solutions that, with a desired probability $s < 1$ are guaranteed to be as good as the unknown ground-truth with respect to the value of the objective function. This is achieved under minimal assumptions on the true model parameters. The method is based on a probabilistic branch-and-bound approach that exploits the distribution of the sum of squared residuals to compute good probabilistic bounds. In addition, we show, through experimental tests on synthetic and real-world data, that our approach can also be employed as a competitive heuristic to find good solutions while keeping the computational costs under control.

The remainder of the paper is structured as follows. Section 2 describes the model and its assumptions and formulates the problem. Section 3 introduces our approach and presents some theoretical results on its performance. Section 4 reports on experiments carried on synthetic datasets. Section 5 presents a real-world use-case scenario that highlights the advantages of our approach. Section 6 concludes the paper and presents some possible avenues for future research.

2 Problem Formulation

Consider a set of n points x_i, y_i with $x_i \in \mathbb{R}^p$ and $y_i \in \mathbb{R}$. The goal of clusterwise linear regression is to fit the data with k regression hyperplanes in order to maximize the log-likelihood under appropriate conditions that allow to avoid spurious/degenerate solutions. We denote with $C = \{S_1(C), S_2(C), \dots, S_k(C)\}$ a partition of the indexes $\{1, 2, \dots, n\}$ of the points into k groups. Let $\sigma = [\sigma_1, \sigma_2, \dots, \sigma_k]^T \in \mathbb{R}^k$ and $\beta = [\beta_1, \beta_2, \dots, \beta_k] \in \mathbb{R}^{p \times k}$. We assume that, for some unknown $\overline{C}, \overline{\beta}, \overline{\sigma}$ (the ground truth – which we consider unknown),

$$y_i = x_i^T \overline{\beta}_j + \epsilon_i,\ \epsilon_i \sim \mathcal{N}(0, \overline{\sigma}_j^2)\ \ \forall j, \forall i \in S_j(\overline{C}).$$

We assume that $\forall j \in \{1, 2, \dots, k\} : |S_j(\overline{C})| \geq p + 1$. Moreover, we assume that the matrix of the independent variables X_j comprised of the points $x_i, i \in S_j(\overline{C})$ is full rank for all $j \in \{1, 2, \dots, k\}$. Moreover, we assume $\overline{\sigma}_j > 0, \forall j \in \{1, 2, \dots, k\}$. We refer to $\overline{C}$ as the ground-truth assignment (of points to clusters).

In a partial assignment some points are left unassigned, i.e., it is a partition of a subset of $\{1, 2, \dots, n\}$. For any (possibly partial) assignment C of points to clusters, we denote with $S_j(C)$ the set of indexes of the points assigned to cluster $j \in \{1, 2, \dots, k\}$ according to C. We denote with $n_j(C)$ the cardinality of cluster j according to C, i.e., $n_j(C) = |S_j(C)|$.

The likelihood of point x, y belonging to cluster j for certain values of the parameters β_j, σ_j is $f_j(x, y) = \frac{1}{\sigma_j \sqrt{2\pi}} \exp\left(-\frac{1}{2}(x^T \beta_j - y)^2 / \sigma_j^2\right)$.

Now, let us assume that a full assignment C of points to clusters is given. Then, the likelihood of the data is $L(C) = \prod_{j=1}^k \prod_{i \in S_j(C)} \pi_j f_j(x_i, y_i)$, where π_j ($\sum_{j=1}^k \pi_j = 1$) are the mixing proportions of the clusters.

We focus on the following optimization problem:

$$\max_{C,\beta,\sigma,\pi} \log(L(C)) = \max_{C,\beta,\sigma,\pi} \sum_{j=1}^{k} \sum_{i \in S_j(C)} \{\log(\pi_j) + \log(f_j(x_i, y_i))\}.$$

Where $\log(L(C))$ is the log-likelihood and with the restriction that C is a partition such that $\log(L(C))$ is bounded and $\sigma_j > 0 \wedge \pi_j > 0\ \forall j \in \{1, 2, \dots, k\}$. In particular, we have that C must be such that $|S_j(C)| \geq p + 1\ \forall j \in \{1, 2, \dots, k\}$. After substituting the value of $f_j(x_i, y_i)$ in $\log(L(C))$, we can obtain

$$\log(L(C)) = -n \log\left(\sqrt{2\pi}\right) + \sum_{j=1}^k \left\{ n_j \log(\pi_j) - n_j \log(\sigma_j) - \frac{\sum_{i \in S_j(C)} (x_i^T \beta_j - y_i)^2}{2\sigma_j^2} \right\}.$$

Therefore, for the given assignment C, the values of π_j, β_j, σ_j which maximize the log-likelihood, are the following (see [6, 9])

$$\hat{\pi}_j(C) = \frac{n_j(C)}{n},\ \hat{\beta}_j(C) = X_j^+ Y_j,\ \hat{\sigma}_j(C) = \sqrt{\frac{\mathrm{SSR}_j(C)}{n_j(C)}},$$

where X_j, Y_j are the matrices of points assigned to cluster j; $X_j^+ = (X_j^T X_j)^{-1} X_j^T$ is the left pseudoinverse of X_j and

$$\mathrm{SSR}_j(C) = \sum_{i \in S_j(C)} (x_i^T \hat{\beta}_j(C) - y_i)^2. \tag{1}$$

We substitute these values into $\log(L(C))$ and drop the explicit dependency of n_j, β_j, SSR_j on C for ease of readability. Thus, we obtain

$$\log(\hat{L}(C)) = -\tfrac{n}{2}\left[\log(2\pi) + 1\right] - n\log(n) + \tfrac{1}{2}\textstyle\sum_{j=1}^{k} 3n_j \log(n_j) - n_j \log\left(\mathrm{SSR}_j\right).$$

Hence, the computationally hard part of the optimization problem is the discrete part, i.e., finding the best assignment of points to clusters $\hat{C}$.

The full optimization problem is the following:

$$\max_C \log(\hat{L}(C)). \tag{2}$$

It is crucial to notice that the (unknown) ground-truth, which is what the model aims to reconstruct, is not necessarily the global optimum of Problem (2), as it happens in cases with overlapping densities. For this reason, the goal of this work is not to find the global optimal solution of Problem (2), rather we are satisfied with the detection of any solution that is at least as good as the ground-truth in terms of log-likelihood. Theorem 1 proves that the proposed approach is able to return such solution with a desired probability $s < 1$.

3 Our Approach: `pclustreg`

The approach we propose is called `pclustreg` (probabilistic clusterwise regression). In addition to the data matrices X (independent variables) and y (dependent variables) and the number of regression hyperplanes k, the routine takes the following parameters as input.

- A desired probability of success $s < 1$ (for example, 0.999).
- An underestimate of the minimum cluster cardinality $\widetilde{n}_{\mathrm{MIN}} \leq \min_j n_j(\overline{C})$.
- An underestimate of the minimum standard deviation of the underlying Gaussian noises $\widetilde{\sigma}_{\mathrm{MIN}} \leq \min_j \overline{\sigma}_j$.

The value of $\widetilde{n}_{\mathrm{MIN}}$ can be set to $p + 1$. However, higher underestimates are preferred if available. The methods for obtaining $\widetilde{\sigma}_{\mathrm{MIN}}$ are not discussed in this work and are instead left for future studies. In the rest of this work we assume that the values $\widetilde{\sigma}_{\mathrm{MIN}}$, $\widetilde{n}_{\mathrm{MIN}}$ are obtained by either previous knowledge of the domain of the problem or by other means. In Sect. 6 we suggest a possible way to compute $\widetilde{\sigma}_{\mathrm{MIN}}$ from the data and a possible way to get rid of $\widetilde{\sigma}_{\mathrm{MIN}}$ entirely, although the development of such ideas is left for future research. Analogously, we do not discuss the selection of k in this work but we mention some possibilities for its informed choice in Sect. 6 as additional future research avenues.

We will show that `pclustreg` finds a solution with log-likelihood at least as high as that of the (unknown) ground truth, with probability at least s. This is obtained by a probabilistic branch-and-bound method.

The proposed routine `pclustreg` also supports setting a limit to the number of nodes expanded. This feature allows the algorithm to be employed as a heuristic method that finds good solutions while keeping under control the execution time. Of course, in this case the aforementioned guarantees on the value of the objective function found by `pclustreg` do not hold.

3.1 Outline of the Algorithm

Firstly, `pclustreg` performs a random permutation of the indexes of the points x, y. The ordering of points defined by this permutation is the one relative to which the points will be indexed in the rest of this paper. In particular, the sets of indexes $S_j(C)$ are defined relative to it. After that, the probabilistic branch-and-bound algorithm starts. The branching happens on the points to clusters assignment variables. In the root node all points are unassigned. In every intermediate node the first l points in the order given by the random shuffling are assigned, the remaining $n - l$ points are unassigned. Note that in branch-and-bound approaches fixing in advance the sequence of branching operations is not customary, rather branching operations are usually chosen adaptively. However, this choice is crucial for the proof of correctness of the upper bound with a known probability, as it will be explained in detail in Sect. 3.3. In particular, it guarantees that, in every intermediate node with partial assignment $\widetilde{C}$ such that $\forall j \in \{1, 2, \ldots, k\} : S_j(\widetilde{C}) \subseteq S_j(\overline{C})$, the sets $S_j(\widetilde{C})$ and $S_j(\overline{C}) \setminus S_j(\widetilde{C})$ are both random subsamples of points taken from $S_j(\overline{C})$.

When an intermediate node is expanded, it creates k new nodes. The children nodes share the assignments of the first l points with the parent node. For every children $j \in \{1, 2, \ldots, k\}$, the point with index $l + 1$ is assigned to cluster j. The remaining (last) $n - l - 1$ points are left unassigned. The nodes are kept in a priority queue ordered by decreasing upper bounds. At every iteration the node with the highest upper bound is extracted from the queue. If its upper bound is lower than or equal to the current lower bound, the algorithm terminates and returns the best solution found. If the upper bound is greater than the current lower bound, then it is expanded into k children as explained above. For each child, the partial assignment is used to compute a new feasible solution according to the procedure described in Sect. 3.2. If the objective function value of the feasible solution is better than the current lower bound, the latter is updated. The computation of the probabilistic upper bound for each child is discussed in detail in Sect. 3.3. New nodes are cut from the search space when the current lower bound is higher than their upper bound.

3.2 Lower Bound Computation

Given a partial assignment C, the lower bound is computed as follows.

1. Ensure that each cluster has at least $p + 1$ points assigned to it by adding random unassigned points to the clusters with less than $p + 1$ points.
2. Fit the points in each cluster with standard least squares linear regression to obtain an estimate of the parameters $\hat{\beta}_j$ and $\hat{\sigma}_j$ for each cluster j.
3. Perform a modified Classification Expectation Maximization (CEM) algorithm to convergence.

The modification applied to the CEM consists of updating the values of $\hat{\pi}_j$ while the points are reassigned instead of keeping them fixed. This change has yielded better lower bounds in practice in the conducted experiments.

3.3 Upper Bound Computation

We first recall the definition (1) of $\mathrm{SSR}_j(C)$, the sum of the squared residuals for the points assigned to cluster j by the (possibly partial) assignment C when they are fitted with standard linear regression. We denote with $\mathcal{X}^2_\epsilon(T)$ the quantile of order ϵ of the distribution $\mathcal{X}^2(T)$ when $T > 0$ and $\mathcal{X}^2_\epsilon(T) = 0$ when $T \leq 0$.

For $\widetilde{n}_j \geq \max\{p+1, n_j(\widetilde{C})\}$, we define $\widetilde{f}_j(\widetilde{n}_j) = 3\widetilde{n}_j \log(\widetilde{n}_j) - \widetilde{n}_j \log(g_j(\widetilde{n}_j))$, where:

$$g_j(\widetilde{n}_j) := \begin{cases} \widetilde{\sigma}^2_{\mathrm{MIN}} \mathcal{X}^2_\epsilon(\widetilde{n}_j - p) & \text{if } n_j(\widetilde{C}) < p+1 \\ \mathrm{SSR}_j(\widetilde{C}) + \widetilde{\sigma}^2_{\mathrm{MIN}} \mathcal{X}^2_\epsilon(\widetilde{n}_j - n_j(\widetilde{C}) - p) & \text{otherwise.} \end{cases}$$

We consider the following problem, with $c_n = -\frac{n}{2}\left[\log(2\pi) + 1\right] - n\log(n)$:

$$\begin{aligned} &\max_{\widetilde{n} \in \mathbb{Z}^k} \; c_n + \tfrac{1}{2}\textstyle\sum_{j=1}^k \widetilde{f}_j(\widetilde{n}_j) \\ &s.t. \quad \widetilde{n}_j \geq \max(n_j(\widetilde{C}), \widetilde{n}_{\mathrm{MIN}}) \quad \forall j \in \{1, 2, \ldots, k\} \\ &\qquad\; \textstyle\sum_{j=1}^k \widetilde{n}_j = n \end{aligned} \tag{3}$$

We will adopt the optimal value of (3) as upper bound in our probabilistic branch-and-bound approach. Now, let us consider an intermediate node that is on the path from the root node to the leaf node representing the ground-truth assignment $\overline{C}$. We will denote the (partial) assignment in such node with $\widetilde{C}$. It is important to notice that the points that are assigned in $\widetilde{C}$ are assigned to the same clusters as they are assigned to in $\overline{C}$ ($\forall j \in \{1, 2, \ldots, k\} : S_j(\widetilde{C}) \subseteq S_j(\overline{C})$). In what follows we prove that at such node, the optimal value of (3) is at least as large as

$$\log(\hat{L}) = c_n + \frac{1}{2}\sum_{j=1}^{k} f_j(\overline{C}),$$

the estimated log-likelihood of the ground-truth, at least with a given probability.

We first state the following lemma.

Lemma 1 *Let $\widetilde{C}$ be the (partial) assignment at a node along the path from the root node to the leaf corresponding to the ground-truth $\overline{C}$. Then, for any $j \in \{1, \ldots, k\}$ it holds that $\widetilde{f}_j(n_j(\overline{C})) \geq f_j(\overline{C})$ with probability at least $1 - \epsilon$.*

Proof See appendix.

As a consequence of the previous result, we can also prove the following.

Lemma 2 *Let $\widetilde{C}$ be the (partial) assignment at the node N of level t along the path from the root node to the leaf corresponding to the ground-truth $\overline{C}$. Then, for node N and all its ancestors,*

$$\widetilde{f}_j(n_j(\overline{C})) \geq f_j(\overline{C}) \;\; \forall\, j \in \{1, \ldots, k\}, \tag{4}$$

with probability not lower than $1 - (k + t)\epsilon$.

Proof See appendix.

An immediate consequence of this result is the following.

Lemma 3 *Let $\widetilde{C}$ be the (partial) assignment at the node N of level t along the path from the root node to the leaf corresponding to the ground-truth $\overline{C}$. Then, for node N and all its ancestors, the optimal value of (3) is at least as large as the estimated log-likelihood $\log(\hat{L})$ of the ground-truth with probability not lower than $1 - (k + t)\epsilon$.*

Proof See appendix.

We are now ready for the main result.

Theorem 1 *Let $\epsilon = \frac{1-s}{n+k}$. Then, the proposed probabilistic branch-and-bound approach is able to detect a solution with log-likelihood at least as good as $\log(\hat{L}(\overline{C}))$, the log-likelihood of the ground-truth, with probability at least s.*

Proof See appendix.

Regarding the solution of Problem (3), notice that since the variables of Problem (3) are all integers, then only the values assumed by $\widetilde{f}_j$ on integer points are relevant. We substitute $\widetilde{f}_j$ with $\widetilde{f}_j^*$, a piecewise-linear function that over-estimates $\widetilde{f}_j$ in integer points. The function $\widetilde{f}_j^*$ is obtained exploiting the concavity of the logarithm in $\widetilde{f}_j$ and then taking the piecewise-linear interpolation of the integer points. We do not have a formal theoretical proof that $\widetilde{f}_j^*$ is convex but `pclustreg` numerically checks its convexity by testing that the slopes are increasing before computing each upper bound. Now, if the function is convex, since the feasible region of Problem (3) is a polytope (more precisely, a simplex) with integer vertices, its maximum value is attained at a vertex of the feasible region (see, e.g., [10]). The number of such vertices is equal to k. In our tests $\widetilde{f}_j^*$ was always found to be convex. However, if the check fails we can still solve Problem (3) through a dynamic programming algorithm that runs in $\mathcal{O}(kn^2)$. We do not discuss such algorithm since it was never used in practice because $\widetilde{f}_j^*$ was always checked to be convex by `pclustreg` in our tests.

4 Numerical Experiments on Synthetic Data

In this section we present and discuss the numerical experiments carried out on synthetic data. Notice that we employed `pclustreg` as a heuristic by limiting the number of nodes expanded (see Sect. 3). We compare three methods.

1. Our proposed approach `pclustreg`.
2. The implementation of `tclustreg` [9, 11] provided in the FSDA robust statistical package [12] without trimming and without restriction factors on the ratios of standard deviations.
3. The `oracle`, i.e., the ground-truth. This is an unfeasible method that we employ as a benchmark for the other methods.

We adjusted the node expansion limit of `pclustreg` to obtain similar computation times as the ones obtained by `tclustreg` to perform a fair comparison. In particular, we used a limit to the number of nodes expanded by `pclustreg` of 100. We used 3000 as the number of subsamples used by `tclustreg`. We ran `pclustreg` with $\widetilde{n}_{\text{MIN}} = p + 1 = 3$ and with $\widetilde{\sigma}_{\text{MIN}} = 1$.

We compare the methods on the following metrics.

1. The value of the log-likelihood $\log(\hat{L})$ obtained. It is worth reiterating that higher values mean better fitness of the model to the data.
2. The computational time $t[s]$, expressed in seconds.

The synthetic data is obtained through the use of the `MixSimreg` [9, 13] and `simdatasetreg` [13, 14] functions of the MATLAB FSDA [12] package. Given the number of regression hyperplanes k and a desired average overlap the function `MixSimreg` creates the corresponding distribution with the values for $\overline{\pi}_j, \overline{\sigma}_j^2, \overline{\beta}_j$. Then, `MixSimreg` uses these values to generate the data and returns the $X, Y, \overline{C}$. We ran the algorithms on 100 randomly generated datasets with $k = 5$, $p = 10$, $n = 1000$ and average overlap of 0.05.

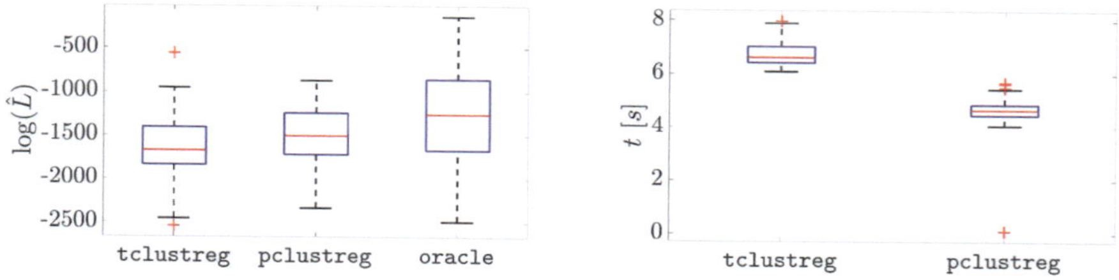

Fig. 1 Boxplots for the log-likelihood (left) and the computation times (right) over 100 runs on synthetic datasets with $k = 5$, $p = 10$, $n = 1000$ and average overlap of 0.05. The computation time of the `oracle` is meaningless

4.1 Simulation Results and Discussion

Results of our experiments are shown in Fig. 1. The left boxplot shows that `pclustreg` outperforms `tclustreg` in terms of log-likelihood of the results. The `oracle` is shown as a benchmark. The right boxplot shows that the computation times of `pclustreg` are comparable to those of `tclustreg`.

5 Real-World Application

In this section we present the tests we conducted on the FSDA [12] Girdles dataset. The dataset consists of real trade data. In particular, it contains the price and quantity of 153 imports of girdles and panty girdles from Israel to Austria. Since we allow an intercept term and there is only one independent variable, $p = 2$.

Since the dataset does not contain the ground-truth the `oracle` method is not feasible, thus we only compare `pclustreg` against `tclustreg`.

We tested the dataset using $k = 3$ and $k = 4$. We repeated the tests increasing the `pclustreg` node limit and the `tclustreg` number of subsamples to keep the computation times approximately equal in the most challenging scenario of $k = 4$. This was done in order to have a fair comparison. In particular, we used a node limit of 100, 500, 10000 and corresponding `tclustreg` number of subsamples of 300, 3000, 60000. We ran `pclustreg` with $\widetilde{n}_{\mathrm{MIN}} = p + 1 = 3$ and with $\widetilde{\sigma}_{\mathrm{MIN}} = 1$. Since we are using `pclustreg` as a heuristic, one could also perform a search on the parameter $\widetilde{\sigma}_{\mathrm{MIN}}$. Such search is not being performed in this experiment.

5.1 Results and Discussion

The results are shown in Table 1. We can see that for $k = 3$ both `tclustreg` and `pclustreg` find the same solution regardless of their execution time. When $k = 4$ `pclustreg` consistently finds solutions that are better than `tclustreg`

Table 1 Log-likelihood and computation time for the Girdles dataset with $k = 3$ and $k = 4$ for `tclustreg` and `pclustreg`. The log-likelihood is expressed in thousands. The computation time is expressed in seconds in the last column

Method	k	$\widetilde{\sigma}_{\text{MIN}}$	node limit	number of samples	$\log(\hat{L})$	t [s]
`tclustreg`	3	–	–	300	−1.1462	0.7457
`pclustreg`	3	1	100	–	−1.1462	0.1456
`tclustreg`	3	–	–	3000	−1.1462	2.7457
`pclustreg`	3	1	500	–	−1.1462	0.5892
`tclustreg`	3	–	–	60000	−1.1462	41.6869
`pclustreg`	3	1	10000	–	−1.1462	7.0427
`tclustreg`	4	–	–	300	−1.1153	0.2702
`pclustreg`	4	1	100	–	−1.0949	0.2415
`tclustreg`	4	–	–	3000	−1.0898	1.4381
`pclustreg`	4	1	500	–	−1.0799	1.2396
`tclustreg`	4	–	–	60000	−1.0801	28.6471
`pclustreg`	4	1	10000	–	−1.0799	9.2513

in comparable times. In general, the execution times of `pclustreg` are lower than those of `tclustreg` while achieving comparable or better results.

6 Conclusion and Future Work

We have provided a reformulation of the maximization of the log-likelihood for the clusterwise regression setting. We then proposed a new algorithm, `pclustreg`. This algorithm, given two underestimates $\widetilde{\sigma}^2_{\text{MIN}}$ and $\widetilde{n}_{\text{MIN}}$ of the minimum variance and minimum cluster cardinality of the ground truth, returns a solution that, with a desired probability $s < 1$, is at least as good as the (unknown) ground-truth in terms of log-likelihood. Moreover, we showed with numerical experiments that `pclustreg` can be effectively used as a heuristic method by limiting the number of nodes expanded. Our tests showed that `pclustreg` is capable of outperforming, as a heuristic, the state-of-the-art implementation `tclustreg` in terms of quality of the result with comparable computation times.

There are several avenues for further research on the approach described in this work. We are currently investigating: (i) the possibility of dropping the necessity to have $\widetilde{\sigma}_{\text{MIN}}$ by using an F distribution with the estimated $\hat{\sigma}(\widetilde{C})$ at intermediate nodes instead of the $\mathcal{X}^2$ distribution. (ii) Estimating the value of $\widetilde{\sigma}_{\text{MIN}}$ by employing a robust trimming-based algorithm for regression or multivariate analysis [15] which aims to identify a single component and $n - \widetilde{n}_{\text{MIN}}$ outliers. (iii) Finding the optimal number of regression hyperplanes k with a linear search using some information criterion [16].

Acknowledgements The author Andrea Fois is grateful to RetailTune S.r.l for sponsoring his Ph.D. and for partially sponsoring the work presented in this paper.

Appendix

Proof of Lemma 1

We consider the two cases $n_j(\widetilde{C}) < p+1$ and $n_j(\widetilde{C}) \geq p+1$.

Case $n_j(\widetilde{C}) < p+1$. First, notice that $\mathrm{SSR}_j(\overline{C}) \sim \overline{\sigma}_j^2 \mathcal{X}^2(n_j(\overline{C}) - p)$ (see, for example, [17]). In this case cluster j does not have enough points to fit with nonzero errors. Since $\mathrm{SSR}_j(\overline{C}) \sim \overline{\sigma}_j^2 \mathcal{X}^2(n_j(\overline{C}) - p)$, the value $\overline{\sigma}_j^2 \mathcal{X}_\epsilon^2(n_j(\overline{C}) - p)$ underestimates $\mathrm{SSR}_j(\overline{C})$ with probability at least $1-\epsilon$. Since $\widetilde{\sigma}_{\mathrm{MIN}} \leq \overline{\sigma}_j$, then $\widetilde{\sigma}_{\mathrm{MIN}}^2 \mathcal{X}_\epsilon^2(n_j(\overline{C}) - p) \leq \mathrm{SSR}_j(\overline{C})$ with probability at least $1-\epsilon$. Since $-n_j(\overline{C})\log(.)$ is decreasing with respect to the argument of the logarithm we have that $\widetilde{f}_j(n_j(\widetilde{C})) \geq f_j(\overline{C})$ with probability at least $1-\epsilon$.

Case $n_j(\widetilde{C}) \geq p+1$. In this case cluster j has enough points to fit with nonzero errors. We have

$$\begin{aligned}\mathrm{SSR}_j(\overline{C}) &= \textstyle\sum_{i\in S_j(\overline{C})} (y_i - x_i^T \hat{\beta}_j(\overline{C}))^2 = \\ &= \textstyle\sum_{i\in S_j(\widetilde{C})} (y_i - x_i^T \hat{\beta}_j(\overline{C}))^2 + \sum_{i\in S_j(\overline{C})\setminus S_j(\widetilde{C})} (y_i - x_i^T \hat{\beta}_j(\overline{C}))^2 =\end{aligned}$$

Firstly, we notice that

$$\sum_{i\in S_j(\widetilde{C})} (y_i - x_i^T \hat{\beta}_j(\overline{C}))^2 \geq \mathrm{SSR}_j(\widetilde{C}),$$

by definition of standard linear regression. Thus,

$$\mathrm{SSR}_j(\overline{C}) \geq \mathrm{SSR}_j(\widetilde{C}) + \textstyle\sum_{i\in S_j(\overline{C})\setminus S_j(\widetilde{C})} (y_i - x_i^T \hat{\beta}_j(\overline{C}))^2 .$$

Secondly, we notice that $S_j(\overline{C}) \setminus S_j(\widetilde{C})$ is a random subsample of $n_j(\overline{C}) - n_j(\widetilde{C})$ points taken from $S_j(\overline{C})$ because the points have been randomly shuffled at the beginning and the points in $S_j(\widetilde{C})$ are simply the first $n_j(\widetilde{C})$ points assigned to cluster j in the ground-truth assignment $\overline{C}$ according to that random order. We denote with $\widetilde{C}'$ the complementary partial assignment with respect to $\overline{C}$, i.e., $S_j(\widetilde{C}') = S_j(\overline{C}) \setminus S_j(\widetilde{C})$ and so $n_j(\widetilde{C}') = n_j(\overline{C}) - n_j(\widetilde{C})$ for all j. We have that, again, by definition of standard linear regression,

$$\sum_{i\in S_j(\overline{C})\setminus S_j(\widetilde{C})} (y_i - x_i^T \hat{\beta}_j(\overline{C}))^2 \geq \mathrm{SSR}_j(\widetilde{C}').$$

Moreover, recall that, when $n_j(\widetilde{C}') \geq p+1$, since $n_j(\widetilde{C}') = n_j(\overline{C}) - n_j(\widetilde{C})$,

$$\mathrm{SSR}_j(\widetilde{C}') \sim \overline{\sigma}_j^2 \mathcal{X}^2(n_j(\overline{C}) - n_j(\widetilde{C}) - p).$$

So we have that, since $\widetilde{\sigma}_{\text{MIN}} \leq \overline{\sigma}_j$ and $n_j(\widetilde{C}') = n_j(\overline{C}) - n_j(\widetilde{C})$,

$$\sum_{i \in S_j(\overline{C}) \setminus S_j(\widetilde{C})} (y_i - x_i^T \hat{\beta}_j(\overline{C}))^2 \geq \widetilde{\sigma}_{\text{MIN}}^2 \mathcal{X}_\epsilon^2(n_j(\widetilde{C}') - p)$$

with probability at least $1 - \epsilon$. When $n_j(\widetilde{C}') \leq p$ the inequality is trivially true since, by our definition, $\mathcal{X}_\epsilon^2(T) = 0$ when $T \leq 0$. Since the logarithm is an increasing function, we have that $\widetilde{f}_j(n_j(\overline{C})) \geq f_j(\overline{C})$ with probability at least $1 - \epsilon$ or, in other terms, $\widetilde{f}_j(n_j(\overline{C})) \geq f_j(\overline{C})$ with probability at least $1 - \epsilon$.

Proof of Lemma 2

We prove the result by induction. It is true at the root node, i.e., when $t = 0$. Indeed, by Lemma 1 it holds that $\widetilde{f}_j(n_j(\overline{C})) \geq f_j(\overline{C})$ at least with probability $1 - \epsilon$ for a single value j, so that it holds for all $j \in \{1, \ldots, k\}$ at least with probability $1 - k\epsilon$.

Now, let us assume that the statement is true for the node at level t and we prove it also holds for its child at level $t + 1$. Only one of the clusters changes when moving from a parent node to a child node because only one new point is assigned to one of the clusters. If such cluster is the r-th one, then only the value $\widetilde{f}_r(n_r(\overline{C}))$ can change. By the inductive assumption,

$$\widetilde{f}_j(n_j(\overline{C})) \geq f_j(\overline{C}) \ \ \forall\, j \neq r,$$

at least with probability $1 - (k + t)\epsilon$. Moreover, by Lemma 1 with probability at least $1 - \epsilon$ it holds that

$$\widetilde{f}_r(n_r(\overline{C})) \geq f_r(\overline{C}).$$

Then, by the Boole-Bonferroni inequality [18] for the two events, the probability of their intersection is at least $1 - (k + t + 1)\epsilon$.

Proof of Lemma 3

It is enough to observe that $[n_1(\overline{C}), n_2(\overline{C}), \ldots, n_k(\overline{C})]$ is a point inside the feasible region of the optimization problem (3), since $\widetilde{C}$ is a partial assignment of $\overline{C}$. Since, by Lemma 2, (4) holds with probability not lower than $1 - (k + t)\epsilon$, then the optimal value of (3) is at least as large as the log-likelihood $\log(\hat{L}(\overline{C}))$ of the ground-truth at least with the same probability.

Proof of Theorem 1

If a solution with log-likelihood at least as good as the one of the ground-truth is detected before reaching the leaf node corresponding to the ground-truth, then we are done. Otherwise, by Lemma 3 no node along the path from the root node to the ground-truth leaf is cut with probability $1 - (n + k)\epsilon = s$, and the ground-truth is directly evaluated at the leaf node. Our method cuts nodes without exploring them when the lower bound is higher than their (probabilistic) upper bound.

References

1. McLachlan, G.J., Lee, S.X., Rathnayake, S.I.: Finite mixture models. Ann. Rev. Stat. Appl. **6**, 355–378 (2019)
2. Veeramisti, N., Paz, A., Khadka, M., Arteaga, C.: A clusterwise regression approach for the estimation of crash frequencies. J. Transp. Saf. & Secur. **13**(3), 247–277 (2021)
3. Skrondal, A., Rabe-Hesketh, S.: Generalized Latent Variable Modeling: Multilevel, Longitudinal, and Structural Equation Models. CRC Press (2004)
4. Poggi, J.M., Portier, B.: PM10 forecasting using clusterwise regression. Atmos. Environ. **45**(38), 7005–7014 (2011)
5. Wedel, M., Kamakura, W.A.: Market Segmentation: Conceptual and Methodological Foundations. Springer Science & Business Media (2000)
6. García-Escudero, L.A., Gordaliza, A., Mayo-Iscar, A., San Martín, R.: Robust clusterwise linear regression through trimming. Comput. Stat. & Data Anal. **54**(12), 3057–3069 (2010)
7. Dempster, A.P., Laird, N.M., Rubin, D.B.: Maximum likelihood from incomplete data via the EM algorithm. J. Roy. Stat. Soc.: Ser. B (Methodol.) **39**(1), 1–22 (1977)
8. Long, Q., Bagirov, A., Taheri, S., Sultanova, N., Wu, X.: Methods and applications of clusterwise linear regression: a survey and comparison. ACM Trans. Knowl. Discov. Data **17**(3), 1–54 (2023)
9. Torti, F., Perrotta, D., Riani, M., Cerioli, A.: Assessing trimming methodologies for clustering linear regression data. Adv. Data Anal. Class. **13**(1), 227–257 (2019)
10. Horst, R., Tuy, H.: Global Optimization: Deterministic Approaches. Springer Science & Business Media (2013)
11. García-Escudero, L.A., Gordaliza, A., Greselin, F., Ingrassia, S., Mayo-Iscar, A.: The joint role of trimming and constraints in robust estimation for mixtures of Gaussian factor analyzers. Comput. Stat. & Data Anal. **99**, 131–147 (2016)
12. Riani, M., Perrotta, D., Torti, F.: FSDA: A MATLAB toolbox for robust analysis and interactive data exploration. Chemom. Intell. Lab. Syst. **116**, 17–32 (2012)
13. Maitra, R., Melnykov, V.: Simulating data to study performance of finite mixture modeling and clustering algorithms. J. Comput. Graph. Stat. **19**(2), 354–376 (2010)
14. Davies, R.: The distribution of a linear combination of chi-square random variables. Appl. Stat. **29**, 323–333 (1980)
15. Atkinson, A.C., Riani, M., Cerioli, A.: Cluster detection and clustering with random start forward searches. J. Appl. Stat. **45**(5), 777–798 (2018)
16. Torti, F., Riani, M., Morelli, G.: Semiautomatic robust regression clustering of international trade data. Stat. Methods & Appl. **30**, 863–894 (2021)
17. Seber, G.A., Lee, A.J.: Linear Regression Analysis. Wiley (2012)
18. Prékopa, András: Boole-Bonferroni inequalities and linear programming. Oper. Res. **36**(1), 145–162 (1988)

Lower Bounds and a Constraint Programming Approach for Scheduling an Assembly Problem with Precedence and Resource Constraints

Andrea Corsini and Mauro Dell'Amico

Abstract In this work, we delve into a real-life scheduling problem involving the production of off-road vehicles, subject to precedence, machines, and resource constraints. This scheduling problem shares characteristics with many others, but also comprises less investigated aspects like a specific precedence structure and a resource constraint limiting the number of simultaneously active machines. We formally present the problem and devise two lower bounds. To solve the problem, we propose a Constraint Programming model and compare it with a Mathematical Programming one. Our findings reveal the superiority of the Constraint Programming approach in both real-life and synthetic instances, and validate the quality of the proposed lower bounds.

Keywords Scheduling · Constraint programming · Mathematical programming · Lower bounds

1 Introduction

This work tackles a scheduling problem encountered by a company that manufactures off-road vehicles according to a *three-level* assembly process. The assembly process occurs in specific areas of the shop floor, known as *workcenters*, each of which houses multiple parallel machines. Each level may contain one or more workcenters. At the 3rd and uppermost level, components like wheels or engines are assembled within a unique workcenter, and sometimes, additional preparations are carried out in intermediate workcenters of the 2nd-level. Finally, these components are installed

A. Corsini (✉) · M. Dell'Amico
University of Modena and Reggio Emilia, Modena, Italy
e-mail: andrea.corsini@unimore.it

M. Dell'Amico
e-mail: mauro.dellamico@unimore.it

M. Di Francesco et al. (eds.), *Operations Research: Closing the Gap Between Research and Practice*, AIRO Springer Series 15, https://doi.org/10.1007/978-3-031-90095-2_16

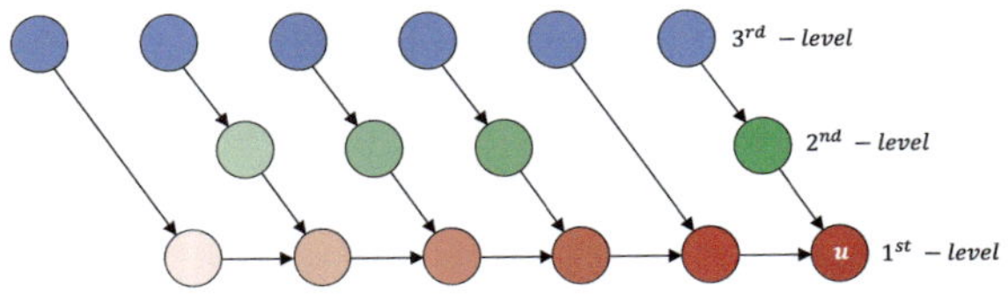

Fig. 1 The job precedences forming a rooted in-tree. 3rd-level operations (blue circles) are executed on the same workcenter; 2nd and 1st-level operations (green and red circles, respectively) are executed on distinct workcenters denoted with different color shades

on vehicles in workcenters of the 1st-level. The vehicle moves through the 1st-level workcenters in a fixed sequence that is identical for all types of vehicles. To perform any assembly operation, both a machine and a human worker are required. The objective is to assign assembly operations to machines and workers by minimizing the total completion time of vehicles.

This manufacturing process forms a distinctive precedence structure among assembly operations resulting in a *directed rooted in-tree* (see Fig. 1), a special kind of directed tree characterized by a root vertex where all paths end.

Although this problem has distinctive characteristics, it also shares aspects with other scheduling problems. The sub-problem stemming from the identical sequence of workcenters at the 1st-level is a *Flexible Flow Shop Problem*, where operations at each stage can be processed on one of several parallel machines (see [1] for a recent survey). However, our problem introduces additional complexity, with each 1st-level operation having one or two extra predecessors at the 2nd and 3rd-level. Whereas, the unique workcenter at the 3rd-level is a *Parallel Machine Scheduling Problem* with identical machines (see e.g., [2]). Additionally, the human workers can be viewed as a *renewable resource* in a *Resource-Constrained Project Scheduling Problem* (RCPSP) or a multi-project RCPSP [3]. However, both the RCPSP and the multi-project variant schedule project activities that often require multiple resources and have arbitrary precedences, see e.g. [4]. We also found connections with *Dual-Resource Constrained Problems* treated in [5, 6], which schedule operations if a machine and a worker are simultaneously available. However, the specific precedence structure of our problem is not covered by any of these dual-resource works.

In a prior study [7], we explored in detail the interrelations between this scheduling problem and others in the literature, and developed heuristic algorithms. In this work, our primary focus lies in devising methods capable of proving the optimality of solutions, thus addressing the limitations of our previous work [7]. Due to its proven quality in traditional scheduling problems (as evidenced in [8]), we propose a Constraint Programming (CP) formulation of the problem. Additionally, we introduce two lower-bounding procedures tailored to leverage key bottlenecks within the problem, namely human resources and workcenters. We systematically evaluate the effectiveness of the CP approach and contrast it with a Mixed Integer Linear Programming (MILP) counterpart across instances of increasing complexity, including eight real-life scenarios and a diverse set of synthetic instances. Our findings unequivocally demonstrate the superiority of the CP approach in solving the problem, while also highlighting the quality of the proposed lower bounds.

In summary, the contributions of this work are:

- We introduce two effective lower bounds for benchmarking algorithms.
- We propose a superior Constraint Programming approach to optimally solve the problem and validate its quality under different scenarios.

The remainder of the paper is organized as follows: Sect. 2 formalizes the problem and introduces the lower bounds; Sect. 3 presents the MILP and CP formulation of the problem; Sect. 4 evaluates the quality of the produced lower and upper bounds; Sect. 5 closes with final considerations.

2 Problem Formulation and Lower Bounds

The problem tackled in this work asks to schedule a set of jobs J onto a set of workcenters W. Each workcenter $w \in W$ contains m_w identical parallel machines that can handle one operation at a time. Every job $j \in J$ comprises a set of operations O_j, where an operation $o \in O_j$ must be executed on one machine of workcenter $w(o) \in W$. All the jobs have a similar precedence structure, which is well expressed through a directed rooted in-tree (see Fig. 1). As one can see, this tree is characterized by a main path (in red) composed of 1st-level workcenters, where enters in each vertex an additional path formed by the 2nd and 3rd-level workcenters (green and blue circles).

Let $G_j = (V_j, A_j)$ be the in-tree representing job j, where V_j contains one vertex for each operation in O_j, and A_j is the set of precedences among operations. The arcs of A_j have no cost while each vertex $v \in V_j$ is given a cost $p_v \in \mathbb{Z}_{\geq 0}$ equal to the processing time of the operation associated to v. Moreover, let $u_j \in V_j$ be the root of G_j and $w_v = w(o)$ denote the workcenter that must execute the operation of v. For each vertex $v \in V_j$, we use $tail(v)$ to refer to the length of the unique path starting at v and ending in u_j, computed as the sum of the costs of the vertices in the path but p_v. Similarly, let $head(v)$ denote the length of the longest path ending in v, again excluding p_v.

A set of $R^{\max} < \sum_{w \in W} m_w$ workers with identical skills is available to execute operations. To start the execution of an operation, a machine and a worker must be available, and the preceding operations must be completed. Preemption is not allowed, i.e., an operation associated with $v \in V$ and starting at time t must be uninterruptedly executed in the time interval $[t, t + p_v)$.

Summarizing, a problem instance is described by the set of workcenters W, the number of machines in each workcenter, the number of workers $R^{\max}$, and the digraph $G = (V := \bigcup_{j \in J} V_j; A := \bigcup_{j \in J} A_j)$. The task is to assign the operation associated with each vertex $v \in V$ to a machine of its workcenter w_v and to a time interval $[t_v, t_v + p_v)$ by respecting precedences in A and without simultaneously executing more than $R^{\max}$ operations. The goal is to minimize the makespan.

Finally, in all our experiments, we adopted the shop floor of the company that inspired this work, which is divided into 11 workcenters as in Fig. 1. At the 1st-level, there are 6 workcenters with a number of machines given by the sequence (4, 3, 2, 2, 3, 2), starting from the beginning to the root node of the main path. At the 2nd-level, there are 4 workcenters with 2 machines each, except for the last one, which has 4 machines. At the 3rd-level there is a single workcenter with 10 machines. Lastly, a crew of $R^{\max} = 20$ workers execute operations.

2.1 Lower Bounds

Herein, we introduce two lower bounds that consider key aspects of the problem.

The first bound focuses on the human resources needed to execute operations. This bound relaxes the precedence constraint and assumes that all the operations are assigned to a unique workcenter with $R^{\max}$ parallel machines. Thus, it solves an identical parallel machine scheduling problem, where each operation has a release date $head(\cdot)$ and a tail $tail(\cdot)$ that is executed immediately after its completion. We propose a CP model for solving this scheduling problem, as utilizing a MILP approach requires disjunctive constraints implemented with a big-M formulation. To present the CP model, we adopt the notation of Table 1.

$$\min \quad LB_R \tag{1}$$

$$\text{s.t. } T_v = \text{IntervalVar}(head(v),\ p_v,\ H) \qquad \forall v \in V \tag{2}$$

$$\text{Cumulative}(\{T_v : v \in V\}) \leq R^{\max} \tag{3}$$

$$LB_R = \max_{v \in V} \{\text{EndOf}(T_v) + tail(v)\} \tag{4}$$

The model minimizes the completion of the last tail (LB_R). Constraints (2) construct an interval variable for each vertex v, where H is a large number. Constraint (3) imposes that no more than $R^{\max}$ interval variables can overlap in each time instant and Constraint (4) computes LB_R.

The other lower bound considers the assignment of operations to workcenter machines by relaxing precedences and disregarding the worker constraint. This bound independently considers each workcenter and derives a bound by observing that the makespan of a machine is at least equal to the sum of the processing times plus the minimum $head(\cdot)$ and $tail(\cdot)$ of the assigned operations, see e.g., [9]. For each workcenter $w \in W$, we find the assignment of operations on the machines resulting

Table 1 Variables and constraints used in the presented CP models

IntervalVar(A, B, C)	It defines an interval variable of length B that cannot start before A and must finish before C
EndOf(·)	It returns the completion time of an interval variable
EndBefore(v1, v2)	It constrains variable v1 to end before the start of v2
Cumulative(·)	It counts how many variables overlap in each time instant

in the minimum makespan. Let $\mathcal{V}_w = \{v \in V : w_v = w\}$ denote the set of operations on workcenter w. We propose a MILP formulation that works slightly faster than a CP equivalent, see also [10]. This MILP model uses a continuous variable $LB_w \geq 0$ and binary decision variables x_{vk} that take value 1 if the operation of vertex $v \in V_w$ is assigned to machine k, 0 otherwise.

$$\min LB_w \tag{5}$$

$$\text{s.t.} \sum_{k=1}^{m_w} x_{vk} = 1 \qquad \forall v \in V_w \tag{6}$$

$$LB_w \geq head(a) + tail(b) + \sum_{v \in I_{ab}} p_v x_{vk} \qquad \forall (a,b) \in \mathcal{V}_w \times \mathcal{V}_w, k = 1, \ldots, m_w \tag{7}$$

Objective (5) minimizes the makespan (LB_w), and Constraints (6) impose that every operation must be assigned to a machine. Constraints (7) compute the lower bound given by the assignment of operations to each machine m. Note that in these constraints we adopt the same reduction of [10], that for every pair of operations (a, b) sums up the processing time of only those operations with a larger head and tail, i.e., $I_{ab} = \{v \in V_w : head(v) \geq head(a), tail(v) \geq tail(b)\}$. We refer to [10] for more details on this reduction.

Finally, an overall lower bound can be easily obtained as:

$$LB_W = \max_{w \in W} LB_w \tag{8}$$

3 Proposed Models

This section first introduces a Mixed Integer Linear Programming model, which serves to exemplify better the problem outlined in Sect. 2 and as a baseline for comparison, and then presents the proposed Constraint Programming model.

3.1 Mixed Integer Linear Programming Model

We present our MILP formulation of the problem introduced in [7]. Let $1, \ldots, H$ be the discrete planning horizon. For each $v \in V$, we define the interval where the operation associated to v can start as: $H_v = \{head(v), \ldots, H - tail(v) - p_v\}$. Our model uses a continuous variable $z \geq 0$ to define the makespan and binary decision variables x_{vt} that take value 1 if $v \in V$ starts at time $t \in H_v$, 0 otherwise.

$$\min z \tag{9}$$

$$\text{s.t.} \sum_{t \in H_{u_j}} t\, x_{u_j t} + p_{u_j} \leq z \qquad \forall j \in J \tag{10}$$

$$\sum_{t \in H_v} x_{vt} = 1 \qquad \forall v \in V \tag{11}$$

$$\sum_{v \in \mathcal{V}_w} \sum_{h=t-p_v+1}^{t} x_{vh} \leq m_w \qquad \forall w \in W, t = 0, \ldots, H \tag{12}$$

$$\sum_{t \in H_a} t x_{at} + p_a \leq \sum_{t \in H_b} t x_{bt} \qquad \forall (a, b) \in A \tag{13}$$

$$\sum_{v \in V} \sum_{h=t-p_v+1}^{t} x_{vh} \leq R^{\max} \qquad \forall t = 0, \ldots, H \tag{14}$$

$$x_{vt} \in \{0, 1\} \qquad \forall v \in V, t \in H_v \tag{15}$$

The model minimizes the makespan (9)–(10). Constraints (11) impose that each operation is executed once, Constraints (12) ensure that no more than m_w machines are used in parallel in a workcenter w. Constraints (13) impose the precedence constraints for each pair $(a, b) \in A$, while constraints (14) impose that no more than $R^{\max}$ operations are executed simultaneously. Model (9)–(15) contains a pseudo-polynomial number of binary variables amounting to $O(|V|H)$.

To define the horizon H, we use a constructive algorithm that prioritizes operations with the earliest start time. The solution produced by this algorithm is also loaded as the initial incumbent before starting the Branch & Bound.

Finally, inequalities (14) are added as lazy constraints. Specifically, the solver starts by disregarding the inequalities declared lazy, and once a feasible integer solution is found, it invokes a user-defined separation procedure. At each invocation, this procedure adds inequality (14) to the smallest time instant violating the $R^{\max}$ constraint. We avoid applying this strategy to constraints (12) as they are often violated when modeled as lazy constraints, lowering performance.

3.2 Constraint Programming Model

Due to its proven effectiveness on many scheduling problems (see, e.g., [8]), we propose a CP formulation that follows the same logic as the MILP counterpart. Our CP model defines an interval variable for each operation and leverages the function Cumulative for handling parallel resources as in the seminal work [11].

$$\min \max_{j \in J} \{\text{EndOf}(T_{u_j})\} \tag{16}$$

$$\text{s.t. } T_v = \text{IntervalVar}(head(v),\ p_v,\ H) \qquad \forall v \in V \tag{17}$$

$$\text{EndBefore}(a,\ b) \qquad \forall (a, b) \in A \tag{18}$$

$$\text{Cumulative}(\{T_v : v \in \mathcal{V}_w\}) \leq m_w \qquad \forall w \in W \tag{19}$$

$$\text{Cumulative}(\{T_v : v \in V\}) \leq R^{\max} \tag{20}$$

The model minimizes the makespan (16). Constraints (17) define an interval variable for each operation. Constraints (18) ensure that the precedences among operations are respected. Constraints (19) guarantee that no more than m_w operations execute simultaneously in each workcenter w, and Constraint (20) ensures that no more than $R^{\max}$ operations execute in parallel. To define the horizon H and to provide a hint to the solver, we adopt the same constructive heuristic used with the MILP model of the previous section.

4 Results

This section evaluates the proposed models on eight real-life scenarios and synthetic instances emulating industrial use cases. The MILP models were solved with Gurobi 10.1, while we used the CP-SAT of Google ORTools for solving the CP ones. All the models have been written in Python 3.10, and executed with a time limit of 3600s on an Ubuntu machine with an Intel Core i9-11900K.

4.1 MILP Versus CP: Performance for Increasing Problem Sizes

We begin the experimental campaign by considering the MILP and CP models of Sect. 3 and compare their performance for increasing sizes of a real-life scenario (instance I8 in the next Sect. 4.2). Specifically, we evaluate the timing and bounds of the models by scheduling only the initial k jobs (in order of memorization), where $k \in \{5, 6, \dots, 15, 20, 25\}$. For each model and k, we save the best bounds obtained, the time taken to find the best upper bound, and the total time used. Figure 2 compares the produced lower and upper bounds on the left, and the corresponding times on the right.

Figure 2 shows that both models rapidly close instances with up to 11 jobs. As observed in [7], with more jobs, the MILP starts to fail in proving the optimality of solutions and requires a lot of time ($\geq$15 min) to retrieve the best UB. Whereas the CP still rapidly retrieves solutions and proves their optimality for instances with up to 15 jobs. However, in scenarios with 20 and 25 jobs, the CP model also requires a lot of time to find good solutions and hardly proves optimality, although the gap between the LB and UB is fairly small (see the small yellow bars of CP for 20 and 25 jobs on the left of Fig. 2).

Consequently, we conclude that the MILP model approaches its natural limit on instances with roughly 10 jobs, while the CP on instances with 20–25 jobs.

4.2 Performance on Real-Life Instances

To further assess the performance of the MILP and CP models, we extend our analysis to eight real-life instances (I1, ..., I8), which involve the assembly of 17–25 vehicles. Specifically, we compare the best solutions of the models with those of the company and contrast the lower bounds with the best obtained in [7] and with the procedures of Sect. 2.1. Note that the company constructs manual solutions by scheduling one

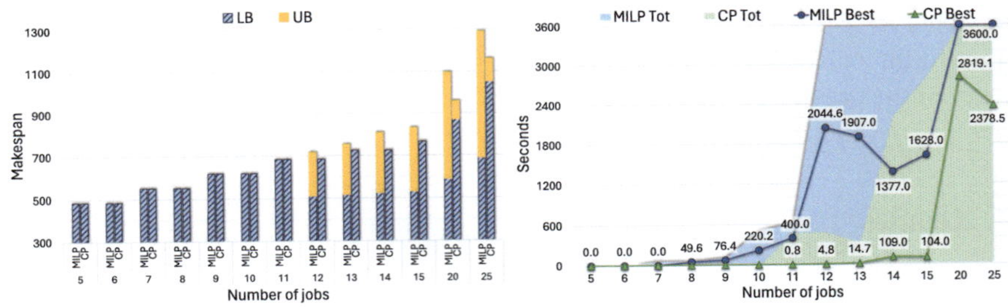

Fig. 2 The comparison between the MILP and CP models on a real-life scenario for increasing numbers of jobs. On the left, the best upper (UB) and lower (LB) bounds produced by each model. On the right, the time for retrieving the best upper bound (Best) and the total time used (Tot)

Table 2 Lower and upper bounds for the eight real-life instances. Best bounds in **bold** and instances marked with * are closed to optimality for the first time

Name	#jobs	LB[7]	LB_R	LB_W	Manual	MILP		CP	
						[LB, UB]	GAP(%)	[LB, UB]	GAP(%)
I1*	17	761	716	**869**	952	[574, 947]	39.4	[761, **869**]	12.4
I2*	21	905	877	**1013**	1116	[688, 1140]	39.6	[905, **1013**]	10.7
I3	18	797	756	**869**	988	[685, 997]	31.3	[797, **877**]	9.1
I4	24	1013	1001	**1013**	1239	[665, 1286]	48.3	[**1013**, **1125**]	10.0
I5	23	977	960	**1013**	1200	[681, 1254]	45.7	[977, **1080**]	9.5
I6	25	1049	1036	**1157**	1268	[833, 1326]	37.2	[1049, **1163**]	9.8
I7	22	941	902	**1013**	1143	[709, 1182]	40.0	[941, **1028**]	8.5
I8	25	1049	1030	**1157**	1267	[687, 1289]	46.7	[1049, **1159**]	9.5
Avg		936.5	909.8	1013.0	1146.6	[690.3, 1177.6]	41.0	[936.5, 1039.3]	9.9

job at a time, in order of their arrival time, and occasionally job exchanges may be done by an expert without any algorithmic or reproducible logic.

Table 2 presents the results for these instances. Each row reports the instance name; the number of jobs (# jobs); the lower bounds of [7], LB_R (1)–(4), and LB_W (8); the makespan of the company solution (Manual); the bounds and gap (GAP = (UB − LB)/UB) of the MILP and CP models. The last row (Avg) averages results across all instances for an overall comparison.

The MILP confirms to be a poor approach for instances featuring over a dozen jobs, as evidenced by its LBs being outperformed by all others and its UBs improving upon manual solutions only once. Therefore, in the next evaluation, we will exclude the MILP. In contrast, the CP consistently outperforms manual solutions and provides reliable LBs aligned with those in [7], maintaining steady optimality gaps of around 10%. Notably, LB_W emerges as the best lower bound in these instances, largely outperforming the bounds in [7]. Lastly, while LB_R gives bounds inferior to LB_W and LB[7], they still largely surpass those of the MILP. This demonstrates the effectiveness of both the CP approach and the proposed lower-bounding procedures, and suggests that these instances seem more constrained by the machines of workcenters than by the human workers.

4.3 Performance on Synthetic Instances

We conclude the experimental campaign by presenting results on synthetic instances. As the company generally schedules orders with 15–30 vehicles, we created instances with a total of 30 jobs. To generate jobs, we defined three processing time intervals: Small = [5, 50], Medium = [30, 100], and Large = [80, 150]; and we combined these intervals into 27 job types in collaboration with stakeholders. From these types, we constructed the following instances: (i) one instance for *each type* (T1, …, T27), comprising 30 jobs created by randomly sampling the processing times at the 1st, 2nd, and 3rd-level in the intervals of the type; (ii) ten instances (U1, …, U10) created by *independently* selecting the type of each job and randomly generating the processing times accordingly; (iii) ten *batched* instances (B1, …, B10) created by randomly selecting a type and an integer $b \in [1, 10]$, and generating a batch of b jobs according to the selected type.

With these diverse instances, we evaluate again the CP model (16)–(20) and the proposed lower bounds by considering only the first 15 jobs and all of them. Table 3 presents results in two vertical sections: the first focuses on instances with 15 jobs, while the second on instances with all the 30 jobs. Each row reports the instance name; the lower bounds in [7], LB_R (1)–(4), and LB_W (8); the bounds and gap of the CP model. The last row (Avg) takes the average of each column.

Table 3 confirms that the CP approach is also effective on instances embodying random characteristics, being able to close a significant portion of them. In addition, the CP model is consistent in terms of gaps, as they are generally below 10% with both 15 and 30 jobs. Focusing on the lower bounds, we observe that LB_W consistently outperforms LB_R in instances with 15 jobs, while their disparity diminishes with 30 jobs. This suggests that the worker constraint becomes a bottleneck when the number of jobs is around 30. Notably, we also observe that LB_R and LB_W rapidly produce lower bounds (matter of minutes) that closely align with those produced by the CP model within one hour of computation, remarking their quality. Finally, we remark that LB [7] obtains better average results as it is the maximum of two different procedures, but it never produces bounds better than the LB_W and LB_R. This empirically confirms the dominance of our proposed lower bounds on those presented in [7].

5 Conclusions

In conclusion, we have discussed a real-life scheduling problem encompassing characteristics of many others and we have introduced two effective lower bounding procedures, which serve as valuable benchmarks for heuristic algorithms. Additionally, we have proposed a Constraint Programming formulation and contrasted it with a Mixed Integer Linear Programming one. Our findings highlight the superiority of the Constraint Programming approach, although there is room for further

Table 3 Lower and upper bounds for synthetic instances with 15 and 30 jobs. Best bounds in **bold** and instances marked with * are closed to optimality for the first time

15 Jobs					
Name	LB [7]	LB_R	LB_W	CP	
				[LB, UB]	GAP(%)
T1	332	319	338	**[355, 370]**	4.1
T2	779	525	783	**[791, 804]**	1.6
T3	1417	806	**1419**	**[1419, 1425]**	0.4
T4*	604	435	**615**	**[615, 615]**	0.0
T5*	820	598	836	**[847, 847]**	0.0
T6	1403	928	1406	**[1414, 1424]**	0.7
T7*	913	566	921	**[922, 922]**	0.0
T8*	1187	746	**1197**	**[1197, 1197]**	0.0
T9	1400	968	1410	**[1454, 1487]**	2.2
T10	602	496	616	**[623, 631]**	1.3
T11	829	666	835	**[838, 846]**	0.9
T12	1418	896	1432	**[1445, 1457]**	0.8
T13	608	593	614	**[615, 677]**	9.2
T14	819	767	820	**[829, 913]**	9.2
T15	1410	1019	1425	**[1440, 1467]**	1.8
T16*	1022	755	**1024**	**[1024, 1024]**	0.0
T17*	1124	930	**1139**	**[1139, 1139]**	0.0
T18	1417	1139	1484	**[1531, 1576]**	2.9
T19	1034	714	1046	**[1080, 1138]**	5.1
T20	1054	865	1074	**[1079, 1148]**	6.0
T21	1474	1103	1483	**[1487, 1491]**	0.3
T22	1074	851	1090	**[1093, 1104]**	1.0
T23	1091	991	1120	**[1145, 1235]**	7.3
T24	1488	1208	1498	**[1498, 1543]**	2.9
T25	1092	1014	1122	**[1125, 1165]**	3.4
T26	1147	1137	1158	**[1164, 1337]**	12.9
T27	1488	1372	1495	**[1548, 1656]**	6.5
U1*	724	802	897	**[936, 936]**	0.0
U2*	883	905	914	**[1040, 1040]**	0.0
U3*	696	858	858	**[896, 896]**	0.0
U4	791	816	816	**[858, 865]**	0.8
U5*	865	933	1018	**[1125, 1125]**	0.0
U6	811	**820**	**820**	**[820, 877]**	6.5
U7*	836	991	992	**[1036, 1036]**	0.0
U8	773	864	864	**[867, 874]**	0.8
U9	741	786	917	**[930, 940]**	1.1
U10	868	877	983	**[1040, 1059]**	1.8
B1	1034	1009	1233	**[1290, 1306]**	1.2
B2	1347	1030	1419	**[1439, 1456]**	1.2
B3	1007	1007	1155	**[1216, 1257]**	3.3
B4	790	724	**797**	**[797, 806]**	1.1
B5	929	905	999	**[1057, 1113]**	5.0
B6	1281	1281	1494	**[1507, 1612]**	6.5
B7	494	512	528	**[566, 614]**	7.8
B8*	639	814	888	**[905, 905]**	0.0
B9	1036	1036	1214	**[1251, 1277]**	2.0
B10	**869**	**869**	856	**[869, 964]**	9.9
Avg	988.5	856.3	1014.2	[1067.3, 1097.8]	2.8

30 Jobs					
Name	LB [7]	LB_R	LB_W	CP	
				[LB, UB]	GAP(%)
T1	**631**	**631**	562	**[631, 701]**	10.0
T2	1309	1309	**1311**	**[1311, 1398]**	6.2
T3	2242	1450	2264	**[2267, 2362]**	4.0
T4*	1123	886	1124	**[1126, 1126]**	0.0
T5	1342	1198	1350	**[1354, 1442]**	6.1
T6*	2304	1642	2306	**[2314, 2314]**	0.0
T7*	1788	1155	**1792**	**[1792, 1792]**	0.0
T8	1957	1498	1974	**[1977, 2015]**	1.9
T9	2266	1958	2266	**[2284, 2444]**	6.5
T10	1154	996	**1179**	[1173, **1212**]	3.2
T11	**1353**	**1353**	**1353**	**[1353, 1551]**	12.8
T12	2351	1780	2369	**[2375, 2425]**	2.1
T13	**1203**	**1203**	**1203**	**[1203, 1284]**	6.3
T14	**1551**	**1551**	**1551**	**[1551, 1750]**	11.4
T15	2263	2000	2268	**[2270, 2441]**	7.0
T16	1829	1507	1836	**[1837, 1865]**	1.5
T17	1963	1832	**1966**	**[1966, 2027]**	3.0
T18	2287	2278	2315	**[2392, 2631]**	9.1
T19	2073	1428	**2082**	[2073, **2097**]	1.1
T20	2115	1760	**2130**	[2115, **2203**]	4.0
T21	2387	2225	**2392**	**[2392, 2576]**	7.1
T22	2090	1657	**2105**	[2090, **2129**]	1.8
T23	2118	1990	**2138**	[2120, **2255**]	6.0
T24	**2463**	**2463**	**2463**	**[2463, 2850]**	13.6
T25	2103	1981	**2122**	[2103, **2170**]	3.1
T26	**2272**	**2272**	**2272**	**[2272, 2465]**	7.8
T27	**2758**	**2758**	**2758**	**[2758, 3090]**	10.7
U1	**1511**	**1511**	1200	**[1511, 1605]**	5.9
U2	**1729**	**1729**	1503	**[1729, 1817]**	4.8
U3	**1565**	**1565**	1330	**[1565, 1651]**	5.2
U4	**1669**	**1669**	1374	**[1669, 1820]**	8.3
U5	**1683**	**1683**	1421	**[1683, 1800]**	6.5
U6	**1609**	**1609**	1262	**[1609, 1683]**	4.4
U7	**1653**	**1653**	1393	**[1653, 1746]**	5.3
U8	**1644**	**1644**	1384	**[1644,1705]**	3.6
U9	**1408**	**1408**	1152	**[1408, 1491]**	5.6
U10	**1684**	**1684**	1371	**[1684, 1784]**	5.6
B1	**2101**	**2101**	1842	**[2101, 2332]**	9.9
B2	**1875**	**1875**	1504	**[1875, 1996]**	6.1
B3	**2001**	**2001**	1611	**[2001, 2108]**	5.1
B4	**1593**	**1593**	1292	**[1593, 1651]**	3.5
B5	**1792**	**1792**	1753	**[1792, 2032]**	11.8
B6	**2036**	**2036**	1734	**[2036, 2274]**	10.5
B7	**1331**	**1331**	1108	**[1331, 1409]**	5.5
B8	**1496**	**1496**	1443	**[1496, 1575]**	5.0
B9	**1817**	**1817**	1744	**[1817, 2135]**	14.9
B10	**1822**	**1822**	1474	**[1822, 1950]**	6.6
Avg	1815.2	1669.6	1691.2	[1820.9, 1940.0]	6.0

improvement. In the future, we will extensively compare the presented approach with various heuristic algorithms and investigate decomposition approaches for tackling this complex scheduling problem.

References

1. Rossit, D.A., Tohmé, F., Frutos, M.: The non-permutation flow-shop scheduling problem: a literature review. Omega **77** (2018)
2. Dell'Amico, M., Martello, S.: Optimal scheduling of tasks on identical parallel processors. J. Comput. **7** (1995)
3. Bredael, D., Vanhoucke, M.: Multi-project scheduling: a benchmark analysis of metaheuristic algorithms on various optimisation criteria and due dates. Eur. J. Oper. Res. **308** (2023)
4. Hartmann, S., Briskorn, D.: An updated survey of variants and extensions of the resource-constrained project scheduling problem. Eur. J. Oper. Res. **297** (2022)
5. Dhiflaoui, M., Nouri, H.E., Driss, O.B.: Dual-resource constraints in classical and flexible job shop problems: a state-of-the-art review. Proc. Comput. Sci. **126** (2018)
6. Xiong, H., Shi, S., Ren, D., Hu, J.: A survey of job shop scheduling problem: the types and models. Comput. & Oper. Res. **142** (2022)
7. Dario, B., Andrea, C., Mauro, D.: Lower and upper bounds for scheduling a real-life assembly problem with precedences and resource constraints. Comput. Oper. Res. **173** (2025)
8. Naderi, B., Ruiz, R., Roshanaei, V.: Mixed-integer programming vs. constraint programming for shop scheduling problems: new results and outlook. J. Comput. **35** (2023)
9. Carlier, J., Pinson, E.: An algorithm for solving the job-shop problem. Manag. Sci. **35** (1989)
10. Juvin, C., Houssin, L., Lopez, P.: Logic-based benders decomposition for the preemptive flexible job-shop scheduling problem. Comput. & Oper. Res. **152** (2023)
11. Laborie, P., Rogerie, J., Shaw, P., Vilím, P.: IBM ILOG CP optimizer for scheduling. Constraints **23** (2018)

Complexity Results on the Bi-Level Interdiction Knapsack Problem with Unit Interdiction Costs or Unit Weights

Alberto Boggio Tomasaz, Margarida Carvalho, Roberto Cordone, and Pierre Hosteins

Abstract In this contribution we pursue the study of the Bi-level Interdiction Knapsack Problem, where a leader interdicts some knapsack items subject to a budget constraint in order to maximally reduce the total profit of the Knapsack Problem that the follower solves on the remaining items. We study the computational complexity of the problem when the leader's interdiction costs are unitary and prove that the problem remains Σ_2^p-complete using a reduction from the bi-level 3-SAT problem. Additionally, we prove that the version with unitary lower level weights but arbitrary interdiction costs is NP-complete.

Keywords Bi-level Knapsack · Polynomial hierarchy · Interdiction · Stackelberg game

1 Introduction

Interdiction games, which are a subclass of *Stackelberg games*, represent the sequential action of two players who seek to optimise the same objective function but in opposite directions. These problems can be naturally modelled as bi-level optimisation programs, with one agent reacting to the decision of the other. In their basic version, an *attacker* has a limited amount of resources to disrupt or disable some elements of the problem (item to pack, edge or node of a network, etc...), after which

A. Boggio Tomasaz · R. Cordone
Università degli Studi di Milano, Milan, Italy
e-mail: alberto.boggio@unimi.it

R. Cordone
e-mail: roberto.cordone@unimi.it

M. Carvalho
Université de Montréal, Montréal, QC, Canada

P. Hosteins (✉)
Università degli Studi di Torino, Turin, Italy
e-mail: hosteins@di.unito.it

M. Di Francesco et al. (eds.), *Operations Research: Closing the Gap Between Research and Practice*, AIRO Springer Series 15, https://doi.org/10.1007/978-3-031-90095-2_17

a *defender* reacts by trying to perform a certain action as best they can, e.g., maximising a flow or finding a shortest path between an origin and a destination node on a network, selecting a set of items given some capacity constraint, etc...

The most classical interdiction problems are based on the attack of network elements to disrupt classic graph problems, such as maximising a quantity of flow (Maximum Flow Interdiction Problem or MFIP [17]) or minimising the length of the shortest path (Shortest Path Interdiction Problem or SPIP [9]) between an origin and a destination node. Such problems help uncover the weak points of, e.g., critical infrastructures and to assess the robustness of such infrastructures to worst-case failures or malevolent attacks. They have a wide range of possible applications, from the analysis of electric power grids [15] to the one of water distribution networks [20]. While these problems usually consider attacks on network arcs, other interdiction problems focus more on identifying the most vital nodes whose removal has the largest impact on network connectivity, such as the Critical Node Problem (CNP) [11]. Even though those problems are bi-level in nature, they are based on a lower level problem which is polynomially solvable. Hence, the decision version of these problems belongs to the class NP and is usually NP-complete in the general case [1, 9, 17].

More recently, harder interdiction problems have been considered, which have an NP-hard lower level and, therefore, do not necessarily belong to NP. This is the case of an interdiction version of the Knapsack Problem (KP), called Bi-level Interdiction Knapsack Problem (BIKP), where a leader can interdict a subset of items, subject to an attack budget, which cannot be packed by the follower when maximising the profit of his/her knapsack [3, 5, 6, 16]. A version of this problem where the lower level KP weights are equal to the items profits is proved to be Σ_2^p-complete, using a bi-level version of the Subset Sum Problem, along with other bi-level KPs, in [2]. Very few interdiction problems with an NP-hard lower level have been shown to be Σ_2^p-complete, e.g., the Maximum Clique Interdiction Problem [8] and the Facility Location Interdiction Problem [7]. Given its simple structure and its higher level extensions, the KP provides a useful basis for clarifying the complexity of further interdiction problems, which are often tackled with unit attack weights. Moreover, since in the single level KP, forcing either the profits or the weights to be unitary makes the problem polynomial, it is interesting to investigate what happens when some sets of costs, weights or profits are unitary in the BIKP. In the following, we propose to clarify the complexity status of the BIKP, first when the attack costs of the leader become unitary, and then when the weights of the follower become unitary. We will show that those two cases belong to different levels of the polynomial hierarchy.

2 Complexity of the Bi-Level Interdiction Knapsack Problem with Unit Attack Costs

As already stated earlier, the computational complexity of the BIKP (as well as a tri-level extension where the defender may act first by protecting some items) has been investigated in the literature, for a specific version where each item has an arbitrary attack cost, and in the lower level (classic KP) its packing weight is equal to the profit [2, 14]. We will consider the more general case where the lower level packing weights are not equal to the profits, but we will concentrate on the specific BIKP version where the attack costs are unitary. We stress that this version is interesting from a theoretical perspective, since in the single level KP, computational complexity tends to be strongly affected when a set of coefficients becomes unitary (as the problem goes from being weakly NP-hard to polynomial). It is therefore a legitimate concern to establish the consequences of adopting unit weights for the higher decision levels in interdiction versions of the KP. In this section, we will establish that even with fortification and attack costs, the BIKP is complete for the second level of the polynomial hierarchy. We stress that the Σ_2^p-hardness proof in [2] for the bilevel interdiction knapsack only holds when the interdiction costs are non-unitary. In their proof, the authors consider two sets of items, D (called *padding* and *dummy* items), with unit attack costs, and O (called *ordinary* items), which cannot be interdicted by the leader. By making the interdiction costs of the items in O unitary, the proof of *only if* does not hold: not only can the leader interdict items in O but there are instances where it has incentive to do so.

Our definition of the Bi-level Interdiction Knapsack Problem with unit attack costs is the following:

UNITARY BILEVEL INTERDICTION KNAPSACK (UBIK):
INSTANCE: A finite set of items I such that each $i \in I$ has a positive integer weight w_i and a positive integer profit p_i, a positive integer capacity W and a positive integer profit goal K for the follower, a positive integer attack budget A for the leader.
QUESTION: Is there a subset $I_l \subseteq I$ of items for the leader to select, with cardinality $|I_l| \leq A$, such that every subset $I_f \subseteq I \setminus I_l$ with $\sum_{i \in I_f} w_i \leq W$ that the follower can select has a total profit $\sum_{i \in I_f} p_i < K$?

In order to prove that the above problem is complete for the second level of the polynomial hierarchy, we will use the following problem, known to be Σ_2^p-complete (Corollary 6 of [10, 18]):

2-ALTERNATING QUANTIFIED SATISFIABILITY ($B_2 \cap 3CNF$):
INSTANCE: Disjoint non-empty sets of Boolean variables X and Y, and a Boolean expression E over $U = X \cup Y$ in conjunctive normal form with at most 3 literals in each clause $c \in C$.
QUESTION: Is there a truth assignment for X such that for all truth assignments of Y, E is never satisfied?

In order to prove the completeness of UBIK, we will introduce a reduction from the $B_2 \cap 3CNF$ which draws inspiration from the reduction from 3-SAT to Subset Sum presented in [4, Theorem 34.15]:

- For each variable $u \in U$, we create two items i_u and $i_{\bar{u}}$, one for each possible truth assignment of u. We define $I_X = \{i_x, i_{\bar{x}} : x \in X\}$ and $I_Y = \{i_y, i_{\bar{y}} : y \in Y\}$. For variables $x \in X$, we introduce a third item j_x, which will serve to force the leader to choose one and only one item between i_x and $i_{\bar{x}}$. In order to ensure a sensible mapping of the variable assignment between Y and I_Y, we also introduce a fourth item j'_x for $x \in X$ to help saturate the budget of the follower for X-related digits (leaving therefore only the Y-related digits of the budget W available for selecting items from I_Y in I_f).
- For each clause $c \in C$, we create two items i_c^1 and i_c^2. We designate by I_C the set of items associated with C.
- Weights, profits, capacity and profit goal will be expressed by integer numbers with $|C| + |Y| + 2|X| + 1$ digits in base 10. The least significant $|C|$ digits are labelled by the clauses, the next $|Y|$ digits by the variables Y, the next $2|X|$ digits by the variables X, two for each variable $x \in X$. Finally, the highest position is labelled as M.

  - For each $y \in Y$, the two corresponding items i_y and $i_{\bar{y}}$ have weights and profits (w_{i_y}, p_{i_y}, $w_{i_{\bar{y}}}$ and $p_{i_{\bar{y}}}$) with digit 1 in the position labelled by the variable y and 0 in the positions labelled by other variables.
    If literal y appears in clause $c \in C$, then p_{i_y} and w_{i_y} have digit 1 in the position labelled as c, and 0 otherwise. Similarly, if the literal $\neg y$ appears in clause $c \in C$, $p_{i_{\bar{y}}}$ and $w_{i_{\bar{y}}}$ have digit 1 in the position labelled by c, and 0 otherwise.
  - For each $x \in X$, the two corresponding items i_x and $i_{\bar{x}}$ have weights w_{i_x} and $w_{i_{\bar{x}}}$ and profits p_{i_x} and $p_{i_{\bar{x}}}$ with digit 1 in the highest position labelled by the variable x and 0 in the lowest x position and in the positions labelled by other variables (remember that each X variable is associated to two consecutive digit positions).
    If the literal x appears in clause $c \in C$, then p_{i_x} and w_{i_x} have digit 1 in the position labelled as c, and 0 otherwise. Similarly, if the literal $\neg x$ appears in clause $c \in C$, $p_{i_{\bar{x}}}$ and $w_{i_{\bar{x}}}$ have digit 1 in the position labelled by c, and 0 otherwise. The weights and profits have digit 0 in all remaining positions.
    The item j_x corresponding to x has weight and profit with digit 2 in the highest position labelled as x (and 0 elsewhere).

The item j'_x corresponding to x has weight $w_{j'_x}$ with digit 1 in the highest position labelled as x and profit p_{j_x} with digit 1 in the lowest position labelled as x (and 0 elsewhere).

- For each $c \in C$, the first item has weight and profit with digit 1 in the position labelled as c and 0 elsewhere, while the second item has weight and profit with digit 2 in the position labelled as c and 0 elsewhere.
- The attack budget A is equal to $|X|$.
- The weight budget W has digit 1 in all positions labelled as variables in Y, 2 in all highest positions labelled as variables in X, 4 in all positions labelled as clauses in C, and digit 0 elsewhere. Hence, I_f can contain any item from I_X (as long as not interdicted).
- The profit goal K has digit $|X|$ at label M[1], 1 for all positions with labels in U and 4 for all positions with labels in C.

See Fig. 1 for an illustration of the proposed reduction.

Analysing an instance from the above reduction, we can infer the following properties:

Property 1 If the leader does not interdict exactly one item between i_x and $i_{\bar{x}}$ for each $x \in X$, the follower can easily reach the profit goal K.

Proof First of all, the attack budget forbids to interdict more than $|X|$ items. If the leader interdicts strictly less than $|X|$ items of I_X, the capacity of the follower allows to select all the uninterdicted items from I_X, reaching the profit goal K. Hence, only those solutions where the leader interdicts exactly $|X|$ items from I_X (and no item j_x or j'_x for $x \in X$) will be of interest for the rest of the proof. For such solutions, the follower has anyway a sufficient capacity to select all uninterdicted items from I_X (any other possibility leading to a failure to reach the profit goal K). Suppose that the two items i_x and $i_{\bar{x}}$ associated with the most significant digits whose label is $x \in X$ are taken simultaneously in I_l. In this case, the follower can select item j_x as well as all remaining uninterdicted items of I_X, achieving a profit of $|X|10^M + 2 \times 10^{|C|+|Y|+2|X|} > K$. Suppose instead that the leader selects neither i_x nor $i_{\bar{x}}$: since, as already stated, the follower has enough budget to select all remaining I_X items, it is possible for I_f to contain both i_x and $i_{\bar{x}}$ and the profit goal K is automatically achieved as a digit of 2 is achieved in the highest digit associated with x (as well as number $|X|$ in column M). The leader must therefore interdict one item between i_x and $i_{\bar{x}}$, after which, in order to be able to reach profit K, the follower can and must include item j'_x in I_f in order to reach digit 1 in the lowest digit associated to x (as not enough residual capacity is available to select item j_x). Consequently, the capacity associated to both digits associated to x is saturated (otherwise the follower cannot reach K).

This reasoning can then be recursively applied until we reach the lowest X-related digit positions, ensuring the above property.

[1] Since $|X|$ can be larger than 10, the contribution of this digit to the value of K should be understood as the number $|X|10^M$. In order to facilitate the exposition, we continue to refer to this contribution to K as a "digit" in base 10.

Fig. 1 Example of construction of an instance of UBIK from an instance $B_2 \cap 3CNF$ with $E = (a \vee b \vee \neg c) \wedge (\neg a \vee \neg b \vee d) \wedge (a \vee b \vee c)$, where $X = \{a, b\}$, $Y = \{c, d\}$ and the clauses are labelled from left to right

I			X				Y		C		
		M	a		b		c	d	c_1	c_2	c_3
i_a	w_{i_a}	0	1	0	0	0	0	0	1	0	1
	p_{i_a}	1	1	0	0	0	0	0	1	0	1
$i_{\bar{a}}$	$w_{i_{\bar{a}}}$	0	1	0	0	0	0	0	0	1	0
	$p_{i_{\bar{a}}}$	1	1	0	0	0	0	0	0	1	0
j_a	w_{j_a}	0	2	0	0	0	0	0	0	0	0
	p_{j_a}	0	2	0	0	0	0	0	0	0	0
j'_a	$w_{j'_a}$	0	1	0	0	0	0	0	0	0	0
	$p_{j'_a}$	0	0	1	0	0	0	0	0	0	0
i_b	w_{i_b}	0	0	0	1	0	0	0	1	0	1
	p_{i_b}	1	0	0	1	0	0	0	1	0	1
$i_{\bar{b}}$	$w_{i_{\bar{b}}}$	0	0	0	1	0	0	0	0	1	0
	$p_{i_{\bar{b}}}$	1	0	0	1	0	0	0	0	1	0
j_b	w_{j_b}	0	0	0	2	0	0	0	0	0	0
	p_{j_b}	0	0	0	2	0	0	0	0	0	0
j'_b	$w_{j'_b}$	0	0	0	1	0	0	0	0	0	0
	$p_{j'_b}$	0	0	0	0	1	0	0	0	0	0
i_c	w_{i_c}	0	0	0	0	0	1	0	0	0	1
	p_{i_c}	0	0	0	0	0	1	0	0	0	1
$i_{\bar{c}}$	$w_{i_{\bar{c}}}$	0	0	0	0	0	1	0	1	0	0
	$p_{i_{\bar{c}}}$	0	0	0	0	0	1	0	1	0	0
i_d	w_{i_d}	0	0	0	0	0	0	1	0	1	0
	p_{i_d}	0	0	0	0	0	0	1	0	1	0
$i_{\bar{d}}$	$w_{i_{\bar{d}}}$	0	0	0	0	0	0	1	0	0	0
	$p_{i_{\bar{d}}}$	0	0	0	0	0	0	1	0	0	0
$i^1_{c_1}$	$w_{i^1_{c_1}}$	0	0	0	0	0	0	0	1	0	0
	$p_{i^1_{c_1}}$	0	0	0	0	0	0	0	1	0	0
$i^2_{c_1}$	$w_{i^2_{c_1}}$	0	0	0	0	0	0	0	2	0	0
	$p_{i^2_{c_1}}$	0	0	0	0	0	0	0	2	0	0
$i^1_{c_2}$	$w_{i^1_{c_2}}$	0	0	0	0	0	0	0	0	1	0
	$p_{i^1_{c_2}}$	0	0	0	0	0	0	0	0	1	0
$i^2_{c_2}$	$w_{i^2_{c_2}}$	0	0	0	0	0	0	0	0	2	0
	$p_{i^2_{c_2}}$	0	0	0	0	0	0	0	0	2	0
$i^1_{c_3}$	$w_{i^1_{c_3}}$	0	0	0	0	0	0	0	0	0	1
	$p_{i^1_{c_3}}$	0	0	0	0	0	0	0	0	0	1
$i^2_{c_3}$	$w_{i^2_{c_3}}$	0	0	0	0	0	0	0	0	0	2
	$p_{i^2_{c_3}}$	0	0	0	0	0	0	0	0	0	2
	W	0	2	0	2	0	1	1	4	4	4
	K	2	1	1	1	1	1	1	4	4	4

Property 2 If the leader interdicts exactly one item between i_x and $i_{\bar{x}}$ for each $x \in X$, any solution of the follower which does not select the remaining $|X|$ uninterdicted items of I_X, all j' items and exactly one item between i_y and $i_{\bar{y}}$ for $y \in Y$ cannot reach the profit goal K.

Proof Suppose the leader does interdict exactly one item between i_x and $i_{\bar{x}}$ for each $x \in X$. The follower has enough capacity to select all remaining uninterdicted I_X items and must do so, otherwise it is impossible to reach value $|X|$ at position M. Following the reasoning in the proof of Property 1, the follower must also select all j'_x items, in order to reach digit 1 in all the lowest digit associated to each $x \in X$, therefore saturating the capacity associated to X variables and forbidding to select any j item. In order to be able to reach the profit goal, the follower must also select a subset of items from I_Y:

- The two items associated with the most significant digit whose label is in Y cannot be taken simultaneously in I_f as it would violate the weight capacity W (remember that all non-interdicted items from I_X must be included in I_f, otherwise the profit K cannot be reached). In fact, exactly one of these items must be taken, as otherwise the follower cannot achieve a profit of K.
- The two items associated with the second most significant digit whose label is in Y cannot be taken simultaneously, since we already know that one of the items associated with the most significant digit in Y is taken which would result in a violation of the weight capacity W. Hence, as before, I_f will include exactly one of the items associated with the second most significant digit in Y.
- The reasoning above propagates until the least significant digit labelled in Y. We conclude that the best set I_f will include exactly one of the items i_y or $i_{\bar{y}}$ for $y \in Y$.

Theorem 1 *UBIK is Σ_2^p-complete.*

Proof The statement of UBIK is of the form $\exists I_l \;\; \forall I_f \;\; Q(I_l, I_f)$ where Q can be tested in polynomial time, directly implying that it is in Σ_2^p. As mentioned above, we use the previously introduced reduction from the $B_2 \cap 3CNF$. In order to prove the completeness of UBIK, we need to prove that the $B_2 \cap 3CNF$ instance is a *Yes* instance if and only if the corresponding UBIK instance is a *Yes* instance.

Let $B_2 \cap 3CNF$ be a *Yes* instance. We form a solution to UBIK by packing in I_l the items i_x such that $x \in X$ is 1 and the items $i_{\bar{x}}$, otherwise. Clearly, the cardinality of I_l is equal to $|X|$. By construction, given I_l, the best I_f takes all $|X|$ items associated with X not packed in I_l, plus all j'_x items for $x \in X$, as it is necessary if the follower wants to reach a profit of at least K. Furthermore, following Property 2 the optimal I_f for the follower will also take exactly one of the items i_y or $i_{\bar{y}}$ for $y \in Y$. Finally, I_f can be completed by items from I_C. However, since the $B_2 \cap 3CNF$ instance is a *Yes* instance, there is at least a clause for which no item from I_X and I_Y with a non-zero value in the associated digit can be selected. Therefore, for this specific digit, set I_f cannot reach value exactly 4 with only items of I_C while respecting the budget W and the UBIK instance is a *Yes* instance.

Next, suppose that the UBIK instance is a *Yes* instance. According to Properties 1 and 2, I_l must contain exactly one of the items i_x and $i_{\bar{x}}$ for $x \in X$ and I_f will contain exactly one item between i_y and $i_{\bar{y}}$ for each $y \in Y$ otherwise the follower cannot achieve profit K. However, since the UBIK instance is a *Yes* instance, no choice of the follower allows to reach a value of 4 for each digit associated to C. This means that for at least one digit associated to $c \in C$, none of the items of X and Y with a digit 1 in position c can be packed, otherwise it is still possible to reach exactly value 6 for this specific digit. Assign 1 to $x \in X$ such that $i_{\bar{x}} \in I_l$, and 0 otherwise. For any follower's packing I_f: assign 1 to $y \in Y$ if $i_{\bar{y}} \in I_f$, 0 if $i_y \in I_f$. Since, by hypothesis, the UBIK instance is a *Yes* instance, there is no Y variables assignment such that E is satisfied in the corresponding $B_2 \cap 3CNF$ instance which is therefore is a *Yes* instance.

3 Complexity of the Bi-Level Interdiction Knapsack Problem with Unit Follower's Weights

When considering the BIKP with unit attack costs, should one set of parameters (either the profits or the follower's weights) become unitary, not only the follower's problem becomes polynomial but the bi-level problem itself also becomes polynomial. In this section, we propose to study the case where the attack costs a_i as well as the profits are arbitrary integers while the lower level weights are all unitary. The attack budget constraint now becomes $\sum_{i \in I_l} a_i \leq A$. In order to prove that this version of BIK is NP-complete, we will use a reduction from Subset Sum:

> SUBSET SUM:
> INSTANCE: Given a set $S = \{s_1, ..., s_n\}$ of n integers whose sum is $\sum_i s_i = 2\Sigma$.
> QUESTION: Is there a subset $S' \subseteq S$ such that $\sum_{s \in S'} s = \Sigma$?

Theorem 2 *BIK with unit lower level weights is NP-complete.*

Proof The statement of BIK with lower level weights is of the form $\exists I_l \;\; Q(I_l)$ where Q can be tested in polyomial time, since the lower level KP is solvable in polynomial time (we simply choose the W items with higher profit not interdicted by the leader), directly implying that it is in NP.

We introduce the following reduction from Subset Sum:

- For each integer s_i, we add an item i to BIK with attack cost and profit $a_i = p_i = s_i$ and unitary knapsack weight $w_i = 1$.
- We set $A = \Sigma$, $K = \Sigma + 1$ and $W = n$, so that potentially the follower can pack all items at the lower level.

From the structure of the BIK instance created, we can see that the only way to ensure that this instance is a *Yes* instance is to interdict a set of items with a total attack cost of Σ, since the profit which can be packed at the lower level is exactly $2\Sigma - \sum_{i \in I_l} a_i$.

Suppose first that the Subset Sum instance is a *Yes* instance. Define a solution of BIK by interdicting all the items corresponding to the integers of S', i.e. $I_l = \{i : s_i \in S'\}$. Since the Subset instance Sum is a *Yes* instance, there exists a solution of the leader in BIK with a total attack cost of Σ, therefore the total profit that can be gathered by the follower is $\Sigma < K$ and the BIK instance is a *Yes* instance.

Conversely, suppose that the BIK instance is a *Yes* instance: in order to have a solution that does not allow a total profit of $\Sigma + 1$, the leader needs to interdict at least a profit of Σ and therefore to interdict a set of total attack cost $\sum_{i \in I_l} a_i = \Sigma$. Define a solution of Subset Sum $S' = \{s_i : i \in I_l\}$, since it has $\sum_{s \in S'} s = \Sigma$, the Subset Sum instance is a *Yes* instance, which proves that BIK with unit lower level weights is NP-complete.

4 Conclusion

We have refined existing complexity results about the Bi-level Interdiction Knapsack Problem, which was proven to be Σ_2^p-complete in [2] with arbitrary attack costs. Given the fact that the Knapsack Problem becomes polynomial when one set of parameters becomes unitary, we investigated the case where one set of parameters of BIKP becomes unitary. We proved that this problem is still Σ_2^p-complete with unitary attack costs and arbitrary lower level weights and profits. When the lower level knapsack weights are unitary, we have proved that the problem, which trivially belongs to NP (since the follower's problem becomes polynomial), remains NP-complete. These results may be used in future works to help prove more straightforwardly the computational complexity of interdiction and fortification problems, which otherwise need more convoluted reductions from multi-level 3-SAT problems [7].

The interesting avenues of research to extend our results include the investigation of the complexity of higher level Knapsack Interdiction Problems with alternating fortification (an item becomes immune to interdiction) and interdiction rounds above the final KP round. The Tri-level IKP, for example has been proved to be Σ_3^p-complete with arbitrary fortification and attack costs in [14], but no other weight configuration has been studied yet. Given that almost no result exists for tri-level fortification-interdiction problem (except those of [14]), we could also use the existing complexity results for the BIKP to prove the Σ_2^p-completeness result for many existing fortification problems with polynomial lower level, such as those of [12, 13, 19].

References

1. Addis, B., Aringhieri, R., Grosso, A., Hosteins, P.: Hybrid constructive heuristics for the critical node problem. Ann. Oper. Res. **238**(1), 637–649 (2016)
2. Caprara, A., Carvalho, M., Lodi, A., Woeginger, G.J.: A study on the computational complexity of the bilevel knapsack problem. SIAM J. Optim. **24**, 823–838 (2014)
3. Caprara, A., Carvalho, M., Lodi, A., Woeginger, G.J.: Bilevel knapsack with interdiction constraints. INFORMS J. Comput. **28**, 319–333 (2016)
4. Cormen, T.H., Leiserson, C.E., Rivest, R.L., Stein, C.: Introduction to Algorithms, 3rd edn. The MIT Press (2009)
5. Della Croce, F., Scatamacchia, R.: An exact approach for the bilevel knapsack problem with interdiction constraints and extensions. Math. Progr. **183**, 249–281 (2020)
6. DeNegre, S.: Interdiction and discrete bilevel linear programming. Ph.D. Thesis, Lehigh University (2011)
7. Fröhlich, N., Ruzika, S.: On the hardness of covering-interdiction problems. Theoret. Comput. Sci. **871**, 1–15 (2021)
8. Furini, F., Ljubic, I., Martin, S., San Segundo, P.: The maximum clique interdiction problem. Eur. J. Oper. Res. **277**, 112–127 (2019)
9. Israeli, E., Wood, R.K.: Shortest-path network interdiction. Networks **40**(2), 97–111 (2002)
10. Johannes, B.: New classes of complete problems for the second level of the polynomial hierarchy. Ph.D. Thesis, Technischen Universitat Berlin (2011)
11. Lalou, M., Amin Tahraoui, M., Kheddouci, H.: The critical node detection problem in networks: a survey. Comput. Sci. Rev. **28**, 92–117 (2018)
12. Leitner, M., Ljubić, I., Monaci, M., Sinnl, M., Tanınmış, K.: An exact method for binary fortification games. Eur. J. Oper. Res. **307**(3), 1026–1039 (2023)
13. Lozano, L., Smith, J.C.: A backward sampling framework for interdiction problems with fortification. INFORMS J. Comput. **29**, 123–139 (2017)
14. Nabli, Adel, Carvalho, Margarida, Hosteins, Pierre: Complexity of the multilevel critical node problem. J. Comput. Syst. Sci. **127**, 122–145 (2022)
15. Salmeron, J., Wood, K., Baldick, R.: Worst-case interdiction analysis of large-scale electric power grids. IEEE Trans. Power Syst. **24**, 96–104 (2009)
16. Weninger, N., Fukasawa, R.: A fast combinatorial algorithm for the bilevel knapsack problem with interdiction constraints. In: Del Pia, A., Kaibel, V. (eds.) Integer Programming and Combinatorial Optimization, pp. 438–452. Springer International Publishing, Cham (2023)
17. Wood, R.K.: Deterministic network interdiction. Math. Comput. Modell. **17**(2), 1–18 (1993)
18. Wrathall, C.: Complete sets and the polynomial-time hierarchy. Theor. Comput. Sci. **3**(1), 23–33 (1976)
19. Wu, X., Conejo, A.J.: An efficient tri-level optimization model for electric grid defense planning. IEEE Trans. Power Syst. **32**, 2984–2994 (2017)
20. Wu, Y., Chen, Z., Gong, H., Feng, Q., Chen, Y., Tang, H.: Defender-attacker-operator: Tri-level game-theoretic interdiction analysis of urban water distribution networks. Reliab. Eng. Syst. Saf. **214**, 107703 (2021)

Shutdown Maintenance Scheduling Using Two-Stage Stochastic Programming with Endogenous Uncertainty

Srimali Gunasekara, Elham Mardaneh, and Sarang Kulkarni

Abstract We introduce the problem of maintenance scheduling across two consecutive maintenance plans which require plant shutdown. Due to budget, resource, and time limitations, maintenance of some components needs to be postponed from the first shutdown to the second with the risk of costly failures in between. We minimise the total expected cost of maintenance and scheduling by developing a two-stage stochastic programming model. The first stage determines which components should be postponed, whereas the second stage determines whether these components undergo maintenance in the second shutdown based on the occurrence of a failure event. The probability of failure depends on the first-stage decision leading to a stochastic problem with endogenous uncertainty. We solve the model for randomly generated instances with a limited number of components to illustrate the complexity of the problem. We propose a strategy for the early detection of infeasible solutions and demonstrate its effectiveness through numerical results.

Keywords Maintenance scheduling · Two-stage stochastic programming · Endogenous uncertainty

1 Introduction

The current study is inspired by the maintenance activities in the mining industry. We consider a setting in which the plant requires shutdown to undergo preventive maintenance of multiple components. Due to the restrictions on budget, resources,

S. Gunasekara (✉) · E. Mardaneh · S. Kulkarni
Curtin University, Bentley, WA, Australia
e-mail: c.ekanayak1@postgrad.curtin.edu.au

S. Gunasekara · E. Mardaneh
ARC Training Centre for Transforming Maintenance through Data Science, Bentley, WA, Australia

S. Kulkarni
Centre for Optimisation and Decision Science, Bentley, WA, Australia

M. Di Francesco et al. (eds.), *Operations Research: Closing the Gap Between Research and Practice*, AIRO Springer Series 15, https://doi.org/10.1007/978-3-031-90095-2_18

or shutdown time, it is difficult to accommodate maintenance of all the required components. Hence, the maintenance planner has to decide on excluding some of the components from the maintenance plan. We study a case in which the delayed components can be accommodated in the next shutdown plan provided no failure occurred in between two such shutdowns. Due to the risk of failure of postponed components, we formulate the problem as a two-stage stochastic programming. The first-stage decision is to determine whether to maintain or postpone the components. The probability of failure is higher if we postpone the maintenance than in the case in which we maintain the component. In the second stage, we decide to accommodate the postponed components in the second shutdown. However, the decision is influenced by the realisation of a failure event of the components. If a component fails between the two shutdowns, we perform corrective maintenance at a high cost. Such a component will not be considered in the second shutdown. Note that the decision to postpone the component depends on maintenance cost and the availability of resources, budget, and time. Hence, each maintenance plan is a resource-constrained project scheduling problem which is NP-hard [8]. Also, the decision-dependent uncertainty (endogenous) is less studied in the literature due to the difficulty in modeling and solving the problem. Hence, the paper aims to provide a two-stage stochastic formulation for the problem that handles the endogenous uncertainty and demonstrates the complexity of the problem by solving small-sized instances with complete enumeration. We also provide a simple heuristic (Method III in Sect. 3) by which we can reduce the number of evaluations in the complete enumeration for this specific problem.

Maintenance scheduling in the context of various industries has been studied extensively with the use of exact as well as heuristic approaches [4]. As per Pinciroli et al. [4], only 2 out of 100 papers considered in their review addressed the problem in the mining industry, with Genetic Algorithm [3] and Dynamic Programming [5] being used, respectively. The problem involving the postponement decision in maintenance planning has also been reported in the literature [2]. However, to the best of our knowledge, the simultaneous decision of postponement and scheduling has not been reported in the literature. In this study, we propose a two-stage stochastic programming approach for the problem. The uncertainty in the two-stage stochastic programming is classified as exogenous and endogenous [6]. Exogenous uncertainty involves the disclosure of true parameter values independently from decisions, while endogenous uncertainty involves parameter realizations that are impacted by decisions [1]. Endogenous uncertainty can be categorized into two main types. Type 1 endogenous uncertainty deals with decision-dependent probability. Decisions impacting Type 2 endogenous uncertainties alter parameter outcomes through their influence on the passage of time [7]. The problem in our study considers Type 1 endogenous uncertainty which is less studied in the literature [9]. Hence, the objectives of this paper are (1) to introduce the new problem of simultaneous postponement and maintenance scheduling to the literature, (2) to model the problem as a two-stage stochastic programming problem with endogenous uncertainty of Type 1, and (3) with the help of small-instances, demonstrate the complexity of the problem.

The rest of the paper is organised as follows. In the next section, we present a detailed description and two-stage stochastic formulation of the problem. In Sect. 3, we present the complete enumeration method and a simple procedure to reduce the number of evaluations and solution times of the complete enumeration, specific to our problem. We present the results using small-sized randomly generated instances. We conclude the paper with Sect. 4 by providing further directions for research.

2 Problem Description and Formulation

Consider two consecutive shutdowns S_1 and S_2 separated by an interval (e.g., 3 months) to perform preventive maintenance. Let J_1 and J_2 be the sets of components that are planned to undergo maintenance in S_1 and S_2, respectively. Let A_j be the set of preventive maintenance activities for component $j \in J_1 \cup J_2$. Each activity $i \in A_j$ for component $j \in J_1 \cup J_2$ has a fixed duration d_i and a set of predecessor activities I_{ij}. Let a_s and b_s denote the start and end time period for shutdown $s \in \{S_1, S_2\}$ and $T_{S_1} = \{a_{S_1}, a_{S_1} + 1, \ldots, b_{S_1}\}$ and $T_{S_2} = \{a_{S_2}, a_{S_2} + 1, \ldots, b_{S_2}\}$ be the set of time periods in S_1 and S_2, respectively. We assume only one type of resource, the amount of which, $\bar{R}_s$, will remain the same for each time period. Let r_i denote the required resource per time period for activity $i \in A_j$. Let c_s denote the unit cost of resource per time period for shutdown $s \in \{S_1, S_2\}$. The available budget for shutdown $s \in \{S_1, S_2\}$ is denoted as B_s.

We consider a case in which a set of components $J \subset J_1$ can be postponed to S_2 due to either resource or budget constraints in S_1. Let $J' = J_1 \setminus J$ denote the set of components that should be maintained in S_1. The postponement of a component, $j \in J$, may result in a failure with probability α_j between S_1 and S_2. If we perform the maintenance, the probability of failure will reduce to $\beta_j < \alpha_j$. For a failed component, we carry out corrective maintenance with a high cost m_j due to the unplanned disruption of the production activities. If a component undergoes corrective maintenance between S_1 and S_2, we do not consider that component in S_2. For a postponed and not failed component $j \in J$, we consider two cases: $j \in J_2$ or $j \notin J_2$. For $j \in J_2$, we must maintain the component in S_2. For $j \notin J_2$, the component will be maintained in S_2 subject to the availability of resources and budget. A component $j \in J$ which is not maintained in either S_1 and S_2 and not underwent corrective maintenance between S_1 and S_2, we incur a corrective maintenance cost m_j. The aim is to decide which components should be postponed to S_2, so that the total cost of preventive maintenance in both shutdowns and the corrective maintenance cost of probabilistic failure will be minimised. We assume the failure of each component is independent and the maintenance activities of each component are independent. However, precedence relations among the activities of the same component may exist.

With the above description and assumptions, we model the problem as a two-stage stochastic programming problem. In the first stage, for each component $j \in J$, we decide whether to perform the preventive maintenance in S_1 or not. Let x_j represent the binary decision variable in the first stage, which takes value 1 if component $j \in J$ undergoes maintenance in S_1, and 0 otherwise. Component $j \in J$ can fail with probability α_j if $x_j = 0$ or β_j if $x_j = 1$. Let ξ_j define a random variable that takes value 1 if component $j \in J$ fails, and 0 otherwise. We define a scenario by the failure status of each component in J. Hence, there are $K = 2^{|J|}$ possible scenarios.

Let ξ_j^k denote the realization of ξ_j in scenario $k \in K$. Hence, given $x_j = 0$, the probability of $\xi_j = \xi_j^k$ for component $j \in J$ in scenario $k \in K$ is given by $P(\xi_j = \xi_j^k | x_j = 0) = \alpha_j \xi_j^k + (1 - \alpha_j)(1 - \xi_j^k)$. Similarly, for $x_j = 1$, the probability of $\xi_j = \xi_j^k$ is given by $P(\xi_j = \xi_j^k | x_j = 1) = \beta_j \xi_j^k + (1 - \beta_j)(1 - \xi_j^k), \forall j \in J, k \in K$. Thus, the probability of $\xi_j = \xi_j^k$ for component $j \in J$ in scenario $k \in K$ depending on the first-stage decision x_j is given by

$$\begin{aligned} Pr_k(x_j) = {} & \alpha_j \xi_j^k + (1 - \alpha_j)(1 - \xi_j^k) + \\ & [(\beta_j \xi_j^k + (1 - \beta_j)(1 - \xi_j^k)) - (\alpha_j \xi_j^k + (1 - \alpha_j)(1 - \xi_j^k))] x_j, \\ & \qquad \forall j \in J, \ \forall k \in K \end{aligned} \tag{1}$$

Let $x = \{x_1, x_2, \ldots, x_{|J|}\}$ be the first-stage decision vector denoting the postponement decisions for components in J. Since each x_j, $j \in J$ is binary, x can take $2^{|J|}$ values. Let $X = \{0, 1\}^{|J|}$ denote all possible values of x. Hence, the probability of scenario $k \in K$ given $x \in X$ is given by

$$P_k(x) = \prod_{j \in J} Pr_k(x_j), \qquad \forall x \in X, \ \forall k \in K \tag{2}$$

such that the summation of probabilities of all scenarios for given x is 1, i.e.

$$\sum_{k \in K} P_k(x) = 1, \qquad \forall x \in X \tag{3}$$

Additionally, we define a first stage scheduling variable u_{it}, which equals 1 if activity $i \in A_j$, $j \in J_1$ finishes at the end of time period $t \in T_{S_1}$, and 0 otherwise. For each scenario $k \in K$, we define a second stage scheduling variable u_{it}^k which equals 1 if activity $i \in A_j$, $j \in J \cup J_2$ finishes at the end of time period $t \in T_{S_2}$, and 0 otherwise. We also define an auxiliary binary variable z_j^k which takes value 1 only if component $j \in J$ is maintained in S_1 or S_2 in scenario $k \in K$, and 0 otherwise. Using the above notations, the two-stage stochastic programming formulation of the problem is presented as below.

$$\begin{aligned} Min \ \ C_E = {} & \sum_{j \in J} \sum_{i \in A_j} c_{S_1} r_i d_i x_j + \sum_{k \in K} \sum_{j \in J} \sum_{i \in A_j} \sum_{t \in T_{S_2}} P_k(x) c_{S_2} r_i d_i u_{it}^k + \\ & \sum_{k \in K} \sum_{j \in J} P_k(x) m_j \xi_j^k + \sum_{k \in K} \sum_{j \in J \setminus (J_2 \cap J)} P_k(x) m_j (1 - z_j^k)(1 - \xi_j^k) \end{aligned} \tag{4}$$

Subject to:

$$\sum_{j\in J_1}\sum_{i\in A_j}\sum_{\tau=t}^{\min\{b_{S_1},t+d_i-1\}} r_i u_{i\tau} \le \bar{R}_{S_1},\ \ \forall t \in T_{S_1} \tag{5}$$

$$\sum_{j\in J\cup J_2}\sum_{i\in A_j}\sum_{\tau=t}^{\min\{b_{S_2},t+d_i-1\}} r_i u_{i\tau}^k \le \bar{R}_{S_2},\ \ \forall t \in T_{S_2}, \forall k \in K \tag{6}$$

$$\sum_{j\in J'}\sum_{i\in A_j} c_{S_1} r_i d_i + \sum_{j\in J}\sum_{i\in A_j}\sum_{t\in T_{S_1}} c_{S_1} r_i d_i u_{it} \le B_{S_1} \tag{7}$$

$$\sum_{j\in J_2\setminus J}\sum_{i\in A_j} c_{S_2} r_i d_i + \sum_{j\in J}\sum_{i\in A_j}\sum_{t\in T_{S_2}} c_{S_2} r_i d_i u_{it}^k \le B_{S_2},\ \ \forall k \in K \tag{8}$$

$$\sum_{t=a_{S_1}}^{b_{S_1}} (t - d_i + 1)u_{it} \ge a_{S_1},\ \ \forall j \in J', \forall i \in A_j \tag{9}$$

$$\sum_{t=a_{S_1}}^{b_{S_1}} t u_{it} \le b_{S_1},\ \ \forall j \in J', \forall i \in A_j \tag{10}$$

$$\sum_{t=a_{S_1}}^{b_{S_1}} (t - d_i + 1)u_{it} \ge a_{S_1} x_j,\ \ \forall j \in J, \forall i \in A_j \tag{11}$$

$$\sum_{t=a_{S_1}}^{b_{S_1}} t u_{it} \le b_{S_1} x_j,\ \ \forall j \in J, \forall i \in A_j \tag{12}$$

$$\sum_{t=a_{S_2}}^{b_{S_2}} (t - d_i + 1)u_{it}^k \ge a_{S_2},\ \ \forall j \in J_2 \setminus J, \forall i \in A_j, \forall k \in K \tag{13}$$

$$\sum_{t=a_{S_2}}^{b_{S_2}} t u_{it}^k \le b_{S_2},\ \ \forall j \in J_2 \setminus J, \forall i \in A_j, \forall k \in K \tag{14}$$

$$\sum_{t=a_{S_2}}^{b_{S_2}} (t - d_i + 1)u_{it}^k \ge a_{S_2}(z_j^k - x_j),\ \ \forall j \in J \setminus (J_2 \cap J), \forall i \in A_j, \forall k \in K \tag{15}$$

$$\sum_{t=a_{S_2}}^{b_{S_2}} t u_{it}^k \le b_{S_2}(z_j^k - x_j),\ \ \forall j \in J \setminus (J_2 \cap J), \forall i \in A_j, \forall k \in K \tag{16}$$

$$\sum_{t=a_{S_2}}^{b_{S_2}} (t - d_i + 1)u_{it}^k \ge a_{S_2}(1 - \xi_j^k),\ \ \forall j \in J_2 \cap J, \forall i \in A_j, \forall k \in K \tag{17}$$

$$\sum_{t=a_{S_2}}^{b_{S_2}} t u_{it}^k \le b_{S_2}(1 - \xi_j^k),\ \ \forall j \in J_2 \cap J, \forall i \in A_j, \forall k \in K \tag{18}$$

$$\sum_{t=a_{S_1}}^{b_{S_1}} (t+1)u_{lt} \le \sum_{t=a_{S_1}}^{b_{S_1}} (t-d_i+1)u_{it}, \quad \forall j \in J', \forall i \in A_j, \forall l \in I_{ij} \tag{19}$$

$$\sum_{t=a_{S_1}}^{b_{S_1}} (t+1)u_{lt} \le \sum_{t=a_{S_1}}^{b_{S_1}} (t-d_i+1)u_{it}, \quad \forall j \in J, \forall i \in A_j, \forall l \in I_{ij} \tag{20}$$

$$\sum_{t=a_{S_2}}^{b_{S_2}} (t+1)u_{lt}^k \le \sum_{t=a_{S_2}}^{b_{S_2}} (t-d_i+1)u_{it}^k, \quad \forall j \in J_2 \setminus J, \forall i \in A_j, \forall l \in I_{ij}, \forall k \in K \tag{21}$$

$$\sum_{t=a_{S_2}}^{b_{S_2}} (t+1)u_{lt}^k \le \sum_{t=a_{S_2}}^{b_{S_2}} (t-d_i+1)u_{it}^k, \quad \forall j \in J, \forall i \in A_j, \forall l \in I_{ij}, \forall k \in K \tag{22}$$

$$z_j^k \ge x_j, \quad \forall j \in J, \forall k \in K \tag{23}$$

$$\sum_{t \in T_{S_1}} u_{it} = x_j, \quad \forall j \in J, \forall i \in A_j \tag{24}$$

$$\sum_{t \in T_{S_1}} u_{it} = 1, \quad \forall j \in J', \forall i \in A_j \tag{25}$$

$$\sum_{t \in T_{S_2}} u_{it}^k = 1, \quad \forall j \in J_2 \setminus J, \forall i \in A_j, \forall k \in K \tag{26}$$

$$\sum_{t \in T_{S_2}} u_{it}^k + \xi_j^k = 1, \quad \forall j \in J_2 \cap J, \forall i \in A_j, \forall k \in K \tag{27}$$

$$\sum_{t \in T_{S_1}} u_{it} + \sum_{t \in T_{S_2}} u_{it}^k \le 1, \quad \forall j \in J \setminus (J_2 \cap J), \forall i \in A_j, \forall k \in K \tag{28}$$

$$\sum_{i \in A_j} \sum_{t \in T_{S_1}} u_{it} + \sum_{i \in A_j} \sum_{t \in T_{S_2}} u_{it}^k \le 2|A_j| z_j^k, \quad \forall j \in J, \forall k \in K \tag{29}$$

$$\sum_{i \in A_j} \sum_{t \in T_{S_1}} u_{it} + \sum_{i \in A_j} \sum_{t \in T_{S_2}} u_{it}^k \ge z_j^k, \quad \forall j \in J, \forall k \in K \tag{30}$$

$$x_j \in \{0,1\}, \ \forall j \in J; \quad u_{it} \in \{0,1\}, \ \forall j \in J_1, \forall i \in A_j, \forall t \in T_{S_1} \tag{31}$$

$$z_j^k \in \{0,1\}, \forall j \in J, \forall k \in K; \quad u_{it}^k \in \{0,1\}, \forall j \in J \cup J_2, \forall i \in A_j, \forall t \in T_{S_2}, \forall k \in K \tag{32}$$

Objective function (4) minimises the total cost of scheduling maintenance of all components in J subject to constraints (5)–(30). Since the first-stage decision (x_j) affects the probability measures, the problem has endogenous uncertainty. The endogenous uncertainty is reflected in Eq. (2) that includes first-stage decision variables. The objective function multiplies the probabilities from Eq. (2) to the second stage decisions and hence, non-linear. Note that components $j \in (J_1 \cup J_2) \setminus J$ have to be scheduled in the respective shutdowns and incur a fixed resource cost. Hence, we do not consider the cost of maintaining these components in the objective function. Constraints (5) and (6) ensure that the resource utilisation in any period does not exceed the maximum available resources in that period. The total spending is restricted to the available budget using constraints (7) and (8). Constraints (9)–(12) enforce the scheduling of all the maintenance activities for component $j \in J'$ in S_1. For each scenario $k \in K$, constraints (13) and (14) ensure that components $j \in J_2 \setminus J$ will be scheduled in S_2. For component $j \in J \setminus (J_2 \cap J)$ and for a given scenario $k \in K$, the term $(z_j^k - x_j)$ equals 1 if the component is scheduled in S_2 else 0. Hence, if scheduled in S_2, constraints (15) and (16) state that for each scenario $k \in K$, all the activities of the component should be scheduled in S_2. For any scenario $k \in K$, if component $j \in J_2 \cap J$ fails (i.e. $\xi_j^k = 1$) between S_1 and S_2, constraints (17) and (18) ensure that the component will not be maintained in S_2. Constraints (19)–(22) enforce the precedence relations between the activities. For any scenario $k \in K$, if component $j \in J$ is maintained in S_1, constraint (23) enforces variable z_j^k to be 1. Constraint (24)–(28) enforce that each activity of any component will be scheduled at most once. For each scenario and for component $j \in J$, constraints (29) and (30) ensure that z_j^k will be 1 only if the component is maintained in either S_1 or S_2. Constraints (31) and (32) define the domain of the variables.

3 Results and Discussion

To study the complexity of the problem, we use randomly generated instances and solve using complete enumeration. To solve each instance, we first enumerate all first-stage postponement decisions. For instance with $|J|$ components, $2^{|J|}$ decisions are possible. Consider an instance with 3 components in J, resulting in 8 possible decisions: $\{(0, 0, 0), (0, 0, 1), (0, 1, 0), (1, 0, 0), (0, 1, 1), (1, 0, 1), (1, 1, 0), (1, 1, 1)\}$, where value 1 indicates that the corresponding component is maintained in S_1, e.g. the solution, (0, 0, 1), states that only component 3 will be maintained in S_1. Once the first-stage decision is fixed, the objective function becomes linear. We enumerate all solutions for the fixed first-stage decision and solve the resulting scheduling problem in S_1 and S_2 using CPLEX. As expected, the solution time increases when the number of components and activities increases. Note that the first-stage decision may be infeasible for a given budget and resource availability in S_1. Hence, the solution time can be reduced by detecting the infeasibility early and removing those solutions. We experimented with the following three methods:

Method I—Complete enumeration without early detection of infeasibility: In this method, we generate all possible first-stage solutions and use CPLEX to solve respective scheduling problems. For each solution, CPLEX will either detect infeasibility or provide the optimal expected cost for the solution. The solution with the minimum expected cost is the optimal solution for the given instance.

Method II—Complete enumeration with feasibility assessment: In this method, we first evaluate the first stage solution with respect to budget and resource requirement in S_1. We simply use CPLEX to solve the formulation with only budget and resource constraints for S_1. If the solution is feasible for S_1 then we solve the complete mathematical formulation for scheduling activities within S_1.

Method III—Complete enumeration with grouping: In this method, we first group the solutions based on the number of components in Set J that are maintained in S_1. For example, Group n contains all the solutions in which n components ($n \in \{0, 1, \ldots, |J|\}$) are maintained in S_1. Consider an instance with 3 components in J. The solutions are grouped as: Group 0 = $\{(0, 0, 0)\}$, Group 1 = $\{(0, 0, 1), (0, 1, 0), (1, 0, 0)\}$, Group 2 = $\{(0, 1, 1), (1, 0, 1), (1, 1, 0)\}$, and Group 3 = $\{(1, 1, 1)\}$. We then assess the feasibility of first-stage solutions similar to Method II but proceed with evaluating solutions from one group at a time in the ascending order of group numbers i.e. from Group 0 to Group $|J|$. In case an infeasible solution is detected in Group n, ($n \in \{0, \ldots, |J|\}$), then we inspect all Group $m : m > n$, for the solutions that maintain the same components (in S_1) as that in the infeasible solution. These solutions are removed from the solution space. For the above example with 3 components, let (0, 0, 1) in Group 1 is infeasible i.e. S_1 does not have sufficient resources or budget to maintain component 3. Hence, we can remove all solutions that require component 3 to be maintained in S_1. Hence, we can eliminate $\{(0, 1, 1), (1, 0, 1)\}$ from Group 2, and $\{(1, 1, 1)\}$ from Group 3. This helps in further reducing the total solution time for the complete enumeration.

We present the results for all of the three methods using instances with 4,5,6,7 and 8 components in set J. All the methods are implemented using Python, and CPLEX is used to solve the mathematical formulations. The experiments are performed on the machine with Intel Core i7 2.80GHz processor and 16 GB RAM. For more than 8 components, CPLEX could not complete the solutions due to memory requirements (requires more than 16 GB RAM). We used 5 randomly generated instances for each size i.e. components 4 to 8. The number of components in Set J' and J_2 are kept at 3. For instances with $|J| \in \{4, 5\}$, two common components are considered in Set J and J_2 whereas for instances with $|J| \in \{6, 7, 8\}$, we consider 3 common components in Set J and J_2. For all the instances, in each component five maintenance activities are required. The duration and resource requirement of each activity is generated randomly between 1 and 4 time periods and 1 and 4 units, respectively. We consider each shutdown with 30 time periods. For S_1, the minimum resource requirement is determined by the addition of the resources required for activities for components in $J_1 \setminus J$, and the maximum resource requirement equals the total resources required for all the activities for components in J_1. Similarly, for S_2, the minimum and maximum resource requirements are computed based on activities for components in J_2 and components in $J \cup J_2$, respectively. The resource availability for each shutdown is set randomly between the calculated minimum and maximum resource requirements for the respective shutdowns. The cost per unit of resource for each activity in S_1 is set at 100 units, while for S_2, it is 150 units. The minimum and maximum budget is calculated using the total cost required for the minimum and maximum resource requirement for each shutdown. The available budget is set randomly between the calculated minimum and maximum budget. Finally, the probability of failure (α_j) for component $j \in J$ is generated randomly between 0.4 and 0.9 whereas the probability (β_j) is generated randomly between a much lower range of 0.03 and 0.15. The corrective maintenance cost for any component is set between 15,000 and 40,000 units.

Figure 1 presents the distribution of the percentage of solutions evaluated and removed in Method III. The data is averaged over five instances for each component. With more infeasible solutions in the instances (see for components 6,7 and 8), we are required to evaluate only 17–25 percent of total infeasible solutions. The rest of the infeasible solutions are removed by inspecting similar solutions in higher-order groups. The effect of such removals can be seen in the solution time as depicted in Fig. 2. As seen in the figure, the solution time for Method III is improved by more than

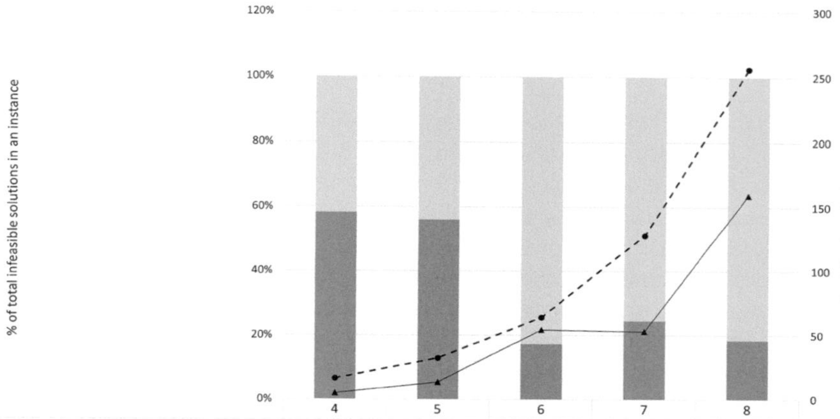

	4	5	6	7	8
% Infeasible solutions removed (Method III)	42%	44%	83%	75%	82%
% Infeasible solutions found (Method III)	58%	56%	17%	25%	18%
No. of infeasible solutions	5	13	54	53	159
No. of Solutions	16	32	64	128	256

Fig. 1 Number of solutions assessed and removed due to infeasibility in Method III

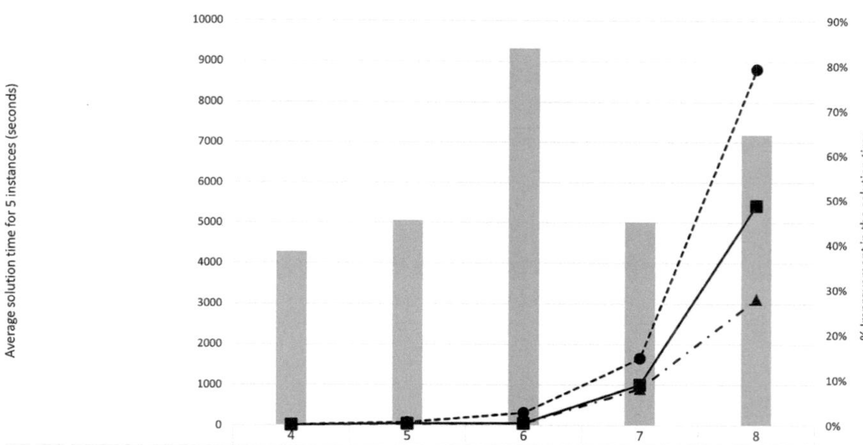

	4	5	6	7	8
% Improvement (Method-I to III)	38%	46%	84%	45%	65%
Method I	17.925	78.833	314.3052	1650.3738	8805.9418
Method II	12.4966	49.3382	56.9414	1003.5246	5419.1968
Method III	11.0258	42.963	50.5826	905.5992	3110.4056

Fig. 2 Solution time comparison

40% from that of Method I. Note that, even with the improvement in the solution time, we could not solve the instances for more than 8 components. For these instances, the mathematical formulation for a given first-stage decision requires more than 16 GB of memory to solve the formulation with all the scenarios. Since the goal of the study is to first model the endogenous uncertainty and understand the complexity of the problem, we did not continue improving the implementation further such as solving each scenario separately.

4 Conclusion and Future Works

This study aims to introduce the problem of scheduling activities across two maintenance plans with postponement decisions. To the best of our knowledge, the problem has not been studied in the literature. Hence, the focus of the study is to introduce the problem to the literature and model the problem as a two-stage stochastic programming problem. For the presented formulation, we are required to handle various complexities. First, the objective function is non-linear due to the endogenous uncertainty. Second, the number of first-stage solutions and the number of scenarios increases exponentially (power of 2) with the increase in the number of components. Third, for any assignment of components to shutdown, we require solving the resource constraint project scheduling problem which itself is NP-hard. Using the simple enumeration for small-sized instances (up to 8 components), we tried to demonstrate and emphasize the complexity of the problem. Hence, there is a need to develop a heuristic approach to solve the practical-sized instances. An approach that explores first-stage decisions heuristically embedded with scenario reduction techniques such as the Sample Average Approximation (SAA) method for each first-stage decision is worth investigating. There is also a possibility of modeling each first-stage decision vector as a variable and using techniques such as column generation to evaluate these decisions iteratively. For complete enumeration, we proposed a simple strategy to use the inherent problem characteristics and reduce the number of evaluations. Such a strategy can reduce the computational time and can be employed in any of the approaches discussed above.

In this study, we used the point estimate of the failure probabilities with the assumption that the components are independent. In reality, we work with the failure distributions and the component maintenance depends on each other. Hence, it is required to extend the formulation to these real cases.

References

1. Apap, R.M., Grossmann, I.E.: Models and computational strategies for multistage stochastic programming under endogenous and exogenous uncertainties. Comput. & Chem. Eng. **103**, 233–274 (2017)
2. Berrade, M.D., Scarf, P.A., Cavalcante, C.A.: A study of postponed replacement in a delay time model. Reliab. Eng. & Syst. Saf. **168**, 70–79 (2017)

3. Javanmard, H., Koraeizadeh, A.A.W.: Optimizing the preventive maintenance scheduling by genetic algorithm based on cost and reliability in National Iranian Drilling Company. J. Ind. Eng. Int. **12**, 509–516 (2016)
4. Pinciroli, L., Baraldi, P., Zio, E.: Maintenance optimization in Industry 4.0. Reliab. Eng. & Syst. Saf. 109204 (2023)
5. Richardson, S., Kefford, A., Hodkiewicz, M.: Optimised asset replacement strategy in the presence of lead time uncertainty. Int. J. Prod. Econ. **141**(2), 659–667 (2013)
6. Şafak, Ö., Çavuş, Ö., Aktürk, M.S.: A two-stage decision dependent stochastic approach for airline flight network expansion. Transp. Res. Part B: Methodol. **158**, 78–101 (2022)
7. Sha, Y., Zhang, J., Cao, H.: Multistage stochastic programming approach for joint optimization of job scheduling and material ordering under endogenous uncertainties. Eur. J. Oper. Res. **290**(3), 886–900 (2021)
8. Ulusoy, G., Hazır, Ö.: Introduction to Project Modeling and Planning. Springer, Switzerland (2021)
9. Zhou, R., Bhuiyan, T.H., Medal, H.R., Sherwin, M.D., Yang, D.: A stochastic programming model with endogenous uncertainty for selecting supplier development programs to proactively mitigate supplier risk. Omega **107**, 102542 (2022)

Scheduling Carbon Dioxide Transfers Between Two Ports via Mixed-Integer Linear Programming and Heuristic Approaches

Mauro Gaggero

Abstract The optimization of CO_2 transfers via ships between two ports is addressed by means of a scheduling approach based on the solution of a mixed-integer linear programming (MILP) problem. The goal is to keep the temporary storage buffers at the ports as empty as possible while satisfying suitable constraints. Since the solution of the optimization problem may be time consuming, a greedy heuristic based on a threshold mechanism is also presented. Owing to its reduced computational requirements, the heuristic can be successfully used to provide an initial solution for the MILP-based approach, thus improving performances. Preliminary simulation results are showcased to assess the effectiveness of the two approaches.

Keywords Scheduling · CO_2 transfer with ships · Mixed-integer linear programming · Greedy heuristic

1 Introduction

In the global pursuit of mitigating climate change and reducing greenhouse gas emissions, the transportation of carbon dioxide (CO_2) via ships from generating plants to treatment facilities has emerged as a crucial component of the carbon capture and storage (CCS) strategy [7, 13]. CCS entails capturing CO_2 emissions from industrial sources, such as power plants and refineries, and transporting them to storage or treatment sites, where they are securely stored or re-used in other industrial processes. In this context, the efficient and reliable transfer of captured CO_2 is crucial. Maritime transport represents a viable and scalable option for CO_2 transfer over long distances [3]. In fact, ships are able to transport large volumes of CO_2 across oceans

This work was supported by the Italian Ministry of University and Research through the Piano Nazionale di Ripresa e Resilienza (PNRR), Centro Nazionale per la Mobilità Sostenibile (CN-MOST), Spoke 3.

M. Gaggero (✉)
National Research Council of Italy, Genoa, Italy
e-mail: mauro.gaggero@cnr.it

M. Di Francesco et al. (eds.), *Operations Research: Closing the Gap Between Research and Practice*, AIRO Springer Series 15, https://doi.org/10.1007/978-3-031-90095-2_19

and waterways. They efficiently connect emission sources to destination sites without requiring the design and construction of complex and expensive ground transportation facilities [9], thus enhancing cost-effectiveness and feasibility. Hence, maritime transport for CO_2 transfer represents a strategic pathway towards a sustainable and resilient future.

The literature on ship-based CCS includes works dealing with cost assessment (see, e.g., [1, 10]) and optimization models to improve performances. For instance, [12] proposed a mixed-integer linear programming (MILP) approach to design a ship-based offshore CCS system. Reference [8] presented a system-wide supply chain optimization model using again a MILP approach. The work [5] evaluated the impact of operational uncertainties and fluctuations on the design and cost of ship-based CO_2 transport. Lastly, [3] investigated operational and investment costs of a CCS logistics chain based on marine vessels.

Besides cost assessment and modeling of the entire logistics chain, another key aspect to ensure efficiency of operations is the optimization and synchronization of transfer activities [14]. In fact, the coordination of the various steps throughout the transfer process is paramount to maximize resource utilization, minimize downtime, and mitigate risks associated with handling CO_2. Effective scheduling encompasses various aspects, including vessel availability, loading and unloading operations, transit routes, and delivery timelines. Optimizing these aspects requires planning and coordination among multiple stakeholders, including generating plants, treatment facilities, shipping companies, and regulatory authorities. By synchronizing these activities, delays and bottlenecks can be minimized, enabling a smooth and uninterrupted flow of CO_2 from source to destination. This paper focuses on the problem of determining the best times for departures and arrivals as well as the optimal load of vessels operating between different ports and carrying CO_2 captured from given industrial plants, with the goal of transferring it to a treatment site for storage or re-use.

CO_2 transfer via ships can be arranged through different transportation strategies that depend on flow rates and geographical location of ports. Two main strategies, named "hub and spokes" and "milk run" can be identified. Figure 1 reports a sketch of the two network topologies. The hub and spokes distribution network is characterized by routes organized as a series of peripheral ports, called spokes, that connect industrial plants generating and/or capturing CO_2 to a central destination hub. A hub and spokes distribution model is a centralized shipment system. The center is the hub and each spoke represents a direction of delivery. The centralization of activities is one of its characteristics, but the organization and synchronization of operations may be difficult, especially in the presence of a large number of spokes. In general, the network is concentrated on a single hub, even if additional hubs can be introduced, thus contributing to make the overall problem much more difficult to deal with. A milk run is a general distribution method that can be used to transport various types of goods from different suppliers to one customer. It gets its name from the dairy industry practice, where one tanker is used to collect milk from several farms for delivery to a milk processing company. All the users are served by the same carrier in a cyclical path. However, in the case of delays or operational gaps in one element of

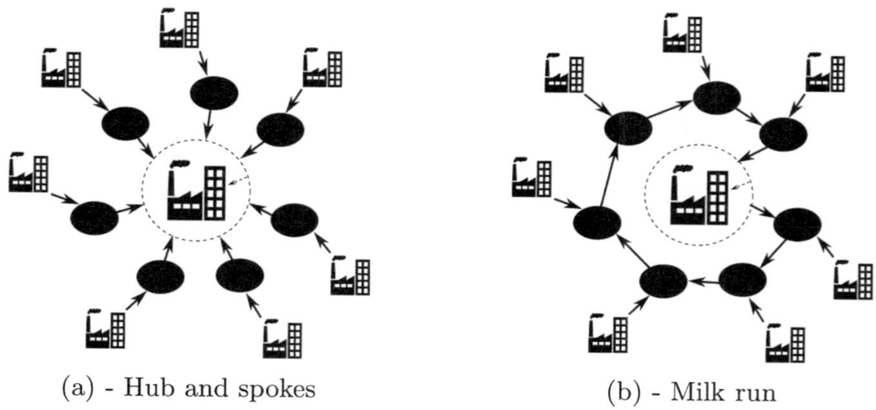

Fig. 1 Network topologies for CO_2 transfer from peripheral manufacturing sites towards a central one for storage or re-use. Black circles represent outgoing ports, while the dashed one is a port connected to a treatment site that collects all incoming flows

the logistic chain, issues can quickly propagate in the entire network, with negative impacts on overall operations.

In this paper, we focus on the milk run case, and in particular on a case study with two manufacturing sites, one that produces CO_2 and another one that collects and safely stocks it. Thus, we assume a one-way transportation of CO_2 from the first manufacturer to the second one. The problem takes into account different aspects, such as distances among the ports, service speed, loading and offloading rates, annual flow rates, and availability of storage at both origin and destination locations. First, a MILP formulation is presented, the solution of which provides optimal schedules for vessels operating among ports. To reduce the computational requirements, which grow with the number of vessels, a heuristic approach based on a greedy mechanism with threshold is presented that is able to compute approximate solutions. The proposed heuristic can be used also to provide a feasible initial solution to the MILP-based optimization, thus allowing an improvement of performance. Preliminary simulation results in two case studies are presented to assess the effectiveness of the proposed approaches.

In the context of CO_2 transportation within CCS systems, the network configuration that is closest to our work is the one of [4], where a MILP model is presented to provide decision support for a CCS supply chain design problem with the goal of minimizing total costs. However, in [4] the scheduling of vessels is not explicitly taken into account, thus making our work innovative and challenging. In fact, works focusing on the scheduling of transportation are scarce in the literature of CCS systems. A parallelism with the much more developed literature of scheduling on parallel machines, where ships performing the transportation of CO_2 play the role of machines executing jobs, is in order. In this context, the problem considered in our work is somehow related to the scheduling on parallel machines with loading and unloading times (see, e.g., [2], where a MILP formulation and some constructive heuristics are presented for operations in a multi-station transhipment terminal) or to the scheduling on parallel machines with limited buffers (see, e.g., [11] and the references therein).

The paper is organized as follows. Section 2 describes the problem as well as the MILP formulation for the optimization of operations and the greedy heuristic approach. Section 3 reports numerical results. Section 4 draws conclusions.

2 Problem Definition, MILP Formulation, and Heuristic Approach

In this section, we define the problem of optimizing the transportation of CO_2 between two ports, according to a milk run strategy. In more detail, we assume that transportation is performed through marine vessels traveling from a source port, named "port of leave" (POL), to a destination port, referred to as "port of destination" (POD). We focus on a set of N vessels to be scheduled within the time interval $[0, T_{fin}]$. Such a time interval is discretized into a number T of sub-intervals, each one of length $\Delta t = T_{fin}/T$. As a result, in the following we will refer to discretized time instants $t = 0, \Delta t, 2\Delta t, \ldots, T\Delta t$, or, with a little abuse of notation, to $t = 0, 1, \ldots, T$.

Both ports are equipped with a single berth dedicated to CO_2 loading and unloading operations. Manufacturer sites are connected to the two ports by means of tubes, and they are supposed to generate/collect CO_2 at given rates, denoted by $r_{1,t}$ and $r_{2,t}$ at time t for the POL and the POD, respectively. The POL is equipped with a temporary stock area with maximum capacity B_1^{max}, simply called buffer, where CO_2 coming from the manufacturer can be stored waiting for being loaded on vessels. Similarly, the POD is equipped with a buffer with maximum capacity B_2^{max}, where CO_2 arriving through incoming cargos can be temporarily stored before being unloaded and transferred to the second manufacturing site as final destination. Figure 2 reports a scheme of the considered network setup.

Each vessel is assumed to have a maximum storage capacity, denoted by X_i^{max}, $i = 1, \ldots, N$. The initial amount of CO_2 stored at the two buffers at $t = 0$ is given by B_1 and B_2. The unitary loading time at the POL is given by T_{load}, while the unitary unloading time at the POD is equal to T_{unload}. The traveling time between ports is denoted by T_{travel}.

Fig. 2 Sketch of the considered network setup with some of the parameters used in the mathematical formulation of the problem (CO_2 flows along with maximum capacity of buffers and vessels)

2.1 MILP Formulation

In this section, a MILP formulation is presented to determine the optimal schedule of vessels. In particular, decision variables are related to berthing and departure times of vessels at the two ports and to the amount of CO_2 to load. In more detail, we define the following variables.

- $x_i \in \mathbb{R}^+$, $i = 1, \ldots, N$, is the amount of CO_2 loaded on vessel i.
- $t^b_{1,i} \in \mathbb{R}^+$, $i = 1, \ldots, N$, is the berthing time of vessel i at the POL, i.e., the time instant when the loading operations begin.
- $t^d_{1,i} \in \mathbb{R}^+$, $i = 1, \ldots, N$, is the departure time of vessel i from the POL, i.e., the time instant when the loading operations end.
- $t^b_{2,i} \in \mathbb{R}^+$, $i = 1, \ldots, N$, is the berthing time of vessel i at the POD, i.e., the time instant when it arrives at the POL and unloading operations start.
- $t^d_{2,i} \in \mathbb{R}^+$, $i = 1, \ldots, N$, is the departure time of vessel i from the POD, i.e., the time instant when the unloading operations end.
- $b_{1,t} \in \mathbb{R}^+$, $t = 0, 1, \ldots, T$, is the amount of CO_2 temporarily stored at the buffer of the POL at time t.
- $b_{2,t} \in \mathbb{R}^+$, $t = 0, 1, \ldots, T$, is the amount of CO_2 temporarily stored at the buffer of the POD at time t.
- $y_{i,j} \in \{0, 1\}$, $i = 1, \ldots, N$, $j = 1, \ldots, N$, is a binary variable equal to 1 if ship i precedes ship j in the scheduling of berthing time at the POL (and consequently at the POD), while it is equal to 0 otherwise.
- $v^{b_1}_{i,t} \in \{0, 1\}, i = 1, \ldots, N, t = 0, 1, \ldots, T$, is a binary variable equal to 1 if vessel i is berthed at time t at the POL, otherwise it is equal to 0.
- $v^{b_2}_{i,t} \in \{0, 1\}, i = 1, \ldots, N, t = 0, 1, \ldots, T$, is a binary variable equal to 1 if vessel i is berthed at time t at the POD, otherwise it is equal to 0.
- $w^{b_1}_{i,t} \in \{0, 1\}$, $i = 1, \ldots, N$, $t = 0, 1, \ldots, T$, is a binary variable equal to 1 if $t \geq t^b_{1,i}$, i.e., if ship i has already berthed at the POL at time t, otherwise it is equal to 0.
- $z^{b_1}_{i,t} \in \{0, 1\}$, $i = 1, \ldots, N$, $t = 0, 1, \ldots, T$, is a binary variable equal to 1 if $t \leq t^d_{1,i}$, i.e., if ship i has not yet left the POL at time t, otherwise it is equal to 0.
- $w^{b_2}_{i,t} \in \{0, 1\}$, $i = 1, \ldots, N$, $t = 0, 1, \ldots, T$, is a binary variable equal to 1 if $t \geq t^b_{2,i}$, i.e., if ship i has already berthed at the POD at time t, otherwise it is equal to 0.
- $z^{b_2}_{i,t} \in \{0, 1\}$, $i = 1, \ldots, N$, $t = 0, 1, \ldots, T$, is a binary variable equal to 1 if $t \leq t^d_{2,i}$, i.e., if ship i has not yet left the POD at time t, otherwise it is equal to 0.

The goal is to organize the transport of CO_2 to minimize the amount of CO_2 temporarily stored at the two ports, i.e., we aim at minimizing, under suitable constraints, a cost function given by the sum over time of $b_{1,t}$ and $b_{2,t}$ (see Eq. (1) below). In more detail, we formulate the following MILP problem.

$$\min \sum_{t=0}^{T} b_{1,t} + b_{2,t} \quad (1)$$

subject to:

$$t^d_{1,i} = t^b_{1,i} + T_{load}\, x_i, \quad i = 1, \ldots, N, \quad (2)$$

$$t^b_{2,i} = t^d_{1,i} + T_{travel}, \quad i = 1, \ldots, N, \quad (3)$$

$$t^d_{2,i} = t^b_{2,i} + T_{unload}\, x_i, \quad i = 1, \ldots, N, \quad (4)$$

$$0 \le x_i \le X^{max}_i, \quad i = 1, \ldots, N, \quad (5)$$

$$0 \le b_{k,t} \le B^{max}_k, \quad k = 1, 2, \quad t = 0, 1, \ldots, T, \quad (6)$$

$$t^b_{1,i} \le t^b_{1,j} + M\left(1 - y_{i,j}\right), \quad i = 1, \ldots, N, \quad j = 1, \ldots, N, \quad (7)$$

$$t^b_{1,j} \le t^b_{1,i} + M y_{i,j}, \quad i = 1, \ldots, N, \quad j = 1, \ldots, N, \quad (8)$$

$$b_{1,t+1} = b_{1,t} + r_{1,t} - \frac{1}{T_{load}} \sum_{i=1}^{N} v^{b_1}_{i,t}, \quad t = 0, 1, \ldots, T-1, \quad (9)$$

$$b_{2,t+1} = b_{2,t} - r_{2,t} + \frac{1}{T_{unload}} \sum_{i=1}^{N} v^{b_2}_{i,t}, \quad t = 0, 1, \ldots, T-1, \quad (10)$$

$$b_{k,0} = B_k, \quad k = 1, 2, \quad (11)$$

$$t \ge t^b_{k,i} - M\left(1 - w^{b_k}_{i,t}\right), \quad i = 1, \ldots, N, \quad k = 1, 2, \quad t = 0, 1, \ldots, T, \quad (12)$$

$$t \le t^b_{k,i} - 1 + M w^{b_k}_{i,t}, \quad i = 1, \ldots, N, \quad k = 1, 2, \quad t = 0, 1, \ldots, T, \quad (13)$$

$$t \le t^d_{k,i} + M\left(1 - z^{b_k}_{i,t}\right), \quad i = 1, \ldots, N, \quad k = 1, 2, \quad t = 0, 1, \ldots, T, \quad (14)$$

$$t \ge t^d_{k,i} + 1 - M z^{b_k}_{i,t}, \quad i = 1, \ldots, N, \quad k = 1, 2, \quad t = 0, 1, \ldots, T, \quad (15)$$

$$w^{b_k}_{i,t} + z^{b_k}_{i,t} \le 2 + M\left(1 - v^{b_k}_{i,t}\right),$$
$$i = 1, \ldots, N, \quad k = 1, 2, \quad t = 0, 1, \ldots, T, \quad (16)$$

$$w^{b_k}_{i,t} + z^{b_k}_{i,t} \ge 2 - M\left(1 - v^{b_k}_{i,t}\right),$$
$$i = 1, \ldots, N, \quad k = 1, 2, \quad t = 0, 1, \ldots, T, \quad (17)$$

$$w^{b_k}_{i,t} + z^{b_k}_{i,t} \le 1 + M v^{b_k}_{i,t},$$
$$i = 1, \ldots, N, \quad k = 1, 2, \quad t = 0, 1, \ldots, T, \quad (18)$$

$$\sum_{i=1}^{N} v^{b_k}_{i,t} \le 1, \quad k = 1, 2, \quad t = 0, 1, \ldots, T, \quad (19)$$

$$t^b_{k,i} \ge 0, \; t^d_{k,i} \ge 0, \quad i = 1, \ldots, N, \quad k = 1, 2, \quad (20)$$

$$x_i \ge 0, \quad i = 1, \ldots, N, \quad (21)$$

$$v^{b_k}_{i,t} \in \{0, 1\}, \quad i = 1, \ldots, N, \quad k = 1, 2, \quad t = 0, 1, \ldots, T, \quad (22)$$

$$w^{b_k}_{i,t} \in \{0, 1\}, \; z^{b_k}_{i,t} \in \{0, 1\}, \quad i = 1, \ldots, N, \quad k = 1, 2, \quad t = 0, 1, \ldots, T. \quad (23)$$

Constraint (2) ensures that vessels depart from the POL after the completion of the loading operations. Equation (3) guarantees that vessels arrive at the POD at the correct time instant, given by the sum of the departure time from the POL and the traveling time, while (4) constrains the departure time from the POD after the completion of the unloading operations. Equations (5) and (6) account for the minimum and maximum capacity of ships and of the buffers at the POL ($k = 1$) and at the POD ($k = 2$), respectively. Constraints (7) and (8) establish the berthing sequence at a port between two vessels i and j according to the value of $y_{i,j}$, where M is a large positive constant that is much bigger than all the other quantities. If $y_{i,j} = 1$, ship i precedes ship j, and therefore (7) holds while (8) is trivially satisfied. On the contrary, if $y_{i,j} = 0$, ship j precedes ship i, and therefore (8) holds, being (7) trivially satisfied. Equations (9) and (10) establish a dynamics for the amount of CO_2 temporarily stored at the POL and the POD, respectively, accounted for by the decision variables $b_{1,t}$ and $b_{2,t}$, respectively, with initialization at $t = 0$ given by (11). Along with (12)–(18), they represent a linear implementation of the following relationships:

$$b_{1,t+1} = \begin{cases} b_{1,t} + r_{1,t} & \text{if no vessel is berthed,} \\ b_{1,t} + r_{1,t} - \frac{1}{T_{load}} & \text{if a vessel is berthed,} \end{cases} \quad t = 0, 1, \ldots, T-1,$$

$$b_{2,t+1} = \begin{cases} b_{2,t} - r_{2,t} & \text{if no vessel is berthed,} \\ b_{2,t} - r_{2,t} + \frac{1}{T_{unload}} & \text{if a vessel is berthed,} \end{cases} \quad t = 0, 1, \ldots, T-1. \quad (24)$$

In fact, (12) and (13) impose that $t \geq t^b_{k,i}$ if $w^{b_k}_{i,t} = 1$, while $t \leq t^b_{k,i}$ if $w^{b_k}_{i,t} = 0$. Similarly, (14) and (15) guarantee that $t \leq t^d_{k,i}$ if $z^{b_k}_{i,t} = 1$, while $t \geq t^d_{k,i}$ if $z^{b_k}_{i,t} = 0$. Lastly, (16), (17), and (18) account for $v^{b_k}_{i,t} = 1$ if $w^{b_k}_{i,t} + z^{b_k}_{i,t} = 2$, i.e., if vessel i is berthed at time t at the POL ($k = 1$) or at the POD ($k = 2$), while $v^{b_k}_{i,t} = 0$ otherwise. Finally, (19) guarantees that only one ship at a time is berthed at each port, while (20)–(23) define the decision variables.

2.2 Heuristic Approach

In this section, we present a heuristic approach to find an approximate solution to problem (1)–(23). It is based on a greedy choice of the amount of CO_2 to transfer between the POL and the POD, and on a thresholding mechanism. The idea is to schedule a vessel berthing at the POL if the amount of CO_2 stored therein is above a given threshold $\theta_1 > 0$. In this way, we avoid that the amount of CO_2 at the POL grows too much, as the goal is that of keeping the buffers of the two ports as empty as possible. Similarly, we schedule a vessel berthing at the POL if the amount of CO_2 stored at the POD is below a given threshold $\theta_2 > 0$. With such a choice, we avoid that the quantity of CO_2 at the POD is too low. In fact, this can generate stockouts for the downstream treatment plant, which requests CO_2 at a rate $r_{2,t}$ for $t = 0, 1, \ldots, T$.

Algorithm 1 Greedy heuristic for the scheduling of vessels

Input: $r_{1,t}, r_{2,t}$ for $t = 0, 1, \ldots, T-1$, B_1, B_2, X_i^{max}, θ_1, θ_2
Output: $t_{1,i}^b, t_{1,i}^d, t_{2,i}^b, t_{2,i}^d, x_i$ for $i = 1, \ldots, N$

1: Let $s_i \leftarrow 0$ for $i = 1, \ldots, N$
2: Let $p_t \leftarrow 0$ for $t = 0, 1, \ldots, T$
3: Let $b_{1,0} \leftarrow B_1$ and $b_{2,0} \leftarrow B_2$
4: **for** $t = 0$ **to** $T - 1$ **do**
5:   **if** $p_t = 0$ **and** $(b_{1,t} \geq \theta_1$ **or** $b_{2,t} \leq \theta_2)$ **then**
6:     **for** $i = 1$ **to** N **do**
7:       **if** $s_i = 0$ **then**
8:         $x_i \leftarrow \min(b_{1,t}, X_i^{max})$
9:         $t_{1,i}^b \leftarrow t$
10:         $t_{1,i}^d \leftarrow t_{1,i}^b + T_{load}\, x_i$
11:         $t_{2,i}^b \leftarrow t_{1,i}^d + T_{travel}$
12:         $t_{2,i}^d \leftarrow t_{2,i}^b + T_{unload}\, x_i$
13:         **if** $i > 1$ **and** $t_{2,i-1}^d < t_{2,i}^b$ **then**
14:           $s_i \leftarrow 1$ and $p_\tau \leftarrow 1$ for $\tau = t, t+1, \ldots, t_{1,i}^d$
15:         **else**
16:           $s_i \leftarrow 0$ and $p_\tau \leftarrow 0$ for $\tau = t, t+1, \ldots, t_{1,i}^d$
17:           $x_i \leftarrow 0$
18:           $t_{1,i}^b \leftarrow 0$
19:           $t_{1,i}^d \leftarrow 0$
20:           $t_{2,i}^b \leftarrow 0$
21:           $t_{2,i}^d \leftarrow 0$
22:         **end if**
23:       **end if**
24:     **end for**
25:   **end if**
26:   Compute $b_{1,t+1}, b_{2,t+1}$ from $b_{1,t}, b_{2,t}$ according to (24)
27: **end for**
28: Compute the cost function $\sum_{t=0}^{T} b_{1,t} + b_{2,t}$
29: Perform a local search around $t_{1,i}^b, t_{1,i}^d, t_{2,i}^b, t_{2,i}^d, x_i$ to further reduce the cost
30: **return** $t_{1,i}^b, t_{1,i}^d, t_{2,i}^b, t_{2,i}^d, x_i$ for $i = 1, \ldots, N$ as updated by local search

Algorithm 1 contains the pseudo-code of the proposed heuristic, where $s_i \in \{0, 1\}$, $i = 1, \ldots, N$, is a binary variable equal to 1 if the ship i has been served, that is, if it has already berthed at the POL (the variable is connected to $w_{i,t}^{b_1}$ of Sect. 2.1), while $p_t \in \{0, 1\}$, $t = 0, 1, \ldots, T$, is a binary variable equal to 1 if the berth at the POL if busy at time t, i.e., if a vessel is performing loading operations (the variable is again connected to $w_{i,t}^{b_1}$ and $z_{i,t}^{b_1}$). At each time $t = 0, 1, \ldots, T-1$, the algorithm verifies whether the berth at the POL is free and the amount of CO_2 at the POL (or at the POD) is above (or below) the threshold θ_1 (or θ_2) (line 5). If yes, the first ship not yet served is selected, and an amount of CO_2 equal to the maximum possible one, given by the minimum between the maximum capacity of the vessel and the amount of CO_2 at the POL, is selected (lines 6–8). Then, decision variables $t_{1,i}^b$, $t_{1,i}^d$, $t_{2,i}^b$, and $t_{2,i}^d$ are updated accordingly (lines 9–12). In more detail, the algorithm checks if a ship can berth at the POD depending on whether the preceding one has completed its unloading operations (lines 13–22). If yes, the ship is marked as "served", and the berth of the POL is set to a busy state from t to $t^d_{1,i}$ (line 14). Otherwise, the starting of loading operations on the vessel at the POL and the consequent unloading operations at the POD are postponed by freeing the berth of the POL and canceling the assignment of x_i, $t^b_{1,i}$, $t^d_{1,i}$, $t^b_{2,i}$, and $t^d_{2,i}$ (lines 16–21). The amount of CO_2 stored at the POL and at the POD is then updated owing to (24) (line 26). Lastly, to further improve the value of the cost function (computed at line 28), a local search is performed around the values of $t^b_{1,i}$, $t^d_{1,i}$, $t^b_{2,i}$, $t^d_{2,i}$, and x_i using a direct search approach (line 29). In particular, the generalized pattern search algorithm is adopted, which is based on the idea of iteratively updating the current solution by sampling the objective function along suitable search directions through the construction of a grid with the goal of reducing the cost [6].

We observe that the thresholds θ_1 and θ_2 play a decisive role in the overall accuracy of Algorithm 1 with respect to the solution of the MILP problem (1)–(23). They may also affect feasibility if the quantities $b_{1,t}$ and $b_{2,t}$ (computed at line 28) become negative for some t. Thus, they require a suitable tuning, which can be performed with a brute-force-like technique owing to the reduced computational requirements of the heuristic. The tuning consists of (i) listing all the values of θ_1 and θ_2 that may be of interest (or randomly sampling a given interval), (ii) computing the overall cost of the solution for each value as in (1), and (iii) selecting the combination of θ_1 and θ_2 that corresponds to the lowest cost.

3 Numerical Results

In this section, we describe the numerical results obtained in two case studies, denoted as "Scenario A" and "Scenario B", designed to evaluate the effectiveness of the proposed approaches. Scenario A is characterized by almost constant input and output rates of CO_2 at the POL and at the POD, respectively, while Scenario B presents some peaks at different time intervals. In both scenarios, we considered a time horizon of 1 week, that is, we chose $T = 720$ and $\Delta t = 1$ h. The number N of vessels was fixed to 15, and each one was assumed to have a maximum capacity $X_i^{max} = 1000$ ton for all $i = 1, \ldots, N$. The maximum capacity of the buffers at the POL and at the POD was taken equal to 3000 ton, while the quantity of CO_2 stored at $t = 0$ was chosen equal to 1000 ton for both ports. Loading and unloading times were both assumed to be equal to 0.05 h/ton, while the traveling time was fixed to 32 h.

The MILP problem (1)–(23) was solved using CPLEX 22.1 on a computer with an Intel 2.5 GHz i7 CPU and 32 GB of RAM. We imposed a maximum time limit of 10 h for the optimization. To evaluate the progress of the optimization, we considered also the results after 1 h, 3 h, 5 h, and 8 h of computation. Concerning the greedy heuristic, we evaluated different combinations of θ_1 and θ_2 with the goal of determining the best configuration. In particular, we varied both thresholds from 100 to 1000 with a step equal to 100, and selected the combination satisfying all the constraints of the problem (it suffices to check whether the amounts of CO_2 stored at the two ports, computed at line 26 of Algorithm 1, always remain non-negative) and characterized by the minimum cost. The selected values were $\theta_1 = 1000$, $\theta_2 = 100$ for Scenario A and $\theta_1 = 1000$, $\theta_2 = 700$ for Scenario B. The heuristic was used also to provide an initialization to the MILP-based solution approach, to be compared with the default initialization generated by CPLEX.

The obtained results are summarized in Table 1 as well as Figs. 3 and 4 for Scenario A and Scenario B, respectively. We observe that the MILP-based approach with initialization provided by the heuristic guaranteed the best results in terms of the cost function in (1). The schedules generated by the MILP formulation with default initialization, instead, were less effective and in line with those computed by the heuristic in the best configuration. In the case of Scenario A, the almost constant

Table 1 Summary of the numerical results

	Scenario A		
	Total cost (ton)	Optimality gap	CPU time (s)
MILP (limit 1 h)	$9.0332 \cdot 10^5$	0.8710	3600
MILP (limit 3 h)	$8.6502 \cdot 10^5$	0.8653	10800
MILP (limit 5 h)	$8.6044 \cdot 10^5$	0.8646	18000
MILP (limit 8 h)	$8.4934 \cdot 10^5$	0.8628	28800
MILP (limit 10 h)	$8.4930 \cdot 10^5$	0.8627	36000
MILP (limit 1 h, with initialization)	$8.3745 \cdot 10^5$	0.8608	3600
MILP (limit 3 h, with initialization)	$8.3232 \cdot 10^5$	0.8600	10800
MILP (limit 5 h, with initialization)	$8.3220 \cdot 10^5$	0.8600	18000
MILP (limit 8 h, with initialization)	$8.3218 \cdot 10^5$	0.8600	28800
MILP (limit 10 h, with initialization)	$8.3148 \cdot 10^5$	0.8598	36000
Greedy heuristic (without local search)	$8.7948 \cdot 10^5$	–	0.0023
Greedy heuristic (with local search)	$8.4146 \cdot 10^5$	–	8.81
	Scenario B		
	Total cost (ton)	Optimality gap	CPU time (s)
MILP (limit 1 h)	$1.0061 \cdot 10^6$	0.8867	3600
MILP (limit 3 h)	$9.9743 \cdot 10^5$	0.8857	10800
MILP (limit 5 h)	$9.8831 \cdot 10^5$	0.8846	18000
MILP (limit 8 h)	$9.7811 \cdot 10^5$	0.8834	28800
MILP (limit 10 h)	$9.5321 \cdot 10^5$	0.8803	36000
MILP (limit 1 h, with initialization)	$9.3823 \cdot 10^5$	0.8785	3600
MILP (limit 3 h, with initialization)	$9.3775 \cdot 10^5$	0.8784	10800
MILP (limit 5 h, with initialization)	$9.3467 \cdot 10^5$	0.8780	18000
MILP (limit 8 h, with initialization)	$9.3169 \cdot 10^5$	0.8776	28800
MILP (limit 10 h, with initialization)	$9.1263 \cdot 10^5$	0.8750	36000
Greedy heuristic (without local search)	$1.0331 \cdot 10^6$	–	0.0019
Greedy heuristic (with local search)	$9.5765 \cdot 10^5$	–	5.87

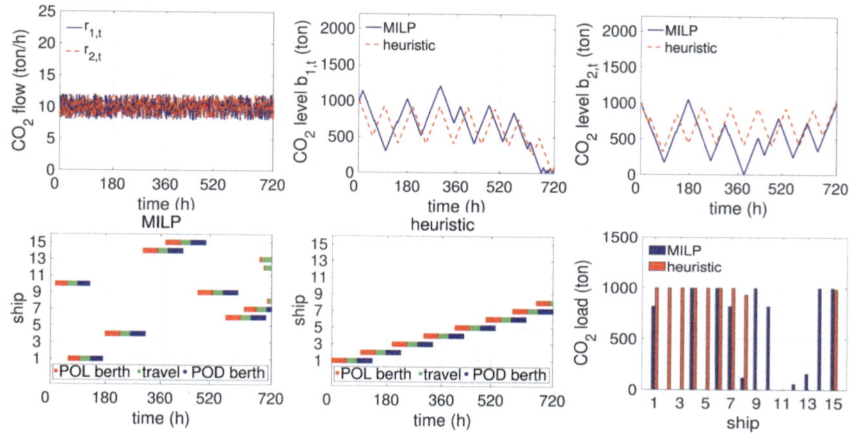

Fig. 3 Numerical results in Scenario A

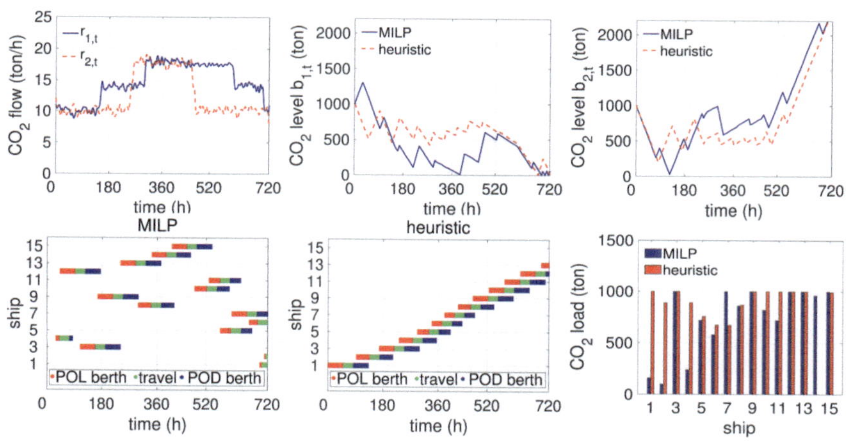

Fig. 4 Numerical results in Scenario B

input and output CO_2 flows $r_{1,t}$ and $r_{2,t}$, respectively, allowed the heuristic with local search to even outperform the MILP-based method with default initialization.

As regards CPU time, the MILP-based approach, either with or without initialization computed by the heuristic, is never able to find an optimal solution with zero optimality gap within the prescribed time limit of 10 h. This is due to the presence of several Big-M-type constraints implementing a large number of disjunctive constraints along with several binary variables, which may generate weak formulations and numerical instabilities. The confirmation is given by the lower bound provided by the relaxation problem, which is almost never updated over 10 h of computation. As a consequence, the values of the optimality gap are quite large (the best value is equal to about 0.8598 for Scenario A and 0.8750 for Scenario B), and the rate of reduction of the total cost in (1) is quite slow, even if always present for all the considered time limits for the optimization. On the contrary, the heuristic approach revealed to be extremely fast, as it provided an approximation of the optimal solution in less than 0.01 s in the version without the final local search step and in less than 10 s in the presence of the refinement through local search. The latter guaranteed a further reduction of the values of the cost. Thus, we can conclude that the local search is worth performing, as it enables to improve the cost at the price of only a reduced increase of the computational effort, which remains much less than the one of the MILP-based approach.

Looking at the plots reported in Figs. 3 and 4, it turns out that the MILP-based and heuristic approaches generated time behaviors of $b_{1,t}$, $b_{2,t}$ characterized by a common pattern, while the schedules of ships are quite different. More specifically, the MILP-based method introduced a larger variability in the scheduling, as regards both berthing/departures times and amount of CO_2 transferred by vessels. On the contrary, the greedy heuristic guaranteed a more "regular" behavior.

4 Conclusions

This paper represents a first step towards the optimization of CO_2 transfers via ships between two ports using a MILP formulation and a heuristic approach. The former suffers from weak relaxations due to the presence of several Big-M constraints, while the latter has revealed very fast in finding approximate solutions and extremely useful in providing an initialization for the MILP-based approach, which has enabled computing better solutions as compared to the case with default initialization.

Future works will account for rolling-horizon optimization strategies rather than considering just a single optimization over the entire timeframe, for which having at disposal faster heuristic approaches will be mandatory. Lastly, we also plan to generalize the proposed methods to more than two ports in complex case studies with both hub and spokes and milk run networks, and to study alternative MILP formulations, possibly with a reduced number of Big-M constraints in order to simplify the search for an optimal solution. In particular, we will investigate the possibility of adapting existing scheduling approaches on parallel machines to our context, where ships performing the transportation of CO_2 can be assimilated to parallel machines where jobs have to be executed.

Acknowledgements M. Gaggero acknowledges the financial support of the Italian Ministry of University and Research through the Piano Nazionale di Ripresa e Resilienza (PNRR), Centro Nazionale per la Mobilità Sostenibile (CN-MOST), Spoke 3 Waterways (project id. CN00000023, CUP n. B43C22000440001). M. Gaggero is also a member of GNAMPA-INdAM (Gruppo Nazionale per l'Analisi Matematica, la Probabilità e le loro Applicazioni—Istituto Nazionale di Alta Matematica).

References

1. Aspelund, A., Molnvik, M., De Koeijer, G.: Ship transport of CO2: Technical solutions and analysis of costs, energy utilization, energy efficiency and CO2 emissions. Chem. Eng. Res. Des. **84**(9), 847–855 (2006)
2. Bazgosha, A., Ranjbar, M., Jamili, N.: Scheduling of loading and unloading operations in a multi stations transshipment terminal with release date and inventory constraints. Comput. & Ind. Eng. **106**, 20–31 (2017)
3. Bennaes, A., Skogset, M., Svorkdal, T., Fagerholt, K., Herlicka, L., Meisel, F., Rickels, W.: Optimization of a ship-based logistics system for carbon capture and storage. In: International Conference on Computational Logistics, pp. 44–59. Springer (2022)
4. Bennaes, A., Skogset, M., Svorkdal, T., Fagerholt, K., Herlicka, L., Meisel, F., Rickels, W.: Modeling a supply chain for carbon capture and offshore storage-a german-norwegian case study. Int. J. Greenhouse Gas Control **132**, 104028 (2024)
5. Bjerketvedt, V.S., Tomasgard, A., Roussanaly, S.: Optimal design and cost of ship-based CO2 transport under uncertainties and fluctuations. Int. J. Greenhouse Gas Control **103**, 103190 (2020)
6. Bogani, C., Gasparo, M.G., Papini, A.: Generalized pattern search methods for a class of nonsmooth optimization problems with structure. J. Comput. Appl. Math. **229**, 283–293 (2009)
7. Burger, J., Nohl, J., Seiler, J., Gabrielli, P., Oeuvray, P., Becattini, V., Reyes-Lua, A., Riboldi, L., Sansavini, G., Bardow, A.: Environmental impacts of carbon capture, transport, and storage supply chains: status and the way forward. Int. J. Greenhouse Gas Control **132**, 104039 (2024)
8. D'Amore, F., Romano, M.C., Bezzo, F.: Optimal design of European supply chains for carbon capture and storage from industrial emission sources including pipe and ship transport. Int. J. Greenhouse Gas Control **109**, 103372 (2021)
9. Jones, E.C., Jr., Yaw, S., Bennett, J.A., Ogland-Hand, J.D., Strahan, C., Middleton, R.S.: Designing multi-phased CO2 capture and storage infrastructure deployments. Renew. Sustain. Energy Trans. **2**, 100023 (2022)
10. Kjarstad, J., Skagestad, R., Eldrup, N.H., Johnsson, F.: Ship transport-a low cost and low risk CO2 transport option in the Nordic countries. Int. J. Greenhouse Gas Control **54**, 168–184 (2016)
11. Liu, S.Q., Kozan, E.: Parallel-identical-machine job-shop scheduling with different stage-dependent buffering requirements. Comput. & Oper. Res. **74**, 31–41 (2016)
12. Nam, H., Lee, T., Lee, J., Chung, H.: Design of carrier-based offshore CCS system: plant location and fleet assignment. Int. J. Greenhouse Gas Control **12**, 220–230 (2013)
13. Rackley, S.: Carbon Capture and Storage. Butterworth-Heinemann (2017)
14. Tapia, J.F., Lee, J.-Y., Ooi, R., Foo, D., Tan, R.: A review of optimization and decision-making models for the planning of CO2 capture, utilization and storage (CCUS) systems. Sustain. Prod. Consump. **13**, 1–15 (2018)

Railway Wheel Impact Force and Alert Prediction Using Machine Learning Models

Gajendra Malviya and Shripad Salsingikar

Abstract The Wheel Impact Load Detector (WILD) sensor measures the force exerted by the railway wheel on the rail track. If the force exerted is beyond a threshold value, the wheel needs inspection. The Problem Solving Competition 2017 organized by INFORMS Railway Application Section, asked participants to predict the force exerted by wheel, when a currently empty rail car would be loaded in the next trip. Further, convert the predicted force value into an alert, if the value is more than the threshold value. We present machine learning models to address the above problem along with results. We label the predicted output into two classes. Records with predicted force greater than or equal to the threshold value as 1 else 0. %True Positives detected were found to be very low (0–6%), which results in poor prediction of critical event. Instead of predicting critical event based on predicted force valued, we label the actual/training records, thereby converting the prediction problem into a classification problem. We apply various classifiers to classify whether the record is critical or not. We apply filtering, data pre-sampling, cross-validation and hyperparameter tuning. As the resultant labeled data is highly imbalance, we apply Synthetic Minority Oversampling Technique (SMOTE) and Cost Sensitive learning techniques to increase True Positives and reduce False Alarms. Random Forest classifier with SMOTE turned out to be the best among all classifiers with %True Positives as 85.76%, %False Alarms as 8.27%.

Keywords Wheel impact load detectors · Kips · Classification · Cost sensitive learning · SMOTE

G. Malviya (✉) · S. Salsingikar
TCS Research, Tata Consultancy Services, Thane, India
e-mail: gajendra.malviya@tcs.com

S. Salsingikar
e-mail: shripad.salsingikar@tcs.com

M. Di Francesco et al. (eds.), *Operations Research: Closing the Gap Between Research and Practice*, AIRO Springer Series 15, https://doi.org/10.1007/978-3-031-90095-2_20

1 Introduction

The Wheel Impact Load Detector (WILD) is a wayside sensor, which measures the magnitude of load exerted by wheels on railway tracks. The unit of force is Kilo Pound-Force (kips) or Kilo Newton. Railroads use WILD sensors to detect defective wheels. The WILD sensor triggers an alarm, if force exerted is above the given threshold value. Rail cars consisting wheels having force above the threshold value are set out for inspection and repair, if required.

The Problem Solving Competition (PSC) 2017 by Railway Application Section (RAS) of INFORMS [1] describes problem in detail. The problem involves prediction of values of peak force exerted by wheels on rail tracks, when a currently empty rail car is loaded in the next trip. The competition requires to generate an alert, if predicted peak load is above the threshold value of 90 kips. The objective of the competition is to enable railroad companies to forecast defective wheels to avoid the disruptions/loss due to on-line failure. The organizers of the problem solving competition [1] provide real life data instance of wheel impact load detectors. The data contains less than 1% records where the peak force value is greater than or equal to the threshold value of 90 kips.

As the given data set is highly imbalance, standard regression models [2] or classification models [3] have some limitations. The accuracy figures are high, in the range of 70–90%, for majority of the classifiers given in [4], however, they produce high false positives or high false negatives. Each false positive alarm forces inspection of the wheel and to decide if that wheel needs repair or replacements. Thus, minimizing false positives is important for reliability and acceptance of prediction model by the practitioners. Given the characteristics of a wheel, when the car is empty, the problem is to predict the peak force exerted by the wheel after certain period of time when the car will be loaded and also to predict whether alarm (Critical Peak) will be raised or not after certain period of time when the car will be loaded. In this work, we present a classification approach to predict alerts when peak force value will go above the threshold value. We use sampling techniques like SMOTE, and also cost sensitive learning techniques like Cost Sensitive Decision Tree (CSDT), Cost Sensitive Logistic Regression (CSLR) and Cost Sensitive Random Forest (CSRF) to overcome the high False positives and False negatives in default classification approach. The application of these sampling techniques and cost sensitive learning techniques along with training data pre-sampling given in Kecman [4] results into satisfactory classification output.

2 Previous Work

Khadilkar [2] in their response to INFORMS RAS PSC 2017 [1] built prediction models using multiple linear regression, regression tree, random forest and neural network techniques to predict the peak force values. Predicted values were then converted into label as ‘defect’, if it was greater than or equal to 90 kips and ‘non-defect’ otherwise. These label were treated as class and accuracy as well as confusion matrix was reported. Although the accuracy was very high, % of True Positives Detected is very small.

Sabnis [3], converted the original prediction problem given in [1] into binary classification problem using the classification logic given in [2]. The resultant labeled data set turns out to be highly imbalanced with more than 99.5% of records fall below 90 kips. They reported that Zero-Inflated Binomial regression with L2 norm model yielded satisfactory results with classification accuracy to 87.00%, % of True Positives Detected as 92.42% and % of False Alarms as 13%.

Kecman [4], also converted the problem [1] into classification problem and added many derived features. They filtered the data and took only two consecutive records, first with car status as 'Empty' and next with car status as 'Loaded' and the difference between two records was less than or equal to 10 days. For wheels that never had a critical peak kip, they randomly selected only one sample for each wheel id. For wheels that had at least one critical peak kip, they selected the first record for each wheel id which had a critical peak kip. They applied twelve different classifiers on this filtered data and compared their results. They found that Random Forest gives best result in terms of accuracy and F1 score. They did not use test data set provided by the organizers but trained their model on 67% of training data and tested it on remaining 33% of records. Further, they did not specifically report figures for False Positives and False Negatives.

3 Methodology

In this work we (a) applied same filters as given in [2, 4], (b) combined features as given in [2, 4], (c) worked on the classification problem as done by [3, 4], (d) applied same pre-processing steps as given in [4], (e) added two more classifiers for comparison and (f) applied sampling and cost sensitive learning techniques, as resultant labeled data is imbalance data-set.

Different approaches used in this work can be seen in flowchart given in Fig. 1. This section further gives detail about data, processing steps applied, classifier as well as sampling and cost sensitive meta-learning techniques used on the data.

3.1 Data Description and Processing

The data-set provide by [1] comprised of various features. Important features in the provided data-set are (a) Type of the car, (b) Location of the car in the train, (c) Axle number of the car to which the wheel is attached, (d) Side of the car on which the wheel is attached (L = left, R = right), (e) Weight of the car when empty (in tons), (f) Date and time of peak force readings, (g) Peak force value (in kips), (h) Speed of the wheel (in mph), (i) Weight of the car (in tons), (j) Loaded or Empty status of

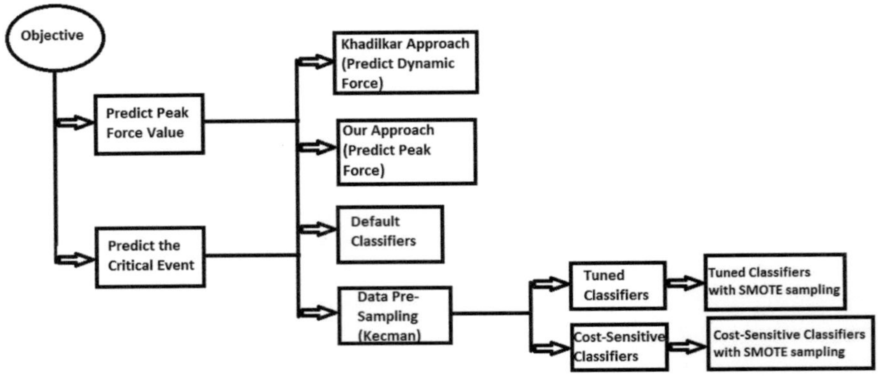

Fig. 1 Approaches flowchart

the car (L = loaded, E = empty), (k) Age of the wheel (in days), (l) Location of the sensor and (m) Vendor/maker of the sensor.

The organizers provided two separate data-sets one each for training and testing. The training data set had 7,032,197 records of WILD readings, corresponding to 113,422 unique wheels. The testing data set had 2,575,499 records.

Based on the provided problem statement, we cleaned and processed the raw data-set (both training and testing subset), according to the steps as listed below. This processing steps are derived from both Khadilkar et al. [2] and Kecman [4].

1. Sort data set by the unique wheel id and the timestamp of readings
2. Add a label named 'Critical'. If the peak force value in the record is greater than or equal to 90 kips then put value '1' ('True') else put value '0' ('False')
3. Add a label named 'Critical Observed', which tells if a wheel ever had a peak force value greater than or equal to 90 kips
4. Add various derived features mainly mean and maximum of some existing features like 'average force', 'peak force', 'dynamic force', 'dynamic ratio','speed of the car', and 'weight of the car'. The mean and maximum values are for the readings in last 10 days from the timestamp of the readings
5. Map the absolute time stamp of each measurement to a 'Season' variable, which corresponds to 1 for Dec, Jan, and Feb (winter), 2 for Mar, Apr, May, Sep, Oct, and Nov (spring and fall), and 3 for Jun, Jul, and Aug (summer)
6. Convert this chronologically successive readings for each wheel into pairs of readings by concatenating successive rows in the tabular raw data
7. Retain only those records where first record has car status of 'Empty' and second record has car status of 'Loaded' (E-L records)
8. Retain only those records where difference between time of reading between two records was less than or equal to 10 days
9. Remove data points where car status is 'Empty' and the peak force reading is already over 90 kips
10. Remove data points where car speed is below 20 mph or above 80 mph

Table 1 Data set summary

Data	Raw data	E-L	Non-critical	Critical
Training	7,032,197	1,274,086	1,267,322	6,764 (0.53%)
Presampled	7,032,197	67,712	62,823	4,898 (7.22%)
Testing	2,575,499	553,062	549,247	3,815 (0.69%)

11. Remove data points where any readings of average force, peak force, dynamic force, speed or weight are missing.

Further, we applied sampling steps given in Kecman [4] for only training set as listed below. This has been called data pre-sampling in report. This pre-sampling is done for better training of our model. For wheels that never had a critical peak kip, randomly select one sample for each wheel id, as we are more interested in finding the critical peak kip. This processing is done for training data set only. For wheels that had at least one critical peak kip, select the first state for each wheel id where next state is with a critical peak kip. This processing is done for training data set only. Convert all categorical variables into one hot encoding both in training as well as testing set.

Table 1 shows summary of the data sets. It shows the total count in raw data, processed and pre-sampled Empty-Loaded (E-L) data for training set and processed Empty-Loaded (E-L) data for testing set, number of records which are Non-Critical (Peak force value $<$ 90 kips), and the number of records which are Critical (Peak force value $\geq$90 kips). In pre-sampling of training data, we have taken one E-L records for each wheel id. It can be seen from Table 1, that after data pre-sampling of training data, percentage of critical records increases which result in better training of our model.

3.2 *Default Classifiers*

We used various classifiers to predict if record is critical peak force or not. Most of the classifiers were also used in [4]. We added two more classifiers namely Decision Tree and Logistic Regression. We used python library Scikit-learn [5] where all above mentioned algorithms are implemented. We trained models for these classifiers as well as tested them using data released by [1] after processing the data as explained in the Sect. 3.1.

We used grid search cross validation method for hyper-parameter tuning. We used hyper-parameter tuning only for classifiers like Decision Tree, Linear SVC with L1 penalty, Logistic Regression and Random Forest, which performed well in default classification after pre-sampling. Table 2 below gives hyper-parameters used for both classifiers, along with their respective range. We also used Voting classifier which trains on an ensemble of numerous models and predicts an output (class) based on their highest probability of chosen class as the output. We used hard-voting in Voting

Table 2 Hyper-parameters and their range

Decision Tree	Min samples leaf	1, 2, 5, 10, 20, 50
	Min samples split	2, 5, 10
Lin. SVC L1	Regularization parameter (C)	0.01, 0.1, 1, 10, 25, 40, 70, 100, 300, 400, 500, 700, 1000, 1200
Logistic Reg	Regularization parameter (C)	0.01, 0.1, 1, 10, 25, 40, 70, 100, 300, 400, 500, 700, 1000, 1200
Random Forest	n estimators	10, 30, 50, 100, 200, 300, 400
	Min samples leaf	1, 5, 10, 50

classifier in which the predicted output class is a class with the highest majority of votes i.e. the class which had the highest probability of being predicted by each of the classifiers.

3.3 Sampling and Cost Sensitive Learning Techniques

As given in Table 1 the input data is highly imbalanced. Standard machine learning algorithms have trouble learning when one class dominates the other. Class imbalance problem is addressed by either sampling of data or cost sensitive learning by tweaking cost function. These sampling techniques are applied on training data only while learning the model. Sampling of data involves over-sampling of minority class or under-sampling of majority of class. Synthetic Minority Oversampling Technique (SMOTE) is oversampling method. It create synthetic samples of the minority class and make the minority class equals to the majority class. It select similar records and alter that record one column at a time by a random amount within the difference to the neighboring records. SMOTE uses k-nearest neighbors to create synthetic examples of the minority class. We used 'imblearn' python library [6] to do SMOTE.

Various classification methods aim at minimizing the misclassification error by assuming that cost for all misclassification errors are same [7]. But, in many real-world applications, cost for misclassification errors are not same i.e. the cost of false positives (FP) is different to the cost of false negatives (FN). Methods that use different misclassification costs are known as cost-sensitive classifiers. Cost sensitive learning involves tweaking the cost function to make misclassification of minority instances more important than misclassification of majority instances. In this case, for each false positive prediction the railroad organization has to spend time and cost for inspection of the wheel to make sure whether repair or replacement is required and thus, minimizing false positive is also important. As in default classification, we were getting high False Positives compared to False Negatives, we took high false positive cost compared to false negative cost while doing cost sensitive learning. Thus, we assumed some cost for false positives and false negatives and zero cost for true positives and true negatives for our cost sensitive learning. We used cost sensitive classifiers such as Cost Sensitive Decision Tree (CSDT), Cost Sensitive Logistic Regression (CSLR) and Cost Sensitive Random Forest (CSRF) in our case. We used 'costcla' python library [8] to do the cost sensitive classification. Sampling techniques and Cost Sensitive classification techniques are mainly used to reduce False Positives and increase True Positives.

4 Results

4.1 Classification Results

We applied data Pre-sampling technique as given in [4] and trained our model. Table 1 tells how our training data changes after applying data pre-sampling. Table 3 shows results after doing Data Pre-sampling and applying default classifiers. Classifiers like Decision Tree, Linear SVC with L1 penalty, Logistic Regression and Random Forest (shown in bold in table have good % of True Positives detected (more than 60%) and less % of False Alarms (less than 9%). Within all classifiers, these classifiers have best results in terms of evaluation measures i.e. TPD (True Positives Detected) % and FA (False Alarms) %. We used these four classifiers for further analysis.

Further, we performed hyper-parameter tuning for both classifiers. We used GridsearchCV method for tuning and used scoring as difference of True Positives and False Positives i.e. (TP-FP) in it. We found that for Decision Tree best parameters are (a) min samples leaf = 50, (b) min samples split = 2. For Linear SVC L1 penalty best parameters are (a) C = 0.01. For Logistic Regression best parameters are (a) C = 700e. For Random Forest best parameters are (a) n estimators = 300, (b) min

Table 3 Default classifiers on pre-sampled data

Classifier	TP #	TN #	FP #	FN #	TPD %	FA %
Decision Tree	2,479	509,237	39,237	1,334	**65.01**	**7.34**
Linear SVC L1	3,193	501,046	47,428	620	**83.74**	**8.69**
Logistic Reg	2,321	524,407	24,067	1,492	**60.87**	**4.63**
Nearest centroid	3,799	11,636	536,838	14	99.63	97.20
Naive Bayes	3,097	336,377	212,097	716	81.22	38.53
Perceptron	2,913	125,659	422,815	900	76.39	76.72
Random Forest	2,722	517,432	31,042	1,091	**71.39**	**5.82**
SGDC L1	2,272	197,495	350,979	1,541	59.58	63.83
SGDC L2	3,414	73,458	475,016	399	89.53	86.08
SGDC Elastic-Net	1,436	320,846	227,628	2,377	37.66	41.64

Table 4 Tuned classifiers

Classifier	TP #	TN #	FP #	FN #	TPD %	FA %
Decision Tree	2,656	516,611	31,863	1,157	69.65	5.98
Linear SVC L1	3,189	500,588	47,886	624	83.63	8.78
Logistic Reg	2,321	524,405	24,069	1,492	60.87	4.63
Random Forest	2,931	516,617	31,857	882	76.87	5.93
Voting classifier	2,671	521,181	27,293	1,142	70.05	5.15

Table 5 Tuned classifiers with SMOTE

Classifier	TP #	TN #	FP #	FN #	TPD %	FA %
Decision Tree	3,338	479,379	69,095	475	87.54	12.59
Linear SVC L1	3,720	426,503	121,971	93	97.56	22.10
Logistic Reg	3,583	462,532	85,942	230	93.97	15.60
Random Forest	3,270	503,323	45,151	543	85.76	8.27
Voting classifier	3,418	486,657	61,817	395	89.64	11.26

samples leaf = 1. Table 4 shows result of Tuned Classifiers. There is slight increase in TPD % and slight decrease in FA %.

4.2 Sampling Results

As TPD % is still less for all the four best classifiers as shown in Table 4. Table 5 shows results after using SMOTE on these tuned classifiers. It can be observed from comparing Tables 4 and 5 that TPD % and FA % changes significantly after use of these sampling techniques.

4.3 Cost Sensitive Learning Results

TP #, TN #, FP #, FN #, TPD % and FA % for cost sensitive classifiers CSDT, CSLR and CSRF can be observed from Table 6. We also applied cost sensitive classifiers with SMOTE sampling and results are shown in Table 7.

Table 6 Cost-sensitive classifiers

Classifier	TP #	TN #	FP #	FN #	TPD %	FA %
CSDT	2,045	529,705	18,769	1,768	53.63	3.72
CSLR	3,813	0	548,474	0	100.00	99.31
CSRF	571	545,810	2,664	3,242	14.98	01.07

Table 7 Cost-sensitive with SMOTE

Classifier	TP #	TN #	FP #	FN #	TPD %	FA %
CSDT	3,391	476,652	71,822	422	88.93	13.08
CSLR	3,813	0	548,474	0	100.00	99.31
CSRF	3,390	479,142	69,332	423	88.91	12.63

Table 8 Comparison of classification approaches for Random Forest classifier

Classifier	TP #	TN #	FP #	FN #	TPD %	FA %
Pre-sampled	2,722	517,432	31,042	1,091	71.39	5.82
Tuned	2,931	516,617	31,856	882	76.87	5.93
Tuned + SMOTE	3,270	503,323	45,151	543	**85.76**	**8.27**
Cost-sensitive	571	545,810	2,664	3,242	14.97	1.07
CS + SMOTE	3,390	479,142	69,332	423	88.91	12.63

Table 9 Comparison of classification approaches for Decision Tree classifier

Classifier	TP #	TN #	FP #	FN #	TPD %	FA %
Tuned	2,656	516,611	31,863	1,157	69.65	5.98
Tuned + SMOTE	3,338	479,379	69,095	475	**87.54**	**12.59**
Cost-sensitive	2,045	529,705	18,769	1,768	53.63	3.72
CS + SMOTE	3,391	476,652	71,822	422	88.93	13.08

4.4 Comparison of Classification Approaches

Tables 8 and 9 shows comparison of different classification approaches for Random Forest and Decision Tree classifier respectively. We have also made bar chart representation of % True Positives detected and % False Alarms for comparison of different classification approaches. The graph of % True Positives detected and % False Alarms for Random Forest classifier is given in Figs. 2 and 3 respectively. The

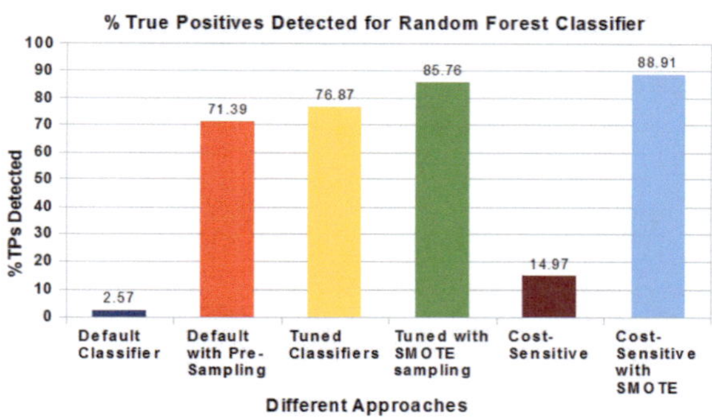

Fig. 2 % of TPs detected for Random Forest classifier

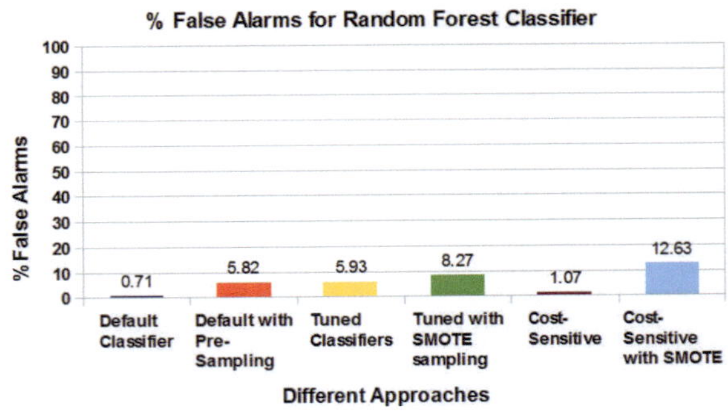

Fig. 3 % of False alarms for Random Forest classifier

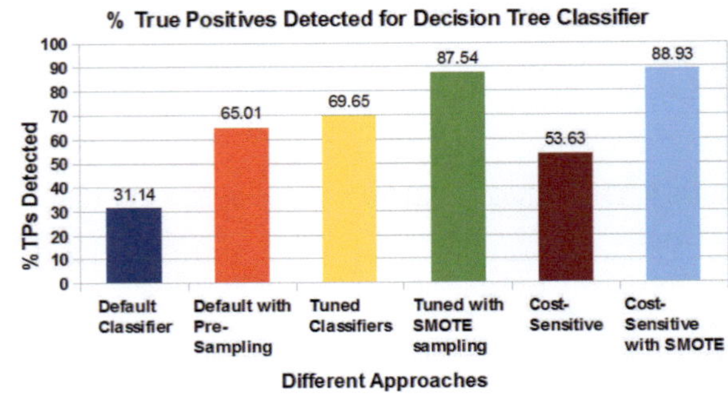

Fig. 4 % of TPs detected for Decision Tree classifier

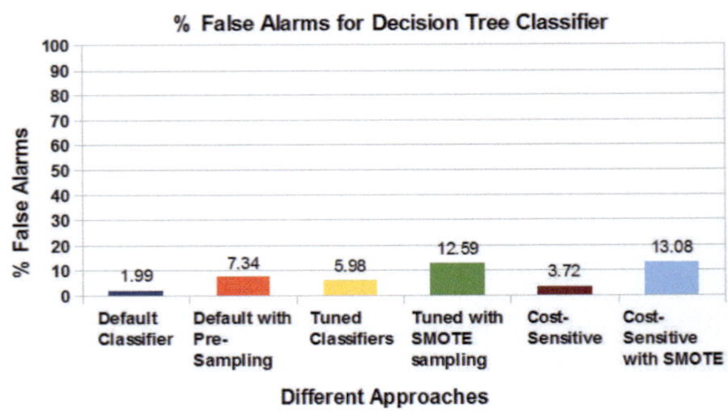

Fig. 5 % of False alarms for Decision Tree classifier

graph of % True Positives detected and % False Alarms for Decision Tree classifier is given in Figs. 4 and 5 respectively.

5 Conclusions

In this paper, we used different classification approaches to solve the wheel impact load alert prediction problem posed in the INFORMS RAS Problem Solving Competition, 2017 [1]. We used multiple classification techniques to identify critical records among the given records i.e. records with peak force above the given threshold values. We found that Decision Tree, Lin. SVC L1, Logistic Regression and Random forest gave comparatively best results in terms of % of TPs detected and % of False Alarms after data pre-sampling on default classifiers as shown in Table 3. However, the False Positives was high for most of them. Higher false positives adversely impact model reliability and acceptance among the practitioner. It also results into loss of time and money for the users. We performed hyper parameters tuning to improve the results further. The parameter tuning was applied on the top 4 classifiers namely Decision Tree, Lin. SVC L1, Logistic Reg and Random forest, resulting into improved % of TPs detected and % of False Alarms compared to the default classification. However, the False Positives were still high, as observed in Table 4.

We used techniques like sampling and cost sensitive learning to reduce the impact of imbalanced data i.e. to increase % of TPs detected and decrease % of False Alarms. We observed that there was significant change in % of TPs detected and % of False Alarms after using above techniques as shown in Tables 5 and 6. After using SMOTE sampling, % of TPs detected increases but % of False Alarms also increases. Cost-sensitive learning applied directly does not give good results in terms of % of TPs detected. That's why we use Cost-Sensitive classifiers with SMOTE sampling and our results were improved as observed in Table 7. In this problem instance, we observed that doing pre-sampling of the training data and using sampling techniques like SMOTE as well cost sensitive learning like CSDT and CSRF resulted into more % of TPs detected and less % of False Alarms. Random Forest with the sampling techniques SMOTE is giving good result, it gives % of TPs detected as 85.76% and % of False Alarms as 8.27% as shown in Table 5. Finally, we concluded that Random Forest with data pre-sampling given in [4], hyperparameters tuning and SMOTE performed the best among all the classifiers.

References

1. INFORMS RAS-PSC: Data analytics for railroad empty-to-load peak kips prediction. In: INFORMS, Railway Application Section, Problem Solving Competition (2017). https://connect.informs.org/railway-applications/new-item3/problem-solving-competition681/new-item11. Last accessed Nov 2019
2. Khadilkar, H., Salsingikar, S., Sinha, S.: A machine learning and big data solution for rail car wheel peak kip prediction. In: INFORMS, Railway Application Section, Problem Solving Competition. Houston, USA (2017)
3. Sabnis, S., Yadav, S., Salsingikar, S.: Identification of defective railway wheels from highly imbalanced wheel impact load detector data. In: Annual Conference of German Operations Research Society (OR2019). Dresden, Germany (2019)
4. Kecman, P.: Tutorial on data analytics for railroad empty-to-load peak kips prediction-INFORMS RAS PSC 2017. In: Annual Conference of German Operations Research Society (OR2019), Germany. http://www.railnorrkoping2019.org/course/Lecture4_Kecman.html. Last accessed Nov 2019
5. Pedregosa, F., Varoquaux, G.: Scikit-learn: machine learning in Python. J. Mach. Learn. Res. **12**, 2825–2830 (2011)

6. Lemaître, G. , Nogueira, F., Aridas, C.: Imbalanced-learn: a Python toolbox to tackle the curse of imbalanced datasets in machine learning. J. Mach. Learn. Res. **18**, 1–5 (2017)
7. Bahnsen, A.C.: Example-dependent cost sensitive classification—applications in financial risk modeling and marketing analytics thesis. Luxembourg (2015)
8. Bahnsen, A.C.: Cost sensitive classification Python library. http://albahnsen.github.io/CostSensitiveClassification. Last accessed Nov 2019
9. Ling, C., Sheng, V.: Cost-sensitive learning. In: Sammut, C., Webb, G., (eds) Encyclopedia of Machine Learning, 1st ed, pp. 231–235. Springer (2011)

Graph Convolutional Neural Network Assisted Monte Carlo Tree Search for the Capacitated Vehicle Routing Problem with Time Windows

Jorin Dornemann and Tobias Klein

Abstract The Capacitated Vehicle Routing Problem with Time Windows (CVRPTW) is a well-known combinatorial optimization problem that extends the classical Vehicle Routing Problem to account for additional real-world constraints such as truck capacity and customer time windows. Recent studies have explored the use of deep graph convolutional networks (GCNs) to predict the arcs that are part of the optimal tour for the Travelling Salesman Problem and related routing problems. We propose a novel context complemented graph convolutional network (CCGCN) which is integrated into a Monte Carlo Tree Search (MCTS) to solve the CVRPTW sequentially. The deep convolutional part of the proposed neural network builds efficient CVRPTW graph representations, whereas the context part processes information of partially built tours to output probabilities on which vertex to add next to the tour, which are used during the expansion of the search tree. For simulating final solution values within the Monte Carlo search tree, we conducted experiments using two different heuristics: LKH3 and a beam search. By leveraging the probabilities provided by our CCGCN and employing a variation of Upper Confidence Bound applied to trees, we effectively manage the tradeoff between exploration and exploitation, thus reducing the optimality gap compared to solutions generated solely by LKH3. Evaluations of the proposed heuristic are performed on benchmark instances of the CVRPTW with up to 100 customers; these show promising results.

Keywords Monte Carlo tree search · Graph convolutional network · Vehicle routing

---

J. Dornemann (✉)
Institute of Mathematics, Hamburg University of Technology, Hamburg, Germany
e-mail: jorin.dornemann@tuhh.de

T. Klein
Institute for Operations Research and Information Systems, Hamburg University of Technology, Hamburg, Germany
e-mail: tobias.klein@tuhh.de

M. Di Francesco et al. (eds.), *Operations Research: Closing the Gap Between Research and Practice*, AIRO Springer Series 15, https://doi.org/10.1007/978-3-031-90095-2_21

1 Introduction

The capacitated vehicle routing problem with time windows (CVRPTW) is a widely studied combinatorial optimization problem first introduced by [17]. In the CVRPTW, a fleet of vehicles is tasked with efficiently delivering goods to a set of customers while respecting capacity and time window constraints. The challenge lies in finding the optimal routes minimizing the total distance traveled. Numerous approaches have been proposed over the years to tackle vehicle routing problems. Exact algorithms such as branch-and-cut and branch-and-price, as outlined by Toth et al. [18], offer optimal solutions but are computationally demanding. Alternatively, heuristics such as genetic algorithms, simulated annealing, and tabu search, as described by [3], provide approximate solutions with varying levels of computational complexity. Advancements in deep learning have recently offered valuable tools for extracting essential graph properties in the context of routing problems [2, 11, 20], which have only recently been utilized to tackle the CVRPTW. Falkner and Schmidt-Thieme [5] adapted the attention model from [14], where a learned policy model was used to construct solutions, which was trained via reinforcement learning, to choose the next node. Falkner and Schmidt-Thieme [5] expanded this approach for the CVRPTW by simultaneously constructing multiple routes and leveraging information from all partially constructed routes to make informed decisions about the next node selection. Falkner et al. [6] built on this approach by using graph neural networks as a replacement for the self-attention layers in their model. They further proposed a Large Neighborhood Search approach that utilizes a learned construction heuristic. Poullet [15] proposed a two-stage algorithm which uses Optimal Classification Trees to build clusters on the set of customers before applying an actor-critic Reinforcement Learning model to solve the CVRPTW on the smaller clusters. Dornemann [4] adapted the constructive approach of [11] for the CVRPTW by modifying the neural network architecture to include directed edges as well as embedding node features given by the capacity and time window constraints, which is embedded in a beam search for constructing a solution. The neural network was further used as a form of learned problem reduction to reduce the size of the representation of the problem as a quadratic binary optimization problem, which is subsequently solved using specialized hardware. Wang et al. [21] introduced a hierarchical reinforcement learning model that relies on pointer networks and operates in two stages: firstly, learning to generate feasible solutions at a lower level, and subsequently using these solutions as input for a second decoder to minimize the total distance. Gao et al. [8] proposed an improvement-based heuristic for the CVRPTW, which uses an improved graph attention network [19] to learn a heuristic for Very Large-scale Neighborhood Search that includes both improvement and destruction operators.

In this work, we aim to combine recent results in deep learning to solve the CVRPTW by adapting a graph convolutional network to extract information from its underlying graph representation and complementing the graph representation with information about the context of partially constructed solutions. We guide the search process for solutions by evaluating the neural network autoregressively within a Monte Carlo Tree Search (MCTS) framework adapted for routing problems.

2 Proposed Approach

Our approach uses a modified graph convolutional network [11], which has been adapted for the CVRPTW and modified to include the context of partially built solutions, to predict promising edges in the graph. The output of the network can be interpreted as assigning probabilities to each directed arc leaving a given vertex, which are subsequently used as decision weights within a MCTS to efficiently search for promising paths. In each iteration of the MCTS procedure one vertex is added to the solution, building the solution sequentially. At each iteration, the context complemented graph convolutional neural network (CCGCN) (see Sect. 2.3) is evaluated given the new context of the partially built solution.

In the subsequent subsections, following the formal problem definition, we provide a comprehensive description of the MCTS procedure and explain the neural network's architecture.

2.1 *Problem Definition*

Given a directed graph $G = (V, E)$ with $|V| = n$, a weight function $c : E \mapsto \mathbb{R}_+$ assigning a cost c_{ij} to each arc $e = (v_i, v_j)$ and a fleet of K vehicles, the goal for the CVRPTW problem is to find a set of up to K routes, such that every vertex except for the depot is included in exactly one route and the capacity and time window constraints are met. The cost c_{ij} for each arc (v_i, v_j) represents the transit time required to travel between nodes v_i and v_j. Each node holds a demand and a time window, which impose additional constraints. For each vehicle the capacity constraint dictates that the sum of demands from each node along the vehicle's route must not exceed its assigned capacity. For each node $v_i \in V$ a time window is provided as $t_i = \left[t_i^0, t_i^1\right]$, representing the permissible time interval within which the start of the service at node v_i is allowed. Moreover, a service duration s_i is associated with each node v_i, denoting the time needed for serving node v_i. The time window constraint states that every route must satisfy the time window requirements at each node along the route. The objective is then to minimize the total costs of all routes combined.

2.2 *Monte Carlo Tree Search*

The Monte Carlo Tree Search is a powerful combination of the Monte Carlo Sampling and a Tree Search, first proposed by [1]. By iteratively simulating decisions and assessing their outcomes, the search tree is gradually constructed while emphasizing promising paths. To balance the tradeoff between exploration and exploitation, the Upper Confidence Bounds applied to Trees (UCT) [13] can be used. The MCTS consists of a four-step process: selection, expansion, simulation and backpropagation.

In the context of CVRPTW, modifications need to be made to all four steps of the traditional MCTS algorithm [1].

In each iteration $i \in [n]$ of the algorithm, one node of the graph is added to the solution. For this, an MCTS search tree T_i is iteratively constructed. Starting from the root node of T_i, which represents the solution built in the first $i - 1$ iteration steps, each node in the search tree represents a continuation of the solution from its parent node. Each rollout of the iteration consists of selecting a leaf node L in the search tree T_i representing a partial solution $L_s = [v_0 \ldots, v_k]$, which is then subsequently expanded by adding child nodes to L for all vertices u not in L_s. A child node L^u in the search tree is added to L, with $L_s^u = [v_0 \ldots, v_k, u]$. Each edge to a child node L^u of L is assigned a decision weight $p_{v_k u}$, which is used to guide the selection phase and can be interpreted as the probability of visiting u after v_k. These decision weights are obtained by the neural network presented in Sect. 2.3, which is evaluated at the start of every iteration i by propagating the graph G as well as the partially built solution S to incorporate the context of S with respect to time window and capacity constraints. It is then simulated which length of a complete solution is to be expected, if we start with the partial solution L_s, by applying a value function. This simulated value is subsequently backpropagated in the search tree up to the root node and the values stored at each node in the search tree, which are used during the selection of a leaf node, are updated accordingly. When the number of rollouts for this iteration is reached, the child node C of the root node R with the minimal tour length found in its subtree is chosen to be the root node for the next iteration $i + 1$. Details to each of the four phases are given in the following.

Selection Phase: The child node is selected that minimizes the following expression:

$$\min \frac{q - q_{\min}}{q_{\max} - q_{\min}} - c_{\text{puct}} \cdot p_{ij} \cdot \sqrt{\frac{n_{\text{parent}}}{n}}. \tag{1}$$

In the first term, $q_{\min}$ and $q_{\max}$ represent the lowest and highest node values found in the subtree starting at the current node. The simulated tour length q is normalized to accommodate variations in distances across different problem instances. In the second term of (1), which represents the PUCT value, c_{puct} denotes a manually chosen constant value that regulates the trade-off between exploration and exploitation, p_{ij} represents a decision weight that the current node v_j is visited by a vehicle after node v_i, n_{parents} indicates the count of how many times the backpropagation has processed the parent node, and n represents the count for the current node. As our objective is to find the set of tours with minimal total length, we select the child node that minimizes (1).

Expansion Phase: We use a trained GCN to obtain a decision policy, denoted as p_{ij}, for each vertex $v_j \neq v_i$, indicating the likelihood of visiting vertex v_j next in the optimal tour given the partial tour represented by the selected node. A higher decision weight p_{ij} implies a higher probability of vertex v_j being visited after vertex v_i. For each unvisited vertex v, that complies with the constraints, a child node is added to the tree with the decision weight p_{ij}.

Simulation Phase: In contrast to the AlphaGo approach [16], where a learned value function is applied to estimate the probability of the player winning from the current position, we apply a heuristic value function to calculate the expected tour length after completing the partial solution from the current position in the search tree. For this, we apply two different value functions. First, we use the LKH3 heuristic [10] by creating a new instance based on the partially given solution, which only consists of the unvisited nodes and the last node v of the partial solution. By adding an auxiliary node with special edge weights, demand and time window we ensure that the LKH3 correctly completes any partial tour included in the partial solution of the initial instance. For the second value function we adapt the beam search proposed in [11] and combine it with our trained CCGCN in a non-autoregressive form to get a good estimation of the possible tour length from the current partial solution. The value function beam search is described in [4]. Determining the invalid expansions of partial solutions during the beam search is computationally demanding, since calculating them with regard to the constraints is unique for every beam in the search, which leads to a linear dependency between the runtime of the simulation and the chosen beam width. To accelerate the simulation process, we alternatively apply a relaxed value function, where the task of filtering invalid expansions with regards to the constraints is delegated to the neural network's decision weights, as the network should naturally assign low weights to invalid next nodes. This translates to relaxing the CVRPTW instances to VRP instances to be solved with the beam search. Given that the value function's determined value serves solely as a guide for the MCTS exploration, any errors resulting from invalid simulations during the simulation phase do not lead to overall invalid solutions, as the feasibility is monitored within the MCTS.

Backpropagation Phase: Our objective is to find a set of tours with minimal total length, therefore, the simulated tour length q is propagated back alongside the minimum $q_{\min}$ and maximum $q_{\max}$ tour lengths using the following formula:

$$
\begin{aligned}
q_{\text{parent}} &= \min(q_{\text{parent}}, q_{\text{child}}) \\
q_{\text{min\_parent}} &= \min(q_{\text{min\_parent}}, q_{\text{min\_child}}) \\
q_{\text{max\_parent}} &= \max(q_{\text{max\_parent}}, q_{\text{max\_child}})
\end{aligned}
$$

This ensures that during the selection phase, the q value of a node always represents the smallest q value among all its descendants.

2.3 Neural Network Architecture

We propose an extension to the deep Graph Convolutional Network (GCN) framework [11] that incorporates two supplementary encoders. These encoders capture contextual information pertaining to previously visited nodes and the current partial tour state representation, respectively.

The GCN presented in [11] allows to efficiently learn to find representations of a graph by computing h_1-dimensional representations of all nodes and edges integrating the node and edge features, respectively, which can be used to produce a 'heatmap' over the edges of a graph, indicating promising edges to be used in a solution. The GCN architecture can be modified for the CVRPTW, incorporating directed arcs and additional node and arc features. The node features consist of the coordinates, the time windows and the demand. The arc features are given by the arc weight. The embedded h_1-dimensional features are then passed through a number of graph convolutional layers and, together with the h_2-dimensional context embedding C, are used to compute the probability of each node to be selected as the next node of the partially built solution.

For the context embedding, define the partial tour state to consist of the current position of the vehicle, given by the coordinates of the last visited node, the current time and the remaining available capacity of the vehicle. The partial tour, consisting of the already visited nodes, gets a separate learned initial embedding parameter. For that, define $y_{\text{tour}} \in \{0, 1\}^{n+1}$ to be the indicator vector with entry one for every already visited node. Let h_s and h_v be the dimensions of the state and tour embeddings, respectively. The partial tour state features are concatenated to the four-dimensional partial tour state feature vector and are then embedded by a feed forward neural network N_s to an h_s-dimensional representation consisting of the $\frac{h_s}{3}$-dimensional representations γ^{loc}, γ^{load}, γ^{time} for the three partial tour state features, respectively. The tour vector y_{tour} is embedded into a h_v-dimensional representation by a second feed forward neural network N_v given as $\gamma^{\text{tour}} = N_v(y_{\text{tour}})$. The context embedding C_{emb} is then given as

$$C_{\text{emb}} = [\gamma^{\text{loc}}, \gamma^{\text{load}}, \gamma^{\text{time}}, \gamma^{\text{tour}}],$$

the concatenated state and tour representations with a dimension of $h_2 = h_s + h_v$, which is then subsequently passed through the context feed forward neural network N_c as follows to derive the h_2-dimensional context representation $C = N_c(C_{\text{emb}})$:

$$C^{\ell+1} = C^{\ell} + \text{ReLU}\left(W^{\ell} C^{\ell}\right),$$

where $W^{\ell} \in \mathbb{R}^{h_2 \times h_2}$ are the learned parameters and ReLU is the rectified linear unit activation function.

3 Computational Experiments

We evaluate our approach on randomly generated instances for three different sizes, ranging from 20 to 100 nodes, and compare the results to baseline models, which include exact commercial solvers as well as state-of-the-art heuristics for routing problems and a neural network assisted beam search heuristic.

3.1 Implementation and Hyperparameters

For the neural network, based on experiments, we set the hidden dimension for the graph convolutional part to $h_1 = 300$ with 30 hidden layers. The context hidden dimension is set to $h_2 = 128$ with 6 layers and the MLP classifier contains 3 layers. In the MCTS we vary the values for the number of playouts per round between 200 and 1000, the beam size for the simulation is set to values between 1 and 10. To find the best configuration of hyperparameters for the MCTS, each combination of (number of playouts, beam size) was tested on datasets containing 50 instances with different c_{puct} values. We report the results for the parameter combinations (200,10), (500,5) and (1000,10) to illustrate the trade-off between computational efficiency and accuracy. We choose a c_{puct} value of 1.4 based on manual experiments. All models are implemented in Python 3.10 and run under Linux, but it is to be mentioned that computational runtimes could be reduced by coding it in C++. The neural network architecture is implemented using PyTorch version 2.0.0 to use GPU computation with Cuda version 11.8.

3.2 Experiment Setup

For the data generation, we follow the approach by [5] to sample the data distribution from the well known Solomon Benchmark set [17]. To generate solutions for the randomly sampled instances for training purposes, we use machines with two CPUs of type Intel Xeon E5-2680v3 @ 2,50GHz with 12 cores. The instances with 20 customers are solved to optimality by Gurobi [9], whereas the instances with 50 and 100 customers are solved using one run of LKH3 [10].

We employ a supervised learning methodology to train the model. The input consists of a graph with the additional node features time windows and demands along with a partial solution. We train the model to generate a probability matrix by minimizing the cross-entropy loss via gradient descent with respect to the adjacency matrix given as the target. To optimize the learning process, we utilize the Adam optimizer [12] and gradually reduce the learning rate to facilitate smoother convergence, starting at a value of 10^{-3}. Each neural network is trained for a maximum of 2000 epochs, each epoch consisting of 500 batches containing 48 pairs of instances and targets. We then select the training level with the lowest validation loss. The training is conducted on machines equipped with the same CPUs as for the data generation. Additionally, the machines are equipped with four NVidia Tesla K80 GPUs, each with 12GB of RAM.

We compare our approach to the commercial exact MILP solver Gurobi version 10.0.0 [9] as well as to the heuristic LKH3 [10] and Google's OR-Tools (GORT) version 9.5.2237 library [7] with a guided local search (GLS), both of which frequently serve as heuristic baselines in related literature. A time limit of 10 s for instances with

20 nodes and 30 s per instance for larger problem sizes is set for GORT. Furthermore, we evaluate the beam search, which is applied as the value function during the simulation phase, by itself as a separate approach to use as additional comparison for neural network assisted heuristics. Gurobi is applied to the 2-index Integer Linear Program formulation of the CVRPTW and is given a time limit of 1200, 3600 and 5400 s per instance for the different problem sizes, respectively.

3.3 Computational Results

We report computational results for the solution quality of our approach as well as an analysis of the impact of our neural network's predictions on the approach. The solution quality is measured by calculating the optimality gap between the solution obtained by our approach and the solution obtained by one run of the LKH3 heuristic, as the solutions by LKH3 were used to train the neural networks. Furthermore, we report results of the LKH3 heuristic, where the number of runs is adjusted to allow for similar runtime as our approach, in order to make the results more comparable, which is abbreviated by $LKH3_{adj}$.

Table 1 shows the results of our approach compared to the baseline models, with the different value function approaches abbreviated as

- CGM^{bs} for the beam search with exact masking,
- CGM^{rs} for the beam search with a relaxed simulation,
- CGM^{LKH3} for the value function LKH3.

Table 1 Mean cost, gap and time per instance over 1000 CVRPTW instances

Model	$n = 20$			$n = 50$			$n = 100$		
	Cost	Gap (%)	Time	Cost	Gap (%)	Time	Cost	Gap (%)	Time
$CGM^{bs}_{(200,10)}$	6.03	1.28	29.48	11.18	10.39	421.05	16.90	12.05	1355.04
$CGM^{bs}_{(500,5)}$	6.01	0.98	50.47	10.93	7.89	766.18	16.53	9.60	2458.87
$CGM^{bs}_{(1000,10)}$	5.97	0.43	112.99	10.79	6.50	1877.02	16.36	8.52	5464.28
CGM^{LKH3}_{200}	5.95	0.00	91.47	10.12	−0.14	1128.64	15.04	−0.24	2641.07
CGM^{LKH3}_{500}	5.95	0.00	187.02	10.11	−0.19	2687.38	15.01	−0.45	5486.11
CGM^{LKH3}_{1000}	5.95	0.00	410.29	10.10	−0.31	4392.90	14.92	−1.04	9542.22
$CGM^{rs}_{(200,10)}$	6.15	3.37	19.62	12.07	19.16	245.62	19.99	32.57	825.47
$CGM^{rs}_{(500,5)}$	6.13	2.96	36.36	11.57	14.22	546.02	19.21	27.35	1639.50
$CGM^{rs}_{(1000,10)}$	6.09	2.39	73.89	11.29	11.48	1107.66	18.92	25.48	3243.79
Beam search 50k	6.12	2.97	240.01	11.31	11.73	1153.01	17.86	18.44	4313.61
MILP (Gurobi)	5.95	0.00	0.60	10.10	−0.31	2092.37	16.98	12.63	5337.80
GORT−GLS	5.95	0.00	10.00	10.21	0.84	30.00	15.60	3.43	30.00
LKH3 1x	5.95	0.00	6.20	10.13	0.00	11.74	15.08	0.00	19.69
$LKH3_{adj}$	5.95	0.00	89.40	10.10	−0.31	453.61	14.88	−1.30	2126.33

Table 2 Mean cost of CGM^{LKH3} with and without predictions given by the GCN

Predictions	Playouts	$n = 20$	$n = 50$	$n = 100$
GCN	200	5.95	10.12	15.04
	500	5.95	10.11	15.01
	1000	5.95	10.10	14.92
Uniformly distributed	200	6.07	10.42	15.58
	500	6.01	10.35	15.49
	1000	5.98	10.32	15.37

Our heuristic CGM^{LKH3} achieves optimality for small instances, while outperforming GORT-GLS, Gurobi as well as LKH3 with one run on medium and large instances, albeit with a significantly longer runtime. It also outperforms the beam search CCGCN heuristic as proposed by [4] for all instance sizes. The negative optimality gap with respect to LKH3 indicates that the solutions calculated by one run of LKH3 are 0.31% worse on average compared to an optimal solution. As our neural network was trained with datasets calculated by one run of LKH3, this negative optimality gap could have an influence on the accuracy. The fact that a better optimality gap can be achieved shows the strength of this approach, that even with imperfect data the neural network manages to learn the important characteristics of the instances and is able to abstract them and thus outperform the heuristic used to create the data. This holds true for large instances with 100 nodes, where an optimality gap of -1.04% is achieved by CGM^{LKH3}_{1000}, outperforming the LKH3 with one run and coming close to the optimality gap of LKH3 with an adjusted number of runs. This comes at the cost of long runtimes, but displays the potential of neural network guided tree searches.

We further conduct experiments where we distribute the p_{ij} values uniformly to obtain insights into the necessity of these predictions for enhancing the performance of the MCTS. Table 2 compares the solutions obtained by CGM^{LKH3} for 200–1000 playouts with predictions p_{ij} obtained by the neural network to the solutions of the MCTS approach without neural network assisted decisions. The GCN predictions enhance the effectiveness of the MCTS, showing that the neural network is able to capture non-trivial instance characteristics and plays a pivotal role in enhancing the performance of our MCTS approach.

4 Conclusion

We presented a Monte Carlo Tree Search approach to solve the CVRPTW efficiently by using a deep neural network to guide the search procedure within the MCTS as well as for the value function during the simulation phase of the MCTS. By using the iteratively built partial solutions during the search, the deep neural network can incorporate the context, specifically addressing the challenging time windows and capacity constraints, to predict promising arcs to guide the search process in the search tree. Encouraging results were achieved, demonstrating the potential of approaches that combine MCTS with AI techniques for routing problems with hard side constraints. The neural network guided search was able to outperform the heuristic used for creating the training data, demonstrating the network's ability to learn non-trivial characteristics from imperfect training data. An ablation study was conducted to show how the network's predictions improve the effectiveness of MCTS, demonstrating the possibility of improvement over classical approaches through neural network support. We further tested the concept of relaxed simulations during the simulation phase of the MCTS to improve computation times. It was demonstrated that our general approach can be used to obtain high quality results, even though the highly specialized heuristics currently still outperform this approach especially regarding the computational runtime. Future research directions include a more efficient implementation as well as employing more time-efficient value functions to reduce the overall computation times.

References

1. Coulom, R.: Efficient selectivity and backup operators in Monte-Carlo tree search. In: Hutchison, D. (ed.) Computers and Games. Lecture Notes in Computer Science, vol. 4630, pp. 72–83. Springer, Berlin Heidelberg (2007)
2. Dai, H., Khalil, E.B., Zhang, Y., Dilkina, B., Le Song: Learning combinatorial optimization algorithms over graphs. In: Advances in Neural Information Processing Systems (NeurIPS) (2017)
3. Desaulniers, G., Lessard, F., Hadjar, A.: Tabu search, partial elementarity, and generalized k-path inequalities for the vehicle routing problem with time windows. Transp. Sci. **42**(3), 387–404 (2008)
4. Dornemann, J.: Solving the capacitated vehicle routing problem with time windows via graph convolutional network assisted tree search and quantum-inspired computing. Front. Appl. Math. Stat. **9** (2023)
5. Falkner, J.K., Schmidt-Thieme, L.: Learning to solve vehicle routing problems with time windows through joint attention. arXiv preprint arXiv:2006.09100 (2020)
6. Falkner, J.K., Thyssens, D., Schmidt-Thieme, L.: Large neighborhood search based on neural construction heuristics (2022)
7. Furnon, V., Perron, L.: Or-tools routing library (2023). https://developers.google.com/optimization/routing/
8. Gao, L., Chen, M., Chen, Q., Luo, G., Zhu, N., Liu, Z.: Learn to design the heuristics for vehicle routing problem (2020)
9. Gurobi Optimization, L.L.: Gurobi optimizer reference manual (2022). https://www.gurobi.com.
10. Helsgaun, K.: An Extension of the Lin-Kernighan-Helsgaun TSP Solver for Constrained Traveling Salesman and Vehicle Routing Problems: Technical report. Roskilde Universitet (2017)
11. Joshi, C.K., Laurent, T., Bresson, X.: On learning paradigms for the travelling salesman problem. arXiv preprint arXiv:1910.07210 (2019)
12. Kingma, D.P., Ba, J.: Adam: a method for stochastic optimization. arXiv preprint arXiv:1412.6980v9 (2014)

13. Kocsis, L., Szepesvári, C.: Bandit based monte-carlo planning. In: Hutchison, D. (ed.) Machine Learning: ECML 2006. Lecture Notes in Computer Science, vol. 4212, pp. 282–293. Springer, Berlin Heidelberg (2006)
14. Kool, W., van Hoof, H., Welling, M.: Attention, learn to solve routing problems! In: International Conference on Learning Representations (2019)
15. Poullet, J.: Leveraging machine learning to solve the vehicle routing problem with time windows. Massachusetts Institute of Technology, Sloan School of Management, Operations Research Center (2020)
16. Silver, D., et al.: Mastering the game of go with deep neural networks and tree search. Nature **529**, 484–489 (2016)
17. Solomon, M.M.: Algorithms for the vehicle routing and scheduling problems with time window constraints. Oper. Res. **35**(2), 254–265 (1987)
18. Toth, P., Vigo, D.: Vehicle Routing. Society for Industrial and Applied Mathematics (2014)
19. Veličković, P., Cucurull, G., Casanova, A., Romero, A., Liò, P., Bengio, Y.: Graph attention networks (2017)
20. Vinyals, O., Fortunato, M., Jaitly, N.: Pointer networks. In: Advances in Neural Information Processing Systems, vol. 28. Curran Associates, Inc (2015)
21. Wang, Y., Sun, S., Li, W.: Hierarchical reinforcement learning for vehicle routing problems with time windows. Proc. Can. Conf. Artif. Intell. (2021)

QoS Estimation in Schedule-Based Vaccination Clinics

Ludovica Adacher, Marta Flamini, and Maurizio Naldi

Abstract Vaccination clinics serve a major role in mass vaccination campaigns, as in the COVID-19 pandemic. The requirement of vaccinating as many people as possible in the shortest possible time calls for lowering the time spent by patients in the clinic. The throughput time is then a major Quality-of-Service (QoS) metric. So far, just the average value has been considered as an overall metric. We show that average values may mask significant differences in the quality received by different patients and advocate the introduction of QoS metrics for different segments of the population of patients, in particular by examining the tail distribution of throughput times.

Keywords Health · Vaccination · Clinics · Hospitals · QoS · Pandemics

1 Introduction

Vaccination clinics serve as central hubs for administering vaccines to individuals across communities, helping to prevent the spread of diseases. Besides offering vaccines according to long-established plans for a wide range of diseases (e.g., influenza, measles, rubella, polio, or hepatitis B), they are employed for emergency situations when a mass vaccination campaign is to be carried out. Their central role stands

L. Adacher · M. Naldi (✉)
Department of Civil, Computer Science and Aeronautical Technologies Engineering, Roma Tre University, Rome, Italy
e-mail: m.naldi@lumsa.it

L. Adacher
e-mail: ludovica.adacher@uniroma3.it

M. Flamini
International Telematic University UNINETTUNO, Rome, Italy

M. Naldi
Department of Law, Economics, Politics, and Modern Languages, LUMSA University, Rome, Italy

M. Di Francesco et al. (eds.), *Operations Research: Closing the Gap Between Research and Practice*, AIRO Springer Series 15, https://doi.org/10.1007/978-3-031-90095-2_22

out in the case of pandemic outbreaks (e.g., the recent COVID-19 one), where they emerge as frontline defences against infectious diseases.

In mass vaccination campaigns, inoculations are typically carried out on a very tight schedule, due to the need to vaccinate as many people as possible in the shortest possible timeframe. Though cost must always be kept as low as possible [3], an efficient vaccination system calls for a low throughput time, the time spent by patients in the system.

The throughput time is recognized as a performance/quality metric for both patients and the public health system. The latter pursues the overall aim of completing the vaccination campaign as fast as possible, as mentioned above. The former wish to stay in the vaccination clinic as little as possible.

Though the throughput time has been provided as a performance metric in the studies of vaccination clinics (see, e.g., [1, 10, 12, 21], which are reviewed in Sect. 2), just its average value is typically considered. The average value is, of course, representative of the whole population of patients, but it may mask significant deviations and the subsequent degradation of service quality for even a large slice of the population.

We claim that more detailed service quality metrics must be introduced for the throughput time in vaccination clinics to highlight different quality levels experienced by different patients.

In this paper, we introduce new quality metrics and propose their classification according to the slice of the population they consider. With respect to the literature on vaccination clinics, we provide the following original contributions:

- we formulate a classification of quality metrics;
- we propose the mean excess function to assess the tail behaviour of the throughput time.

By applying these metrics to the output of a vaccination clinic simulator, we also show how those metrics reveal different features of service quality across the population of patients.

2 Related Literature

In this section, we review the literature concerning the performance of vaccination clinics. We also examine the literature concerning the very similar problem of performance analysis of emergency departments. An exhaustive review of the works on vaccination clinics is reported in [9].

The literature on vaccination clinics is quite unanimous in considering just the average time spent by patients in the system as the performance measure. Aaby et al. simulated a vaccination centre against smallpox using approximation formulas to reduce simulation times [1]. Washington used the commercial tool Arena to simulate the operations of a mass influenza/pneumococcal vaccination clinic with different paths for patients with different medical insurance coverage [19]. Gupta et al. again used Arena and introduced post-vaccination observation in a drive-through vaccination system [10]. Lee et al. developed their own RealOpt©software to simulate a mass dispensing centre of medications against infectious disease outbreaks and bioterrorist attacks [12]. Building on the alert generated by the 2009 H1N1 outbreak, the 2011 European Escherichia coli outbreak, and Hurricane Sandy, Yaylali et al. employed a discrete event simulation model devised through the Simio software to model a mass vaccination clinic [21]. The experience of COVID-19 triggered the building of an R-based simulation tool for vaccination clinics in [3], where, again, the average time was considered as a performance metric, but quality classes were also introduced to account for the differences in service levels experienced by patients.

We can now examine the wider body of literature on emergency departments, for which the problem of assessing time performances has been posed for a longer time. We see a large predominance of average quantities as figures of merit but with some new proposals. The most relevant quantity that was examined is the length of stay in the emergency department. We see first those papers that employed the average length of stay. Arya et al. adopted linear regression to evaluate the impact of a split in the workflow based on the Emergency Score Index [5]. Park et al. also used linear regression to estimate the average length of stay and to examine the effect of race/ethnicity and payment source [16]. In addition to the length of stay, Song et al. estimated the influence of pooled queuing systems on the average waiting time [17]. Wang et al. used the simulation tool ARIS to estimate the mean waiting time before admission to the consultation room [18]. There are also some papers that introduced additional figures of merit in addition to the average value of the length of stay. Approximate formulas from queuing theory and an add-in tool on Excel were employed in [15] to evaluate the probability of waiting for a recovery bed, in addition to the average length of stay. In [11], the additional figure of merit to be estimated was the probability of leaving the department before being seen and before treatment is complete. Finally, Xu et al. again used the queuing theory and System Dynamics simulation to predict the average waiting time and the probability of waiting time beyond some threshold (5, 7, and 9 h) [20].

3 The Vaccination Clinic Simulator

In order to demonstrate the use of QoS metrics in the case of vaccination clinics, we employ the simulator described in [4]. In this section, we describe the vaccination clinic workflow implemented in that simulator.

The vaccination clinic model we employ is shown in Fig. 1.

We assume that patients have booked specific dates and times for their vaccination appointments. Our model accounts for continuous vaccination operations over a period T (typically, eight-hour working day). Patient appointments are scheduled at regular intervals of Δ minutes. However, patients exhibit random arrival behaviour, arriving either earlier or later than their scheduled appointment time. We model

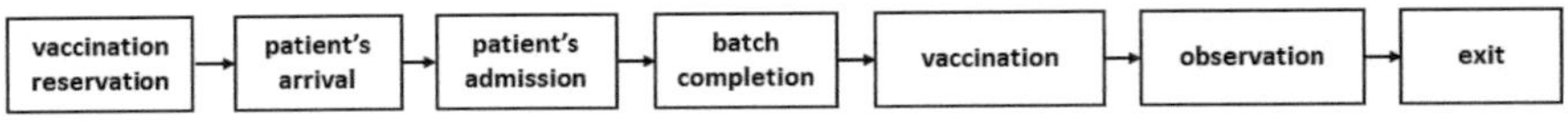

Fig. 1 Service flow in the vaccination clinic

this arrival pattern using a truncated Gaussian distribution with a zero mean and a standard deviation of one minute to represent the perturbation with respect to the orderly arrival according to the schedule. Since patients are served on a FIFO basis, the perturbation may alter the order of service.

Patients enter the clinic through a reception desk and then move to the waiting room, where they join a queue to be served by the first available nurse. Patients are not served individually but in batches. Serving batches allows nurses to pipeline all operations required before the actual inoculation (dose and syringe retrieval and preparation of all needed equipment). After vaccination, patients move to the observation room, where they stay under observation to examine possible adverse reactions to the vaccine [7]. Their length of stay in the observation room is fixed, but it depends on the patient's medical history. Here, we consider observation periods of 15 or 30 min, with percentages equal to 82 and 18%, respectively.

The simulator employs a Monte Carlo approach, where randomness is due to arrival times. Each Monte Carlo instance spans a full working day of arrivals.

4 Quality of Service Metrics

We have established the time needed for the patient to get through the vaccination clinic as the major quality of service indicator. However, since that time may be different for each individual, we must compute an aggregate indicator for the population of patients. In this section, we describe the different ways we can employ to compute the aggregate indicator of QoS.

In its journey through the clinic, the patient experiences several waiting periods in queues: waiting for the batch to be formed, for a nurse to be available, and for actual inoculation. These waiting times are random, since they depend on the clinic occupancy status at the time of arrival. Finally, the patient has to stay in the observation room. However, this last period is of fixed length and is not due to the clinic's performance. Hence, we remove it from the overall computation: all the times considered in the following must be intended as the throughput time minus the time spent in the observation room.

Since the most complete description of the time required of the population of patients to get through the vaccination clinic is the distribution of throughput times, we define the following three classes of QoS metrics based on the portions of the distribution curve that we consider:

- the whole distribution;
- the bulk of the distribution (or segments of it);

- the tail of the distribution.

It is to be noted that we refer to the shape of the distribution curve rather than the mathematical function. Hence, by the whole distribution, we mean that the derived statistics are computed over the whole range of the random variable, which is not the case when we refer to the *bulk* (i.e., when we consider just the range where the probability density takes the higher values, or some interval of values for the variable of interest) and the *tail* (i.e., the distribution curve portion pertaining to the highest values taken by the random variable of interest or, equivalently, those values exceeding some threshold).

The first category of metrics is the one mostly employed in the literature, as shown in Sect. 2. It includes all standard measures of location, i.e., the mean and the median values, hopefully accompanied by a measure of dispersion (e.g., the standard deviation) to describe deviations from the measure of location across the population. These metrics provide the overall view of quality, though exceptionally bad cases of very low quality (i.e., very high waiting times) are somewhat absorbed by the great majority, especially if we opt for robust statistics like the median value.

In the second category, we consider the quality of service experienced by most patients. A typical example of this class of metrics is the percentage of patients being served within a prescribed amount of time. We can also imagine different classes of service, where the time grows progressively. This is the approach considered, e.g., in [3], where the grade classification for airport check-in services was borrowed from [2]. The scale employs five QoS classes from A (Excellent) to E (Inadequate) and is shown in Table 1. This classification allows us to segment the population of patients. Alternatively, we could opt for a binary classification, where we have a single threshold, marking the passage from an acceptable service level to an inadequate one. Keeping track of the fraction of the population that receives a given service level is a step forward with respect to the single overall indicator represented by the mean throughput time. Computing the fraction of patients falling into Class E allows us to see which fraction of the population receives an inadequate service level.

However, we may wish to go deeper and see how inadequate the service is. With reference to Table 1, Class E includes patients who experience a throughput time just above the threshold (i.e., just a bit worse than patients in Class D) but also patients whose throughput time is way worse (e.g., two or three times as long). For the purpose of getting a more detailed view of the actual throughput time when we step into Class E, we need to look into the tail of the distribution of throughput times. Here, we propose to employ the most prominent descriptor of a distribution tail shape [13]: the mean excess function, which outputs the mean value of throughput times exceeding the acceptability threshold. The mean excess function has been employed,

Table 1 QoS classes

QoS class	A	B	C	D	E
Max waiting time [mins]	10	25	40	60	$\geq$61

e.g., to model quality in cloud services by measuring the outage durations [14]. Hereafter, we briefly recall its definition.

For a random variable X, the mean excess function $e(t)$ is defined as $e(t) = \mathbb{E}[X - t | X > t]$, i.e., the mean value in excess of a given threshold t, conditioned to the random variable exceeding that threshold (see Sect. 1.2.2 of [8]).

For many probability distributions, the mean excess function can be easily computed. A special case is the Generalized Pareto distribution (GPD), which is the only distribution whose mean excess function is linear. If we consider the sample $\{x_1, x_2, \ldots, x_n\}$ of throughput times obtained through our simulator, where $n = T/\Delta$, the mean excess function can be estimated through its empirical counterpart $\hat{e}(t) = \sum_{i=1}^{n} x_i \mathbb{I}(x_i > t) / \sum_{i=1}^{n} \mathbb{I}(x_i > t) - t$.

5 Results

In this section, having defined the performance metrics of interest in Sect. 4, we report the indications provided by those metrics for a set of load (attendance rates) and clinic configurations. We wish to stress that the aim here is not to identify the best-performing clinic configuration but rather to assess the different views that performance metrics offer.

The vaccination time has been set equal to 1 min. We have considered two load situations (low attendance and high attendance). Under low attendance, the scheduled frequency of arrivals is one patient per minute (480 patients per working day). Under high attendance, we have instead two patients per minute (960 patients per working day). Also, we have considered the clinic configurations in Table 2, which have been chosen with the aim of showing the clinic under stress, i.e., with the service quality perceived by patients being very far from ideal. For each configuration, we have run the simulator for ten working days. In the following, we refer to the simulation of one working day as one Monte Carlo instance. For arrival times, we have generated Gaussian-distributed random variates, performing truncation by using the acceptance-rejection technique (see Sect. II.2b of [6]). The patient class is a Bernoulli variable, since we consider just two classes.

Table 2 Clinic configurations

Configuration	Attendance	Batch size	No. of nurses	Observation room size
S_1	H	4	4	35
S_2	H	3	3	30
S_3	H	3	4	35
S_4	L	3	2	15
S_5	L	3	2	20
S_6	L	4	2	20

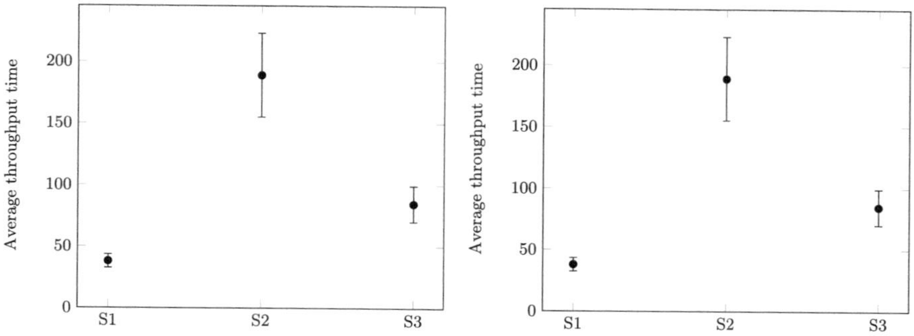

Fig. 2 Throughput time under high attendance (left) and low attendance (right)

For each Monte Carlo instance, we have computed the performance metrics defined in Sect. 4. The quantity of interest is always the time spent by the patient in the vaccination clinic, which we call throughput time but is also known as time-in-the-system, after removing the fixed observation time.

In Fig. 2, we show the metrics pertaining to the overall distribution, i.e., the average throughput time and the associated 1-σ confidence intervals. The three configurations under high attendance originate three distinct performance outcomes. We clearly see that Configuration S_1 is better than Configuration S_3, which in turn is better than Configuration S_2. This statement is well rooted in the observation that the three configurations are well ordered according both to their average waiting time and its standard deviation, better configurations being characterized by lower average waiting times and lower dispersion for all the Monte Carlo instances.

However, if we consider the performance metrics based on the bulk of the distribution, we see in Fig. 3 (high attendance) that for the best-performing class (Class A), the two configurations S_1 and S_3 exhibit a certain degree of overlapping. Though the average waiting time is quite different for those configurations, their performance for the top class is not very dissimilar. The percentage of patients receiving top service quality lies in the range 2–6% for S_1 and 2–4% for S_3. Hence, the two configurations differ most in the performances for quality classes lower than Class A. Instead, Configuration S_2 marks quite a lower presence for classes A to D, i.e., all the classes considered adequate, with most patients receiving an inadequate service level (Class E in Table 1). In the case of low attendance (Fig. 4), we see that the performance for the top class is very similar for all three configurations, but the configurations differ in the lower quality classes, with configuration S_3 achieving higher percentages than S_1 and S_2.

We can finally examine the tail behaviour through the mean excess function, shown in Fig. 5. Under high attendance, the three configurations do not differ just for the average throughput time, but they remain quite distinct for the tail behaviours also, since Configuration S_2 shows much larger values than the other two. Instead, under low attendance, Configurations S_4 and S_5 exhibit similar behaviour in the tail, confirming the overlapping behaviour in the bulk in Fig. 4. However, the mean excess

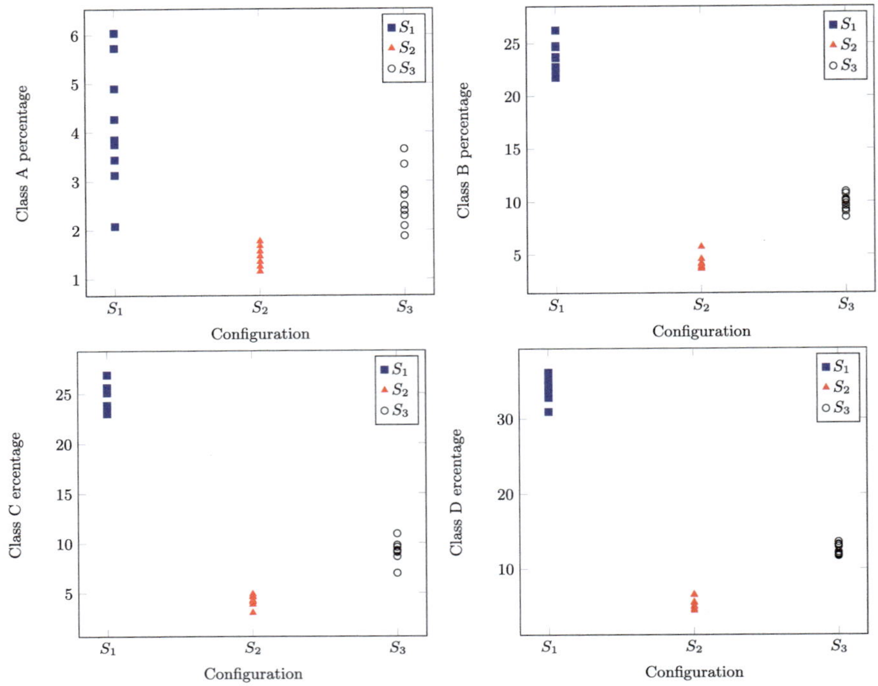

Fig. 3 Performance metrics for quality classes under high attendance

function also gives us a numeric indication of how much the quality degrades for patients in the tail portion of the throughput time distribution. Also, we see that all the mean excess function curves exhibit a linear trend, supporting the hypothesis of a Generalized Pareto distribution to model the shape of the distribution tail of throughput times.

6 Conclusions

We have defined three categories of quality metrics for vaccination clinics based on the time the patient spends in the clinic. The introduction of metrics that look at specific portions of the probability distribution represents a step forward with respect to the current status of performance analysis for vaccination clinics.

We have shown that the average time alone does not allow us to identify the differences (or even similarities) in the quality offered across the population of patients when we consider different clinic configurations. In particular, we have seen that, though the average quality may be quite different for the two configurations, those configurations can provide top quality for roughly the same percentage of patients.

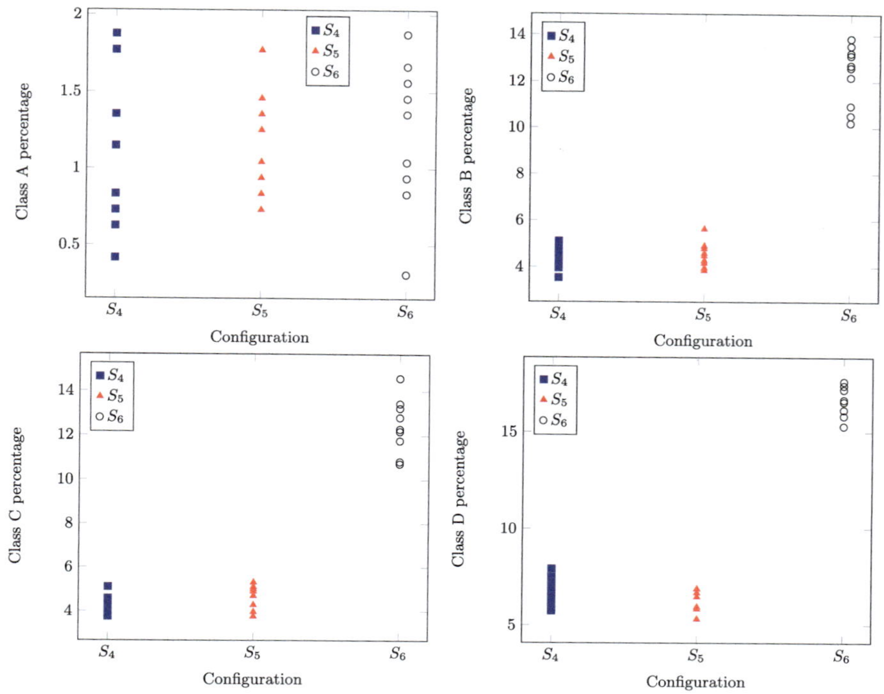

Fig. 4 Performance metrics for quality classes under low attendance

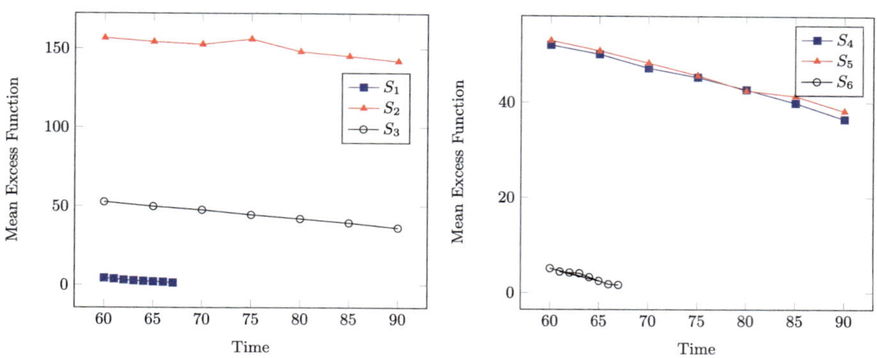

Fig. 5 Mean excess function under high attendance (left) and low attendance (right)

On the other hand, a configuration with similar average metrics as another one may show significant differences in the distribution tails, i.e., it may offer a significantly worse quality for a slice of the population.

These observations make a case for the establishment of a wider range of performance metrics for vaccination clinics than the average throughput time considered so far. Metrics based on quality classes allow us to segment the population and understand the differences in the quality level received by the different segments. Metrics based on the distribution tail, such as the mean excess, allow us to track abnormal phenomena, where a small percentage of the population receives an overly worse treatment than the bulk, and measure the level of performance degradation. The information derived from such a range of quality metrics allows the decision maker to see such inequalities in QoS and better align the service configuration with their policy aims. e.g., by deciding to live with those inequalities and accordingly stating the quality of service it offers, or deciding to reduce inequalities by changing the service configuration.

References

1. Aaby, K., Herrmann, J.W., Jordan, C.S., Treadwell, M., Wood, K.: Montgomery county's public health service uses operations research to plan emergency mass dispensing and vaccination clinics. Interfaces **36**(6), 569–579 (2006)
2. Adacher, L., Flamini, M.: Optimization of airport check-in service quality focused on operational costs and passengers' satisfaction. Plos One **16**(8), e0253586 (2021)
3. Adacher, L., Flamini, M., Naldi, M.: Optimal design of a vaccination clinic: the trade-off between costs and QoS. In: Proceedings of the International Conference on Optimization and Decision Science ODS (2023)
4. Adacher, L., Flamini, M., Naldi, M.: Simulation of a schedule-based mass vaccination clinic. J. Simul. 1–17 (2023)
5. Arya, R., Wei, G., McCoy, J.V., Crane, J., Ohman-Strickland, P., Eisenstein, R.M.: Decreasing length of stay in the emergency department with a split emergency severity index 3 patient flow model. Acad. Emerg. Med. **20**(11), 1171–1179 (2013)
6. Asmussen, S., Glynn, P.W.: Stochastic Simulation: Algorithms and Analysis, vol. 57. Springer (2007)
7. Beatty, A.L., Peyser, N.D., Butcher, X.E., Cocohoba, J.M., Lin, F., Olgin, J.E., Pletcher, M.J., Marcus, G.M.: Analysis of covid-19 vaccine type and adverse effects following vaccination. JAMA Netw. Open **4**(12), e2140364–e2140364 (2021)
8. Beirlant, J., Goegebeur, Y., Segers, J., Teugels, J.L.: Statistics of Extremes: Theory and Applications. John Wiley & Sons (2006)
9. Gianfredi, V., Pennisi, F., Lume, A., Ricciardi, G.E., Minerva, M., Riccò, M., Odone, A., Signorelli, C.: Challenges and opportunities of mass vaccination centers in covid-19 times: a rapid review of literature. Vaccines **9**(6), 574 (2021)
10. Gupta, A., Evans, G.W., Heragu, S.S.: Simulation and optimization modeling for drive-through mass vaccination—a generalized approach. Simul. Model. Pract. Theory **37**, 99–106 (2013)
11. Khalifa, M., Zabani, I.: Utilizing health analytics in improving the performance of healthcare services: a case study on a tertiary care hospital. J. Infect. Public Health **9**(6), 757–765 (2016)
12. Lee, E.K., Maheshwary, S., Mason, J., Glisson, W.: Decision support system for mass dispensing of medications for infectious disease outbreaks and bioterrorist attacks. Ann. Oper. Res. **148**(1), 25–53 (2006)

13. McNeil, A.J., Frey, R., Embrechts, P.: Quantitative Risk Management: Concepts, Techniques and Tools. Princeton University Press (2005)
14. Naldi, M.: The availability of cloud-based services: Is it living up to its promise? In: 2013 9th International Conference on the Design of Reliable Communication Networks (DRCN), pp. 282–289. IEEE (2013)
15. Palvannan, R.K., Teow, K.L.: Queueing for healthcare. J. Med. Syst. **36**, 541–547 (2012)
16. Park, C.Y., Lee, M.A., Epstein, A.J.: Variation in emergency department wait times for children by race/ethnicity and payment source. Health Serv. Res. **44**(6), 2022–2039 (2009)
17. Song, H., Tucker, A.L., Murrell, K.L.: The diseconomies of queue pooling: an empirical investigation of emergency department length of stay. Manage. Sci. **61**(12), 3032–3053 (2015)
18. Wang, T., Guinet, A., Belaidi, A., Besombes, B.: Modelling and simulation of emergency services with ARIS and arena. case study: the emergency department of saint joseph and saint Luc hospital. Prod. Plan. Control **20**(6), 484–495 (2009)
19. Washington, M.L.: Evaluating the capability and cost of a mass influenza and pneumococcal vaccination clinic via computer simulation. Med. Decis. Mak. **29**(4), 414–423 (2009)
20. Xu, K., Chan, C.W.: Using future information to reduce waiting times in the emergency department via diversion. Manuf. Serv. Oper. Manage. **18**(3), 314–331 (2016)
21. Yaylali, E., Ivy, J.S., Taheri, J.: Systems engineering methods for enhancing the value stream in public health preparedness: the role of Markov models, simulation, and optimization. Public Health Rep. **129**(6_suppl4), 145–153 (2014)

Integrated Optimization of Building and Curing in Tires Production

Raniero Bagozzi, Giovanni Righini, and Andrea Taverna

Abstract We address the problem of optimally planning the two main phases in tires production, namely building and curing. They are executed by different machines subject to specific constraints; furthermore, the buffer capacity between the two phases is limited. We present a mathematical model of the optimization problem, that turns out to be a computationally difficult MILP problem. Computational tests show that reducing the symmetries by variable aggregation allows to solve large size instances arising in a real production plant to near-optimality within a reasonable amount of computing time.

Keywords Mixed-integer linear programming · Production planning · Scheduling

1 Introduction

We consider a planning problem arising in a tire manufacturing company that needs to plan its daily production over a monthly horizon, so that it matches given target quantities. The production of tires goes through several phases. The two main phases are called building and curing. In the building phase the semi-finished components are assembled on suitable tire building machines to produce so-called "green tires". In the curing phase, tires are vulcanized in curing presses, that is they are subjected to certain temperature and pressure conditions for a certain amount of time. Different types of tires are required; they are identified as codes and they have different

R. Bagozzi
Pirelli Tires, Milan, Italy
e-mail: raniero.bagozzi@pirelli.com

G. Righini (✉)
Department of Computer Science, University of Milan, Milan, Italy
e-mail: giovanni.righini@unimi.it

A. Taverna
Amazon Development Center Germany GmbH, Berlin, Germany
e-mail: atavern@amazon.de

M. Di Francesco et al. (eds.), *Operations Research: Closing the Gap Between Research and Practice*, AIRO Springer Series 15, https://doi.org/10.1007/978-3-031-90095-2_23

characteristics and dimensions. Since building machines and curing presses are not identical, the assignment of production lots to machines and presses is also subject to compatibility constraints.

The problem of maximizing the production efficiency in curing was introduced by Degraeve and Schrage [1], who reported significant improvements obtained by mathematical optimization techniques compared with manual planning. More recently, Cancela et al. [2] analyzed a scheduling problem to optimize the curing phase, taking into account compatibility restrictions between molds and machines. Lot-sizing and scheduling in curing has been also considered by Koch et al. [3], who developed a MILP model for a scheduling problem with 170 codes, 70 identical parallel curing machines and 7 periods. Planning on longer periods (up tp 42 days) was possible through math-heuristics propoed by Koch et al. [4].

The objective of this study is to simultaneously address building and curing and to define a detailed schedule for both phases, covering a given planning horizon. The schedule must specify the sequence of operations for each building machine and each curing press in each day.

In Sect. 2 we define a mathematical model of the optimization problem in terms of data, variables, constraints and objectives. We present a MILP model able to take into account several details of the production process, such as set-up and mounting/unmounting operations as well as different working modes of some specific machines. Computational results on large-scale real instances are presented and discussed in Sect. 3.

2 Problem Description

The mathematical model of the optimization problem can be decomposed into three sub-problems, connected by some linking constraints:

- planning: determines when each tire code enters production and the amounts produced for each day;
- building: determines the assignment of tire codes to building machines;
- curing: determines the assignment of tire codes to curing presses.

The three sub-problems are separately described in the next three subsections. For each sub-problem, data, variables and constraints are defined, while the main objective to be optimized is outlined at the end of the Section. Since the notation is rich, because of the many details of the problem, each symbol is introduced when needed.

2.1 Planning Sub-problem

An industrial plant produces many types of tires of different size and with different characteristics. To be general, in the remainder we use the term *code* to identify every specific combination of tire characteristics associated with a production order and we indicate by N the set of codes.

The planning horizon consists of a given number of consecutive days (typically thirty days). We indicate the indexed set of days by $D = \{1, \ldots, \overline{d}\}$.

Production.

The main decision variables in the planning sub-problem are variables with two indices corresponding to codes and days: each binary variable w_{id} indicates whether code $i \in N$ is produced in day $d \in D$ or not; each continuous non-negative variable p_{id} indicates the amount of production of code $i \in N$ in day $d \in D$. Since on each specific day the building production and the curing production of a specific code can be slightly different, we assume p_{id} to be the building production.

Let the maximum allowed daily production for each code $i \in N$ be $\bar{P}_i$. Binary variables w and continuous variables p are linked by constraints

$$p_{id} \leq \bar{P}_i w_{id} \quad \forall i \in N, d \in D. \tag{1}$$

There is a given upper limit W to the number of codes that can be produced in a day. The corresponding constraint is

$$\sum_{i \in N} w_{id} \leq W \quad \forall d \in D. \tag{2}$$

Target production and allowed deviations.

It is assumed that a master production plan has been previously defined at a higher planning level. It sets a total production target K_i to be reached in the planning period for each code $i \in N$. However, limited deviations are allowed with respect to such target values. To represent them, we use continuous variables $s_i^- \geq 0$ (slack) and $s_i^+ \geq 0$ (surplus). They are linked with the production variables p by the constraints

$$\sum_{d \in D} p_{id} + s_i^- - s_i^+ = K_i \quad \forall i \in N. \tag{3}$$

Phase-in operations.

A crucial feature of a production plan is the number of phase-in and phase-out operations. A phase-in occurs when a code $i \in N$ not produced in day $d-1$ is produced in day d; phase-out is symmetric. In general, the fewer phase-in/phase-out operations a plan contains, the better it is. To represent phase-in operations, we use binary variables ϕ_{id}, meaning that the production of code $i \in N$ starts on day $d \in D$. We also indicate by w_{i0} whether code $i \in N$ is in production the day before day 1. The following constraints are imposed:

$$\phi_{id} \geq w_{id} - w_{i(d-1)} \quad \forall i \in N, d \in D. \tag{4}$$

Each code can undergo a given limited amount ϕ_i^{TOT} of additional phase-ins over the planning horizon, besides the first one. For those codes that are already in production at the beginning (i.e. those for which $w_{i0} = 1$) all phase-ins count as additional.

$$\sum_{d \in D} \phi_{id} - (1 - w_{i0}) \leq \phi_i^{TOT} \quad \forall i \in N. \tag{5}$$

A daily limit Φ for the total number of phase-ins is also given. The corresponding constraint is

$$\sum_{i \in N} \phi_{id} \leq \Phi \; \forall d \in D. \tag{6}$$

Since phase-ins imply to mount and unmount molds, which is a labor-intensive and time-consuming activity, by default $\phi_i^{TOT} = 0 \quad \forall i \in N$. The generic parameter has been kept in the model, just in case a non-zero value is needed to make an instance feasible, for some exceptional reason.

Attributes.
Each code has some specific attributes. An example of attribute is "Run-Flat", a special feature that helps keeping tires inflated after puncture, allowing drivers to safely keep driving, albeit at reduced speed. Let A be the set of attributes and $N^a \subset N$ the subset of codes with attribute $a \in A$. For each attribute $a \in A$, a maximum daily production limit π_a is imposed by the following constraints:

$$\sum_{i \in N^a} p_{id} \leq \pi_a \quad \forall a \in A, d \in D. \tag{7}$$

2.2 Building Sub-Problem

Data.
The set of building machines is indicated by B. A set of special codes is indicated by $N^{\text{SPECIAL}} \subseteq N$. These codes can be produced only by a specific subset of building machines, indicated by $B_{\text{SPECIAL}} \subseteq B$. The complement of B_{SPECIAL} is B_{NORMAL}. Machines can be further distinguished based on their ability to work in two different production modes. By default, machines can produce only one code at a time, that is they work in MONO production mode. However, some machines may also work in DOUBLE production mode, which means they can produce two different codes in parallel although in a less efficient way. Special codes can be produced only in MONO production mode. Hence, the set of machines B is partitioned into three subsets B_{SPECIAL}, $B_{\text{NORMAL}}^{\text{MONO}}$ and B^{DOUBLE}.

Variables.
The main variables in the building sub-problem should represent the assignment of building machines to codes. However, we used aggregate integer variables x_{id}^m,

indicating the number of machines assigned to a code $i \in N$ on day $d \in D$ and working in mode $m \in M = \{\text{MONO, DOUBLE}\}$. This is allowed because all machines working in the same mode are identical. This variable aggregation was instrumental in making the resulting MILP model solvable in a reasonable amount of computing time.

MONO and DOUBLE mode.
There is a limit $\bar{y}$ to the number of building machines in B^{DOUBLE} that can work in DOUBLE mode in any day $d \in D$:

$$\sum_{i \in N} x_{id}^{\text{DOUBLE}} \leq \bar{y} \quad \forall d \in D. \tag{8}$$

Special codes cannot be built in DOUBLE mode:

$$x_{id}^{\text{DOUBLE}} = 0 \quad \forall i \in N^{\text{SPECIAL}}, d \in D. \tag{9}$$

Special codes cannot be assigned to normal machines.

$$\sum_{i \in N^{\text{SPECIAL}}} x_{id}^{\text{MONO}} \leq |B_{\text{SPECIAL}}| \quad \forall d \in D. \tag{10}$$

Assignment constraints.
The total number of machines used in each day is limited by the number of available machines

$$\sum_{i \in N} (x_{id}^{\text{MONO}} + x_{id}^{\text{DOUBLE}}/2) \leq |B| \quad \forall d \in D \tag{11}$$

Production.
Continuous variables $\beta_{id}^{m} \geq 0$ indicate the total production of code $i \in N$ on machines working in mode $m \in M$ in day $d \in D$. Some specific data are needed to formulate the constraints on production. The coefficient τ_i is the cycle time of code $i \in N$, measured in minutes per cycle, where a "cycle" is a standard sequence of operations of building machines. In each cycle one or more tires are produced. The coefficient ω^{MONO} indicates the efficiency of building machines working in MONO mode. It is measured in tires per cycle. Hence the productivity of a machine producing code $i \in N$ in MONO mode is given by ω^{MONO}/τ_i and it is expressed in tires per minute. The time available in each day $d \in D$ is indicated by Δ_d and it is expressed in minutes (usually $\Delta_d = 1440$ min). Therefore the maximum amount of tires of code $i \in N$ that can be produced in day $d \in D$ by a machine working in MONO mode is $\omega^{MONO}\Delta_d/\tau_i$. This allows to express the constraints on the maximum production of each code in each day in MONO mode.

$$\beta_{id}^{\text{MONO}} \leq \frac{\Delta_d \omega^{\text{MONO}}}{\tau_i} x_{id}^{\text{MONO}} \quad \forall i \in N, d \in D. \tag{12}$$

For machines working in DOUBLE mode, a coefficient ω^{DOUBLE} is given. Since MONO mode is more efficient than DOUBLE mode, the inequality $\omega^{\text{DOUBLE}} < \omega^{\text{MONO}}$ holds. An immediate consequence is that it is never optimal to assign the same code $i \in N$ to more than one machine working in DOUBLE mode i.e. $x_{id}^{\text{DOUBLE}} \leq 1 \, \forall i \in N, d \in D$. It can be safely assumed that the available time on each machine working in DOUBLE mode is equally split among the two codes. Hence, the constraint on the maximum production of each code in each day in DOUBLE mode is:

$$\beta_{id}^{\text{DOUBLE}} \leq \frac{1}{2}\frac{\Delta_d \omega^{\text{DOUBLE}}}{\tau_i} x_{id}^{\text{DOUBLE}} \quad \forall i \in N^{\text{NORMAL}}, d \in D. \tag{13}$$

Stop operations.
Similarly to phase-out in planning, building machines can have "stop operations", when a code $i \in N$ is assigned to fewer machines in day d than in day $d-1$. Stop operations are disruptive and a good plan should contain a minimum number of them. To control the amount of stop operations we impose the following constraints:

$$\sum_{m \in M} (x_{id}^m - x_{i(d-1)}^m) \geq \sum_{m \in M} (x_{i(d+1)}^m - x_{id}^m) \quad \forall i \in N, d \in D \setminus \{\bar{d}\}, \tag{14}$$

where for $d = 1$ x_{i0}^m is the amount of machines in mode $m \in M$ assigned to code $i \in I$ the day before day 1. These conditions imply that the daily number of machines assigned to a code i varies as a (discrete) concave function, i.e. they prevent it to have subsequent "ups" and "downs" over the planning horizon, while still allowing it to increase and decrease during the period in which a code is produced.

Rim assignment.
When processed on a building machine, each tire is contained in a rim. Each product code is associated with a specific rim type, that must be mounted on the building machine. We indicate the set of rim types by T. Some codes may share the same rim type. Building machines can work in DOUBLE mode only if the two assigned codes have the same rim type. $R_t \subset N$ is the subset of codes with rim type $t \in T$. Integer variables $v_{td}^m \geq 0$ indicate the number of rims of type $t \in T$ mounted on machines that work in mode $m \in M$ in day $d \in D$. The relationship between variables v and x in machines working in DOUBLE mode is represented by the following constraints:

$$\sum_{i \in R_t} x_{id}^{DOUBLE} = 2v_{td}^{\text{DOUBLE}} \quad \forall t \in T, d \in D. \tag{15}$$

When the same machine works in MONO mode, it needs to mount a rim for the corresponding code.

$$\sum_{i \in R_t} x_{id}^{\text{MONO}} \leq v_{td}^{\text{MONO}} \quad \forall t \in T, d \in D. \tag{16}$$

Constraints (15) force two different codes to be assigned to each machine working in DOUBLE mode. This is not restrictive, because processing a single code on a machine working in DOUBLE mode cannot be optimal.

Link between planning and building
The production amounts p are linked with the β variables by equations

$$p_{id} = \sum_{m \in M} \beta_{id}^{m} \quad \forall i \in N, d \in D. \tag{17}$$

2.3 Curing Sub-problem

A set of curing presses C can "cure" tires produced in the building phase, by mounting the corresponding molds. Each curing press mounts up to two molds.

Molds usage.
An integer variable $n_{id} \geq 0$ counts the number of molds dedicated to the production of each code $i \in N$ in each day $d \in D$. As each press can mount at most two molds in each day, the total daily number of molds that are used is limited by the following constraint:

$$\sum_{i \in N} n_{id} \leq 2|C| \quad \forall d \in D. \tag{18}$$

In each day, a maximum given number $\bar{n}_i$ of molds can be mounted for each code $i \in N$:

$$n_{id} \leq \bar{n}_i w_{id} \quad \forall i \in N, d \in D. \tag{19}$$

Mold mounting and unmounting.
Integer variables $n_{id}^{up} \geq 0$ and $n_{id}^{dn} \geq 0$ count the number of mold mounting and unmounting operations, respectively, for each code $i \in N$ in each day $d \in D$. Variables n^{up} and n^{dn} are linked to variables n by the following constraints:

$$n_{id}^{up} \geq n_{id} - n_{i(d-1)} \quad \forall i \in N, d \in D \tag{20}$$

$$n_{id}^{dn} \geq n_{i(d-1)} - n_{id} \quad \forall i \in N, d \in D. \tag{21}$$

A datum n_{i0} indicates the number of molds used in the day before day 1. An upper limit U on the total daily number of mounting and unmounting operations is given:

$$\sum_{i \in N} (n_{id}^{up} + n_{id}^{dn}) \leq U \quad \forall d \in D. \tag{22}$$

2.4 Links Between Planning, Building and Curing

The curing capacity allocated to each day should ideally match the number of tires produced in the building phase, to guarantee a continuous flow of semi-finished tires and a smooth run of the two consecutive operations. On the other hand, forcing strict equality between the building output and the curing input could be infeasible, due to the different throughput of the two phases, or unrealistic, due to unplanned delays that may occur in the operations. For these reasons, our model allows the planner to control the maximum allowed difference between building and curing production, using a continuous adimensional parameter Γ, as shown below.

Molds have a given daily yield, which is the number of tires they can cure in a day: for each code $i \in N$, its corresponding mold can cure up to ρ_i tires in a day. For each day $d \in D$ and code $i \in N$, a continuous non-negative variable δ_{id} is the difference between the installed curing capacity and the planned production:

$$\delta_{id} = \rho_i n_{id} - p_{id} \quad \forall i \in N, d \in D. \tag{23}$$

The requirement $\delta_{id} \geq 0$ means that the curing phase must allocate enough molds to cure all tires produced in the building phase in the same day.

Suitable constraints limit the daily curing capacity surplus not to exceed a fraction $0 \leq \Gamma \leq 1$ of the allocated curing capacity for each code $i \in N$.

$$\delta_{id} \leq \Gamma \rho_i n_{id} \quad \forall i \in N, d \in D. \tag{24}$$

2.5 Objective Function

The main objective in our problem is the minimization of production slacks, because its optimization pursues the consistency between the long-term master production plan of the company and the detailed monthly production plan of individual plants.

$$\min g = \sum_{i \in N} s_i^- .$$

3 Computational Results

We ran our tests on a 31-days instance, 37 building machines, 9 of which are for special codes, and 138 curing machines that can cure up two tires at a time. We started from real problem instance with 103 different codes and a total demand of 404 thousands pieces, i.e. 4 thousands pieces per code on average. The corresponding MILP istance has 41779 constraints and 35360 variables, including 15996 integer and 3193 binary.

Table 1 Demand for each instance

Instance	Total demand
1	404140
2	428208
3	391243
4	416428
5	404384
6	436392

Instances generation.
To test our model, we generated 5 more instances from the real one by randomly changing the demand and the number of available molds of each code. In particular, the randomization mechanism preserves, on average, the ratio between demand and available molds of the original instance; this allows reaching a high amount of demand satisfaction (more than 80%), as it would be realistically planned by an experienced human planner.

The demand for each code $i \in N$ is multiplied by a factor $1 + \alpha_i$, where α_i follows a triangular distribution in the range $[-60\%, +60\%]$ and mode $\pm 12\%$, with the sign chosen uniformly at random as well.

The number of daily available molds $\bar{n}_i$ is multiplied by a factor $(1 + \theta_i)(1 + \alpha_i)$, where θ_i follows a triangular distribution in range $[-35\%, 35\%]$ with zero mode, and $1 + \alpha_i$ is the ratio between the new demand and the old one. In particular, for $\theta_1 = 0$ the number of molds scales proportionally with the new demand.

The total demand for each instance is reported in Table 1.

Computational tests.
For each instance we made two tests, one with a 15-days horizon and one with a full 31-days horizon.

We used Gurobi 10.0.1 to solve the MILP model. To avoid long tailing off effects, we set a time limit of two hours, a relative gap limit of 1% and an absolute gap limit of 1% of the total demand.

In Table 2 we report the results with the two planning horizons. Primal and dual bounds are expressed as percentage of total slack over total demand $\frac{\sum_{i \in N} s_i^-}{\sum_{i \in N} K_i}$.

In the experiment with the long horizon, we noticed that the solver could quickly find a solution that is close to the optimum, while it took a long time to improve the dual bound. To illustrate this, we report for both horizons the "time for primal bound+1%" as the time required to find a solution that is 1% larger than the final primal bound and, similarly, the "time for dual bound-1%" as the time required to reach a dual bound that is 1% lower than the final dual bound.

The half-horizon instances were easier to solve: the solver could find good primal and dual bounds within a couple of minutes. The rest of the time was spent in "tailing-off", i.e. in slowly improving both bounds until they reached the desired accuracy.

Table 2 Computational results

Horizon	Instance	Gap (%)	Primal Bound (PB) (%)	Dual Bound (DB) (%)	Time (s)	Time PB + 1% (s)	Time DB − 1% (s)
31 days	1	3.85	22.19	21.33	4806	1509	3745
	2	4.79	21.31	20.89	1935	1502	1916
	3	4.62	22.72	21.67	2663	916	2374
	4	4.41	24.13	23.06	1863	943	1611
	5	10.45	22.79	20.40	7200	2291	4905
	6	5.81	22.19	21.20	4806	1509	3745
15 days	1	1.97	48.09	47.15	160	94	6
	2	1.86	51.01	50.05	110	88	9
	3	2.10	47.46	46.42	264	106	16
	4	1.98	49.85	48.86	131	85	6
	5	2.02	48.49	47.51	156	109	8
	6	1.66	51.67	50.81	145	119	11

For most long-horizon instances, the solver could find an almost optimal solution in less than 30 min. Due to many symmetries in the formulation, it took almost an hour and a half to find an equally good dual bound for five instances out of six. However, given the tactical level of the decision problem at hand, few hours of computation are definitely compatible with the needs of the company.

Sensitivity analysis on parameter Γ.
In an additional set of experiments we solved the same six instances over the 31-days horizon with $\Gamma = 0.25$ instead of $\Gamma = 1$, implying that in each day the difference between building production and curing capacity has to be less than 25% of a mold yield.

We could observe that a lower value of Γ slightly increases the computing time. This is expected due to the increased complexity, as the match between building and curing phases is more demanding. However, we could not observe any significant change in the bound values, as shown in Table 3.

Table 3 Computational results with $\Gamma = 0.25$

Instance	Gap (%)	Primal Bound (PB) (%)	Dual Bound (DB) (%)	Time (s)	Time PB + 1% (s)	Time DB − 1% (s)
1	4.44	22.71	20.32	7200	1953	5403
2	4.38	21.95	20.99	4265	2815	2830
3	10.8	22.99	20.5	7200	1191	5294
4	4.43	24.51	23.45	4982	1893	3837
5	8.08	22.85	21.06	7200	1253	5040
6	4.47	22.06	21.07	3397	2451	3142

4 Conclusions

Synchronization between long-term production plans and short-term schedules as well as coordination between different phases in production are two typical issues in manufacturing. The use of mathematical optimization techniques allowed to solve a complex optimization problem with a significant impact on the productivity of a large tire production company. Usable and explainable solutions were obtained from a mathematical model solved by a state-of-the-art mixed-integer linear programming software solver.

Acknowledgements The contributions of Andrea Taverna to this paper are not related to his role at Amazon.

References

1. Degraeve, Z., Schrage, L.: A tire production scheduling system for Bridgestone/firestone off-the-road. Oper. Res. **45**(6), 789–796 (1997)
2. Cancela, H., Piñeyro, P., Velázquez, J.: A MILP formulation for a tire curing scheduling problem. Electron. Notes Discrete Math. **69**, 61–68 (2018)
3. Koch, C., Ouazene, Y., Arbaoui, T., Yalaoui, F., Jaunet, N., De Wulf, A.: A dedicated lot sizing problem in tire industry. IFAC-PapersOnLine **53**(2), 10664–10669 (2020)
4. Koch, C., Arbaoui, T., Ouazene, Y., Yalaoui, F., De Brunier, H., Jaunet, N., De Wulf, A.: A matheuristic approach for solving a simultaneous lot sizing and scheduling problem with client prioritization in tire industry. Comput. Ind. Eng. **165** (2022)

Extended Formulations for the Electric Asymmetric TSP

Alberto Ceselli, Cristina Ondei, and Giovanni Righini

Abstract The Electric Traveling Salesman Problem (ETSP) is a variant of the traditional TSP in which the vehicle is assumed to be equipped with a battery of limited capacity. Therefore, the vehicle needs to be recharged along the route and, for this purpose, a set of recharge stations is given. Each station has potentially different unit cost for energy and multiple visits to the same station are allowed. The aim is to visit all the customers once, minimizing the overall cost of the recharged energy. We consider the asymmetric version of the ETSP (EATSP). We also consider the possibility that traveling along some arcs of the underlying graph implies a negative consumption of energy. This corresponds to traveling along downhill roads with vehicles that can recharge their batteries. We propose an extended formulation for the problem, having one variable for each possible sequence of customers which can be visited with no recharges from one station to another. Furthermore, we modify such a formulation in a less intuitive form that requires fewer variables and constraints. We compare these two approaches theoretically and experimentally.

Keywords TSP · Electric vehicles · Partial recharges · Extended formulation

1 Introduction

In the Traveling Salesman Problem (TSP) [1] a set of customers must be visited by a vehicle. The distance between customer locations is known and the objective is to minimize the total distance traveled by the vehicle. When the graph representing the

A. Ceselli · C. Ondei (✉) · G. Righini
Dipartimento di Informatica, Università degli Studi di Milano, Milano, Italy
e-mail: cristina.ondei@unimi.it

A. Ceselli
e-mail: alberto.ceselli@unimi.it

G. Righini
e-mail: giovanni.righini@unimi.it

M. Di Francesco et al. (eds.), *Operations Research: Closing the Gap Between Research and Practice*, AIRO Springer Series 15, https://doi.org/10.1007/978-3-031-90095-2_24

road network is directed and asymmetric, the problem is known as the Asymmetric TSP (ATSP). Both the TSP and the ATSP are NP-hard.

A recent stream of research concerns electric routing problems [2]. Electric vehicles are equipped with batteries of relatively small capacity that may be insufficient to travel the whole route. Unlike gasoline stations, electric recharge stations may be rare; furthermore, recharging time may be significant compared to traveling time. Hence, recharge operations need to be planned carefully, as an optimized choice may lead to important savings.

An electric variant of the TSP (ETSP), in which time windows are imposed for visiting each customer, has been studied in [3]. The authors propose mathematical programming formulations, which allow to optimize small-sized instances and a heuristic that can tackle instances with up to 200 customers. Some basic properties of the ETSP (without time windows) have been studied by Ceselli and Righini [4]. At the best of our knowledge, the only study on the electric ATSP (EATSP) is the paper by Roberti and Wen [3], that does not consider negative consumption arcs and multiple recharge technologies and assumes time windows associated with the customers. The authors formulate a MILP model based on recharge paths connecting customers through stations, while paths we use in this paper connect stations through customers. Roberti and Wen also assume to split each station into several copies, each corresponding to a distinct visit.

The multi-vehicle version of the ETSP is the Electric Vehicle Routing Problem (EVRP). For the EVRP several algorithms have been developed, both exact and heuristic, especially for the symmetric problem (see for instance [5–7]). In particular, Ceselli et al. [8] presented a branch-and-price algorithm for the EVRP with partial recharges and multiple technologies that relies upon a path-based formulation, where a path is a sequence of arcs between two stations.

In this paper we consider the EATSP with partial recharges, without assuming the presence of time windows or other restrictions that decrease the size of the solution space. A specific feature of this problem is the possible presence of negative consumption arcs, corresponding to downhill roads along which electric vehicles can recharge their batteries. Our purpose is to adapt the path-based EVRP extended formulation of Ceselli et al. [8] to the EATSP.

In Sect. 2, we formally define the EATSP. In Sect. 3 we propose an extended formulation for the problem. In Sect. 4 we propose a modified version of the formulation and compare it to the former one. In Sect. 5 we experimentally compare the proposed formulations. Conclusions are reported in Sect. 6.

2 Problem Description

Let $\mathcal{G} = (\mathcal{N} \cup \mathcal{R}, \mathcal{A})$ be a directed weighted graph, whose set of nodes is the union of a set $\mathcal{N}$ of n customers and a set $\mathcal{R}$ of r recharge stations. The graph is assumed complete and without self-loops, so that $\mathcal{A} = \{(i, j) : i \in \mathcal{N} \cup \mathcal{R}, j \in \mathcal{N} \cup \mathcal{R}, i \neq j\}$. The customers in $\mathcal{N}$ must be visited exactly once, while the stations can be visited to recharge the vehicle when needed. Multiple visits to the same station as well as partial recharges are allowed.

For each station $j \in \mathcal{R}$ we assume a positive recharge cost per unit of energy, γ_j. A weight e_{ij} is associated with each arc $(i, j) \in \mathcal{A}$, representing the energy consumed to travel from node i to node j. Because of the capability of some electric vehicles to recharge when traveling downhill, negative cost arcs are allowed. However, negative cost cycles cannot exist.

The vehicle is equipped with a battery of given capacity B.

A feasible solution is a closed walk complying with the following constraints:

- The tour starts and ends at a given depot node.
- Every customer node is visited once.
- The battery level is always kept between 0 and B. When the vehicle traverses an arc (i, j) with cost e_{ij}, the battery level is correspondingly decreased if $e_{ij} \geq 0$, increased if $e_{ij} < 0$. If, due to some negative cost arcs, the battery level should exceed B, the solution is considered feasible and the redundant amount of energy is lost. If the battery level decreases below 0 the solution is infeasible.

The objective is to minimize the overall cost of recharged energy, as opposed to the traditional TSP, where one wants to minimize the traveled distance. Although time and distance are not explicitly considered in the objective (nor in constraints), the minimization of total costs is likely to provide space-efficient and time-efficient solutions. Actually, even when distance is minimized, it is because it translates into cost.

3 An Extended MILP Model

Following Ceselli et al. [8], we define a path as a sequence of arcs between two recharge stations, visiting some customers in between. An empty path is a single arc directly connecting two different recharge stations.

Let Λ be the set of all feasible paths. We indicate with $\Lambda_{u,v}$ the set of all feasible paths starting from station $u \in \mathcal{R}$ and ending in station $v \in \mathcal{R}$; the two endpoints can coincide, because it is allowed for the vehicle to leave and return to the same station after visiting some customers. We indicate with l the index of the generic path.

Binary coefficients y_i^l take value 1 if and only if customer $i \in \mathcal{N}$ is visited along path $l \in \Lambda_{u,v}$. Binary coefficients x_{ij}^l take value 1 if and only if arc $(i, j) \in \mathcal{A}$ is traversed along path $l \in \Lambda_{u,v}$. Finally, b_j^l represents the battery level at node j along path l, assuming a starting level of charge equal to B.

With this notation, we can now state the formal definition of feasible paths for each ordered pair of stations (u, v).

$$\Lambda_{u,v} = \{(y^l, x^l, b^l) :$$

$$\sum_{j\in\mathcal{N}\cup\mathcal{R}\setminus\{i\}} x^l_{ij} = \sum_{j\in\mathcal{N}\cup\mathcal{R}\setminus\{i\}} x^l_{ji} = y^l_i \quad \forall i \in \mathcal{N} \tag{1}$$

$$\sum_{j\in\mathcal{N}\cup\mathcal{R}\setminus\{u\}} x^l_{uj} = \sum_{j\in\mathcal{N}\cup\mathcal{R}\setminus\{v\}} x^l_{jv} = 1 \tag{2}$$

$$\sum_{j\in\mathcal{N}\cup\mathcal{R}\setminus\{i\}} x^l_{ij} = \sum_{j\in\mathcal{N}\cup\mathcal{R}\setminus\{i\}} x^l_{ji} = 0 \quad \forall i \in \mathcal{R} \setminus \{u, v\} \tag{3}$$

$$b^l_i - e_{ij}x^l_{ij} \geq b^l_j - B(1 - x^l_{ij}) \qquad \forall (i, j) \in \mathcal{A} \tag{4}$$

$$0 \leq y^l_i \leq 1 \qquad \forall i \in \mathcal{N} \tag{5}$$

$$x^l_{ij} \in \{0, 1\} \qquad \forall (i, j) \in \mathcal{A} \tag{6}$$

$$0 \leq b^l_i \leq By^l_i \qquad \forall i \in \mathcal{N} \cup \mathcal{R}\}. \tag{7}$$

To feasibly connect paths to form complete solutions, it is necessary to associate an amount of energy consumption with each of them. However, owing to negative consumption arcs and to the limited battery capacity, it may occur that some energy is wasted along the path. This is a major difference between the symmetric and the asymmetric versions of the Electric TSP.

In general, the amount of wasted energy depends on the starting level of charge. For example, if the first arc after a recharge has a negative cost, leaving the station with full battery would cause a loss of energy, that would not happen with a lower starting level of battery. It is anyway possible that the waste of energy cannot be prevented. An example is shown in Fig. 1.

It follows that the total energy consumption along a path is not necessarily equal to the total energy consumption along its arcs. Indicating by s_l the minimum amount of energy that must be wasted along path l, and by e_l the total energy consumption along path l, the following conservation constraint holds:

$$e_l = \sum_{(i,j)\in\mathcal{A}} e_{ij}x^l_{ij} + s_l. \tag{8}$$

To avoid an unnecessary loss of energy, we need to introduce meaningful bounds to the initial battery charge. For this purpose, we define ϕ^+_l to be the maximum increase of the battery level, with respect to the starting level, and ϕ^-_l to be the maximum decrease of the battery level, with respect to the starting level, as shown in Fig. 2.

These values can be easily computed for each path, as shown in Algorithm 3.1.

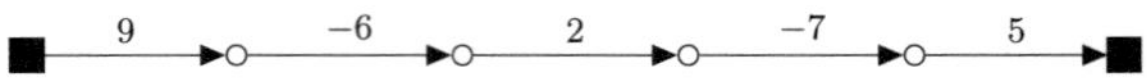

Fig. 1 Example of a path where the vehicle is forced to waste energy. On each arc is displayed the value e, $B = 10$. Regardless of the initial battery level, the vehicle will waste at least one unit of energy when traversing the arc with $e = -7$

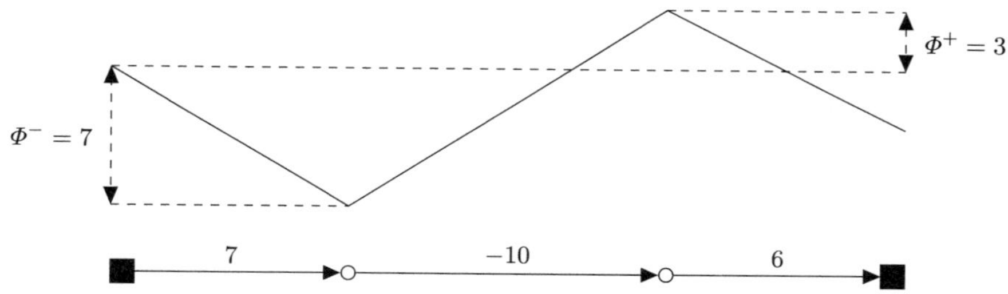

Fig. 2 The values of ϕ^+ and ϕ^- along a path made of three arcs between two recharge stations, assuming $B = 10$. The plot indicates the value of charge: it increases when the vehicle travels downhill along the second arc

---

Algorithm 3.1 An algorithm to compute ϕ^-, ϕ^+ and s along a path L represented as a sequence of arcs, indexed by $k = 1, \ldots, |L|$. The variable z indicates the difference of charge level with respect to the beginning of the path.

```
z ← 0
φ⁺ ← 0
φ⁻ ← 0
s ← 0
for k = 1, ..., |L| do
    (i, j) ← L_k
    z ← z − e_ij
    if z < −φ⁻ then
        φ⁻ ← −z
    if z < φ⁺ − B then
        Infeasible
    if z > φ⁺ then
        if z > B − φ⁻ then
            s ← s + z − (B − φ⁻)
            z ← B − φ⁻
        φ⁺ ← z
```

It is immediate to verify that ϕ^+ and ϕ^- always satisfy the condition $\phi^+ + \phi^- \leq B$. The difference between B and the sum $\phi^+ + \phi^-$ corresponds to the width of the interval of charge level values that are feasible at the beginning of the path. A path can be feasibly traversed if and only if the initial level of charge is not lower than ϕ^-; on the other hand, no additional energy is wasted along a path, besides s, if and only if the initial level of charge is not greater than $B - \phi^+$. These properties allow to formulate the master problem of the extended formulation as shown in the remainder.

3.1 Master Problem

Following the notation used by Ceselli et al. [8], we associate a binary variable θ_l with each feasible path $l \in \Lambda = \bigcup_{u,v} \Lambda_{u,v}$: θ_l takes value 1 if and only if path l is selected to be part of the solution. With each feasible path $l \in \Lambda$, we also associate a continuous variable a_l, representing the battery level at the beginning of path l. We use continuous variables $\delta_j \geq 0$ to indicate the total amount of energy recharged at each station $j \in \mathcal{R}$. Discrete variables ω_j indicate the number of visits to each station $j \in \mathcal{R}$.

For any subset $S \subseteq \mathcal{R}$, Λ_S indicates the set of paths with both endpoints in S; $\mathcal{L}_S^-$ and $\mathcal{L}_S^+$ indicate the set of paths entering and leaving S, respectively. Suffixes e and ne stand for "empty" and "non-empty", respectively: so, Λ^e (see constraints (29)) is the set of empty paths, while Λ_S^{ne} (see constraints (28)) is the set of non-empty paths with both endpoints in S.

An extended MILP model (from now on, MILP1) for the EATSP reads as follows.

$$\min \sum_{j \in \mathcal{R}} \gamma_j \delta_j \tag{9}$$

$$\text{s.t.} \sum_{l \in \mathcal{L}_j^-} \theta_l = \sum_{l \in \mathcal{L}_j^+} \theta_l = \omega_j \qquad \forall j \in \mathcal{R} \tag{10}$$

$$\omega_0 = 1 \tag{11}$$

$$\sum_{l \in \mathcal{L}_S^+} \theta_l \geq \sum_{l' \in \Lambda_S^{ne}} y_i^{l'} \theta_{l'} \qquad \forall S \subseteq \mathcal{R} \setminus \{0\},\ \forall i \in \mathcal{N} \tag{12}$$

$$\sum_{l \in \Lambda} y_i^l \theta_l \geq 1 \qquad \forall i \in \mathcal{N} \tag{13}$$

$$\phi_l^- \theta_l \leq a_l \leq (B - \phi_l^+)\theta_l \qquad \forall l \in \Lambda \tag{14}$$

$$\delta_j = \sum_{l \in \mathcal{L}_j^+} a_l - \sum_{l \in \mathcal{L}_j^-} (a_l - e_l \theta_l) \qquad \forall j \in \mathcal{R} \tag{15}$$

$$\theta_l \in \{0, 1\} \qquad \forall l \in \Lambda^{ne} \tag{16}$$

$$\theta_l \text{ integer} \qquad \forall l \in \Lambda^e \tag{17}$$

$$0 \leq a_l \leq B \qquad \forall l \in \Lambda \tag{18}$$

$$\omega_j \text{ integer} \qquad \forall j \in \mathcal{R} \tag{19}$$

$$\delta_j \geq 0 \qquad \forall j \in \mathcal{R}. \tag{20}$$

Constraints (10) and (11) are degree constraints on the stations and for the depot. Constraints (12) are subtour elimination constraints. Constraints (13) impose that all customer nodes are visited. Constraints (14) set a lower and upper bound to the initial level of battery for each path l. Finally, constraints (15) determine the amount of energy to be recharged in each station $j \in \mathcal{R}$, that is the difference between the energy required at the beginning of the outgoing paths and the residual energy of the incoming paths. Since the possible loss of energy is already accounted in e_l, as shown in (8), the constraints can be formulated as an equality.

Model MILP1 is indeed a relaxation of the EATSP, as shown in the remainder. An exact formulation of the EATSP can be obtained by splitting each station since the beginning into several copies, one for each possible visit, as in [3]. However, our preliminary computational results suggest this pre-processing that enlarges the size of the graph can be avoided. Only in very exceptional cases the optimal solution of our MILP models is not feasible. The way these rare cases can be dealt with is illustrated in [8].

4 An Alternative MILP Model

Model MILP1 presented in Sect. 3 is intuitive, however it requires the addition of variables a_l and constraints (14) for each $l \in \Lambda$, whose number grows exponentially in the number of customers.

We modify model MILP1 to make it simpler. In particular we remove the a_l variables and we reformulate the energy constraints, following Ceselli et al. [8].

Again, we consider binary variables θ_l for each feasible path $l \in \Lambda$, continuous variables δ_j and discrete variables ω_j for each $j \in \mathcal{R}$; variables a_l are not needed.

4.1 Master Problem

A second MILP model (from now on, MILP2) for the EATSP reads as follows.

$$\min \sum_{j \in \mathcal{R}} \gamma_j \delta_j \tag{21}$$

$$\text{s.t.} \sum_{l \in \mathcal{L}_j^-} \theta_l = \sum_{l \in \mathcal{L}_j^+} \theta_l = \omega_j \qquad \forall j \in \mathcal{R} \tag{22}$$

$$\omega_0 = 1 \tag{23}$$

$$\sum_{l \in \mathcal{L}_S^+} \theta_l \geq \sum_{l' \in \Lambda_S^{ne}} y_i^{l'} \theta_{l'} \qquad \forall S \subseteq \mathcal{R} \setminus \{0\},\ \forall i \in \mathcal{N} \tag{24}$$

$$\sum_{l \in \Lambda} y_i^l \theta_l \geq 1 \qquad \forall i \in \mathcal{N} \tag{25}$$

$$\sum_{j \in S} \delta_j \geq \sum_{l \in \Lambda_S} e_l \theta_l + \sum_{l \in \mathcal{L}_S^+} \phi_l^- \theta_l - \sum_{l \in \mathcal{L}_S^-} (B - \phi_l^+ - e_l) \theta_l \qquad \forall S \subseteq \mathcal{R} \tag{26}$$

$$\sum_{j \in S} \delta_j \leq \sum_{l \in \Lambda_S} e_l \theta_l + \sum_{l \in \mathcal{L}_S^+} (B - \phi_l^+) \theta_l - \sum_{l \in \mathcal{L}_S^-} (\phi_l^- - e_l) \theta_l \qquad \forall S \subseteq \mathcal{R} \tag{27}$$

$$\theta_l \in \{0, 1\} \qquad \forall l \in \Lambda^{ne} \qquad (28)$$

$$\theta_l \text{ integer} \qquad \forall l \in \Lambda^{e} \qquad (29)$$

$$\omega_j \text{ integer} \qquad \forall j \in \mathcal{R} \qquad (30)$$

$$\delta_j \geq 0 \qquad \forall j \in \mathcal{R}. \qquad (31)$$

We keep constraints (22)–(25) from model MILP1. Constraints (26) and (27) impose respectively a lower bound and an upper bound to the amount of recharged energy. For each path $l \in \Lambda$, the minimum amount of starting energy that is necessary to traverse l is equal to ϕ_l^-. In general, $e_l \leq \phi_l^-$. On the other hand, $B - \phi_l^+$ is the maximum amount of energy that can be available at the beginning of path l. If more energy is available, the amount exceeding $B - \phi_l^+$ must be wasted. These two bounds allow to compute lower and upper bounds on the amount of energy recharged in any subset of recharge stations. Given a subset of recharge stations $S \subseteq \mathcal{R}$, each constraint in (26) states a lower bound on the total amount of energy recharged in the stations in S. On the supply side,

- the term $\sum_{j \in S} \delta_j$ is the total amount of energy recharged in S;
- the term $\sum_{l \in \mathcal{L}_S^-} (B - \phi_l^+ - e_l)\theta_l$ is the maximum amount of energy at the end of the paths entering S.

On the consumption side,

- the term $\sum_{l \in \Lambda_S} e_l \theta_l$ is the total amount of energy needed to travel within S;
- the term $\sum_{l \in \mathcal{L}_S^+} \phi_l^- \theta_l$ is the minimum amount of energy needed to feasibly traverse the paths outgoing from S.

Constraints (26) state that the minimum total consumption must not exceed the maximum total supply for any S.

Symmetrically, constraints (27) state upper bounds to the amount of recharged energy for each set of recharge stations $S \subseteq \mathcal{R}$. On the supply side,

- the term $\sum_{j \in S} \delta_j$ is the total amount of energy recharged in S, as before;
- the term $\sum_{l \in \mathcal{L}_S^-} (\phi_l^+ - e_l)\theta_l$ is the minimum residual energy at the end of the paths entering S.

On the consumption side,

- the term $\sum_{l \in \Lambda_S} e_l \theta_l$ is the total amount of energy needed to travel within S, as before;
- the term $\sum_{l \in \mathcal{L}_S^+} (B - \phi_l^+)\theta_l$ is the maximum amount of energy at the beginning of the paths outgoing from S.

Constraints (27) state that the minimum total supply must not exceed the maximum total consumption for any S.

4.2 Comparison Between the Two MILP Models

The proposed MILP models represent a relaxation for the problem, meaning they allow for integer solutions, potentially super-optimal, that would be impossible to be actualized. This situation can occur when a station is visited multiple times, due to the aggregation of different recharges into a single variable δ. An example is shown in Fig. 3.

It is worth noticing that the formulations are not equivalent and do not dominate each other.

In particular, constraints (14) and (15) in model MILP1 imply

$$\delta_j \geq \sum_{l \in \mathcal{L}_j^+} \phi_l^- \theta_l - \sum_{l \in \mathcal{L}_j^-} (B - \phi_l^+ - e_l)\theta_l \qquad \forall j \in \mathcal{R} \qquad (32)$$

$$\delta_j \leq \sum_{l \in \mathcal{L}_j^+} (B - \phi_l^+)\theta_l - \sum_{l \in \mathcal{L}_j^-} (\phi_l^- - e_l)\theta_l \qquad \forall j \in \mathcal{R} \qquad (33)$$

These inequalities are included in constraints (26) and (27). However, model MILP2 is then strengthened by adding a pair of constraints for each subset of stations $S \subseteq \mathcal{R}$.

On the other hand, we observe that constraints (26) and (27) in model MILP2 include (32) and (33), imposing a lower and an upper bound on δ_j for each $j \in \mathcal{R}$. Constraints (14) in model MILP1 imply

$$\sum_{l \in \mathcal{L}_j^+} a_l - \sum_{l \in \mathcal{L}_j^-} (a_l - e_l\theta_l) \geq \sum_{l \in \mathcal{L}_j^+} \phi_l^- \theta_l - \sum_{l \in \mathcal{L}_j^-} (B - \phi_l^+ - e_l)\theta_l \qquad \forall j \in \mathcal{R} \qquad (34)$$

$$\sum_{l \in \mathcal{L}_j^+} a_l - \sum_{l \in \mathcal{L}_j^-} (a_l - e_l\theta_l) \leq \sum_{l \in \mathcal{L}_j^+} (B - \phi_l^+)\theta_l - \sum_{l \in \mathcal{L}_j^-} (\phi_l^- - e_l)\theta_l \qquad \forall j \in \mathcal{R} \qquad (35)$$

It follows that δ_j and $\sum_{l \in \mathcal{L}_j^+} a_l - \sum_{l \in \mathcal{L}_j^-} (a_l - e_l\theta_l)$ have the same lower and upper bound for each $j \in \mathcal{R}$. However there is no guarantee that the equality in (15) holds.

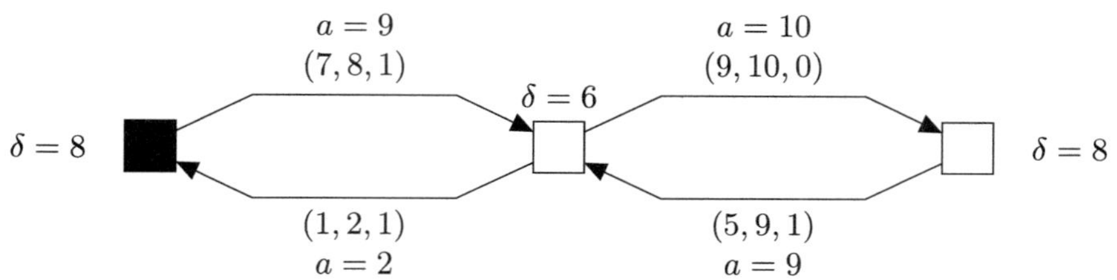

Fig. 3 A solution with multiple visits, which is feasible with respect to both formulations. It is however easy to verify how this solution is impossible to be implemented. On each arc are displayed the values (e, ϕ^-, ϕ^+), $B = 10$

5 Computational Experiments

We performed a preliminary analysis to compare the MILP models presented in Sects. 3 and 4.

In particular, we tested the models on small instances, explicitly enumerating the variables and the constraints at the beginning. This approach would be impossible to use on real size instances. However, it is helpful for our preliminary analysis, to establish which formulation is worth investing on.

We considered instances derived from the Solomon dataset for the VRP. The nodes, customers and recharge stations, are represented as points in a plane, i.e. they are associated with a pair of coordinates (x, y). To create meaningful negative arcs, we randomly generate a third coordinate, namely z, and use the following to determine the cost.

$$e_{ij} = \tilde{e}_{ij} + k(z_j - z_i) \tag{36}$$

where $\tilde{e}_{ij}$ would be the cost of the projected arc on the plane (x, y); k simulates the multiplicative inverse of the friction coefficient. In our experiments we consider $k = 50$.

We implemented the models in C++ and used Gurobi 11.0 to solve them.

We summarize the results in Table 1. We considered instances with 5 customers and up to 4 stations and instances with 10 customers and up to 5 stations. Both algorithms could reach the optimal solution in every instance considered, meaning no infeasible super-optimal solution was found.

Clearly, model MILP2 requires about half the variables than model MILP1. However the execution times are comparable; it seems that model MILP2 is slightly faster to be solved on the instances with 10 customers.

In Table 2 we compare the root node relaxations of the formulations. We see that in general model MILP2 provides better lower bounds. However, on one instance it is model MILP1 that provides the best bound, confirming that the two models do not dominate each other.

Table 1 Comparison between the two MILP models: average results

		MILP1			MILP2		
Customers	N. Inst.	Opt.	N. Var.	Time (s)	Opt.	N. Var.	Time (s)
5	12	12	2443	0.25	12	1226	0.30
10	12	12	85320	23.68	12	42665	22.09

Table 2 Comparison between the root node relaxations. The gap is computed between the lower bound found and the optimal solution

		MILP1		MILP2	
Customers	N. Inst.	Gap (%)	Best bound	Gap (%)	Best bound
5	12	12.06	0	11.57	2
10	12	11.72	1	10.82	9

6 Conclusions

In this paper we investigated two extended formulations for the Electric ATSP with negative consumption arcs. Both represent integer relaxations of the problem, including one variable for each possible feasible path between two recharge stations. Their differences rely on the set of master constraints enforcing consistent recharges, and the set of variables encoding battery levels; the second MILP model requires fewer variables and constraints, under the assumption that the set of station is smaller than the set of customers.

We show by both theoretical arguments and experiments that they do not dominate each other.

Despite being integer relaxations, both formulations could always find the optimal solution on a relevant set of small instances from the literature. From a computational point of view, on the same set of instances, the formulation with fewer variables performs better in terms of execution times and continuous relaxation bounds, thus being a promising starting point for further investigations.

In order to deal with instances of larger size, future work will focus on the development of a column generation algorithm, which dynamically adds path variables to the master problem. Moreover, given the structure of the energy constraints, the second MILP seems more suitable for the development of a column generation algorithm.

References

1. Applegate, D.L., Bixby, R.E., Chvátal, V., Cook, W.J.: The traveling salesman problem: a computational study. In: Princeton Series in Applied Mathematics (2007)
2. Pelletier, S., Jabali, O., Laporte, G.: Goods distribution with electric vehicles: review and research perspectives. Transp. Sci. **50**(1), 3–22 (2016)
3. Roberti, R., Wen, M.: The electric traveling salesman problem with time windows. Transp. Res. **89**, 32–52 (2016)
4. Ceselli, A., Righini, G.: The electric traveling salesman problem: properties and models. Technical Report 2434/789142—University of Milan (2020). http://doi.org/10.13140/RG.2.2.17712.99848
5. Desaulniers, G., Errico, F., Irnich, S., Schneider, M.: Exact algorithms for electric vehicle-routing problems with time windows. Oper. Res. **64**(6), 1388–1405 (2016)

6. Felipe Ortega, A., Ortuno Sanchez, M.T., Righini, G., Tirado Dominguez, G.: A heuristic approach for the green vehicle routing problem with multiple technologies and partial recharges. Transp. Res. **71**, 111–128 (2014)
7. Keskin, M., Catay, B.: A matheuristic method for the electric vehicle routing problem with time windows and fast chargers. Comput. Oper. Res. **100**, 172–188 (2018)
8. Ceselli, Á., Felipe Ortega, M.T., Ortuño Sánchez, G., Righini, G., Domínguez, T.: A branch-and-cut-and-price algorithm for the electric vehicle routing problem with multiple technologies. SN Oper. Res. Forum **2**(8) (2021)

Optimization of Nutrition for Patients in Intensive Care Units

Anita Papetti, Giovanni Righini, and Pier Giorgio Villani

Abstract Nutrition of patients in Intensive Care Units (ICU) is a critical treatment that needs to be optimized according to the specific needs of each patient. This study concerns the problem of optimally mixing commercially available formulas for enteral nutrition to achieve any desired mix of nutrients. The mathematical model has been discussed and improved according to feedback received from physicians in ICUs; it is solved in negligible computing time by a free integer linear programming solver. A software tool has been developed with a user-friendly interface, to allow extending the testing phase to several hospitals; it allows studying the trade-off between different objective functions.

Keywords Integer linear programming · Multi-objective optimization · Personalized medicine

1 Introduction

Nutrition plays a critical role among treatments on intensive care (IC) patients. Several studies have shown how the lack of nutrients (intended as calories, proteins, electrolytes, etc.) tends to worsen the outcome and to lengthen the duration of hospitalization [1, 2]

Each IC patient, according to the pathology for which they are hospitalized, has different needs in nutritional terms [3]; for this reason, different ready-to-use formulas are available on the market. These formulas for enteral nutrition (EN) are designed for

A. Papetti · G. Righini (✉)
Department of Computer Science, University of Milan, Milan, Italy
e-mail: giovanni.righini@unimi.it

A. Papetti
e-mail: anita.papetti@studenti.unimi.it

P. G. Villani
Department of Emergency and Critical Care Medicine, Hospital of Cremona, Cremona, Italy
e-mail: piergiorgio.villani@asst-cremona.it

M. Di Francesco et al. (eds.), *Operations Research: Closing the Gap Between Research and Practice*, AIRO Springer Series 15, https://doi.org/10.1007/978-3-031-90095-2_25

standard classes of patients and they cover a wide range of needs: from nephropathic patients to patients with chronic bronchitis [4]. However, they are not optimized for patients with specific nutritional needs (e.g. polypathological patients or patients with specific deficiencies). Therefore, the need arises to optimize the composition of EN mix of formulas adapting them to specific patients' needs in a personalized fashion [5]; However, hospitals do not have the means to self-produce the needed individual formulas for their patients. On the contrary they must rely on a clever use of the available formulas.

The problem of finding the optimal composition of EN mix of formulas starting from available formulas and complying with several restrictions and objectives was studied in collaboration by an IC doctor and an academic operations research lab. The problem was formulated in mixed-integer linear programming terms in order to allow non-O.R.-expert users computing optimal solutions by a standard MILP solver.

The purpose of this short article is to illustrate and discuss the use of mathematical optimization models for decision-making support in such a real context. It illustrates the mathematical model, that turns out to be a variation of the classical diet problem that requires discrete variables and a multi-objective optimization method. Section 2 illustrates the mathematical model; Sect. 3 presents computational results; Sect. 4 outlines some conclusions and directions for possible future developments.

2 Problem Formulation

Given a set of formulas for EN, each one characterized by a percentage for each nutrient in a given set of interest (proteins, carbohydrates, glucides, lipids, etc.) it is required to compute a mix of some of them that optimally approximates a user-defined target nutrient content, based on specific nutritional needs for each individual patient. To achieve better approximations, one can also use oral solutions, containing only one nutrient each. They are also available in units (such as bottles) of known volume. In addition, the model should allow to assign a relative importance to nutrient target values, by a weight between 0 and 1.

Data. The problem data are the following:

- a set N of nutrients; the nutrients and their units are as follows: energy [kcal], protein [g], fat [g], carbohydrates [g], sodium [mg], potassium [mg], calcium [mg], magnesium [mg], phosphorus [mg], iron [mg];
- the set M of available EN formulas;
- the set C of available oral solutions;
- the set $O \subseteq N$ of target nutrients;
- the maximum number $M_{\max} \in \mathbb{Z}$ of EN formulas to use;
- the maximum number $C_{\max} \in \mathbb{Z}$ of oral solutions to use;
- the amount $a_{ij} \in \Re_+$ of each nutrient $j \in N$ contained in 100 mL of each formula or oral solution $i \in M \cup C$, expressed in the units listed above for each nutrient;

- the volume $b_i \in \Re_+$ of a unit of formula or oral solution $i \in M \cup C$, expressed in milliliters;
- the maximum number $f_i^{\max} \in \mathbb{Z}_+$ of units available for each formula or oral solution $i \in M \cup C$;
- the target value o_j of each nutrient $j \in O$, expressed in the units listed above for each nutrient;
- the priority $0 \leq p_i \leq 1$ of each formula $i \in M$;
- the nutritional importance $0 \leq w_j \leq 1$ of each nutrient $j \in O$.

Variables. The variables of the problem are the following:

- $x_i \in \Re_+$: volume used for each formula or oral solution $i \in M \cup C$, expressed in milliliters;
- $y_i \in \{0, 1\}$: use of each formula or oral solution: $y_i = 1$ if and only if formula or oral solution $i \in M \cup C$ is used in the mix;
- $f_i \in \mathbb{Z}_+$: number of units that are used for each formula or oral solution $i \in M \cup C$;
- $\delta \in \Re_+$: maximum percentage deviation between each nutrient's target and the obtained value;
- $s \in \Re_+$: maximum percentage waste of a formula;
- $d \in \Re_+$: penalty due to the priority of the formulas used;
- $n_j \in \Re_+$: total amount obtained in the mix for each nutrient $j \in N$, expressed in the units listed above for each nutrient;
- $V \in \Re_+$: total volume of the mix, expressed in milliliters.

Constraints. The following constraints are imposed.

- For each formula or oral solution, the number of units that can be used is upper bounded by the corresponding maximum available amount if the formula or oral solution is used.

$$f_i \leq y_i \cdot f_i^{\max} \quad \forall i \in M \cup C.$$

- For each formula or oral solution, the number of units that are used must be enough to provide the requested amount.

$$x_i \leq b_i \cdot f_i \quad \forall i \in M \cup C.$$

- The number of formulas that can be used is upper bounded by a parameter $M_{\max}$.

$$\sum_{i \in M} y_i \leq M_{\max}.$$

- The number of oral solutions that can be used is upper bounded by a parameter $C_{\max}$.

$$\sum_{i \in C} y_i \leq C_{\max}.$$

- The amount of each nutrient in the mix is assigned to variable n_j. As the amounts a_{ij} of nutrients are given for 100 mL of formula or oral solution, they are multiplied by 0.01.

$$n_j = \sum_{i \in M \cup C} a_{ij} \cdot 0.01 \cdot x_i \qquad \forall j \in N.$$

- The volume of the obtained mix, expressed in milliliters, is assigned to variable V.

$$V = \sum_{i \in M \cup C} x_i.$$

Objectives. The main objective is to meet the patient's nutrient requirements as closely as possible. For this purpose the variable δ is defined as the maximum relative deviation between each nutrient's target and the obtained amount. Each deviation is weighted by the nutrient importance:

$$\text{minimize } \delta = \max_{k \in O} \left\{ w_k \cdot \left| \frac{n_k - o_k}{o_k} \right| \right\}.$$

The min-max objective is linearized through the constraints

$$\begin{cases} \delta \geq w_k \cdot \left(\dfrac{n_k}{o_k} - 1 \right) & \forall k \in O \\ \delta \geq w_k \cdot \left(-\dfrac{n_k}{o_k} + 1 \right) & \forall k \in O \end{cases}$$

Two secondary objectives were also considered to minimize waste and to prioritize the use of formulas.

One of them is the minimization of wasted amounts of the formulas used. For this purpose a variable s is defined as the maximum percentage waste of a formula, i.e. the ratio between the residual amount of a formula in the last opened unit divided by the volume of the unit.

$$\text{minimize } s = \max_{i \in M} \left\{ \frac{b_i f_i - x_i}{b_i} \right\}.$$

Another secondary objective aims at prioritizing the use of formulas, on the basis of their availability, cost or expiration date. For this purpose a variable d is defined as a penalty, depending on the priority of formulas used in the mix.

$$\text{minimize } d = \sum_{i \in M} \frac{x_i}{p_i}.$$

Classification. The problem turns out to be a variation of the classical Stigler diet problem [6], but with additional requirements that make the model a discrete optimization one. For an introduction to linear programming and mixed-integer linear programming, the interested reader is referred to textbooks like [7]. Furthermore, the presence of multiple and conflicting objective functions requires a tailored approach, illustrated hereafter.

3 Computational Results

Hardware and software. Glpsol [8] was initially selected as an ILP solver, because it is free and easy to use. The prototype model was written in MathProg, that is a simplified version of the AMPL modelling language [9].

The interface between the nutrients and formulas database and the solver was implemented by a Python script. The output was converted in an interpretable text file through available MathProg commands.

As there was need for a more user-friendly final solution, a second implementation was carried out, enriched by a Graphic User Interface (GUI) written in Python. Also the MILP model was rewritten in Python to ease its integration with the GUI, using the Python-MIP package, a collection of Python tools for the modeling and solution of Mixed-Integer Linear programs (MIPs) [10]. CBC MILP Solver was selected as a solver in this implementation, as it is integrated in the Python-MIP package.

The tests have been carried out on a PC equipped with CPU AMD Ryzen 5 4600H with Radeon Graphics 3.00 GHz, 64 bit, 16GB RAM, Windows 10. Computing times were negligible, as reported in the remainder.

In order to validate the model on realistic problem instances, target nutritional values were calculated as follows. Basal Metabolic Rate (BMR) was estimated from each patient's weight, height, age and gender using Harris-Benedict equations revised by Roza and Shizgal in 1984 [11]. Non-protein energy was estimated to be equal to $1.2 \cdot BMR$ for a sedentary lifestyle [1]. Total energy, protein, fat and carbohydrates were calculated from patient's weight and non-protein energy [1]. Micronutrients values were fixed for each patient as follows: Sodium = 2000 mg, Potassium = 3500 mg, Calcium = 800 mg, Magnesium = 3750 mg, Phosphorus = 800 mg, Iron = 14 mg [12].

The main objective δ and the two secondary objectives s and d were optimized in three different ways, illustrated hereafter.

Hierarchical optimization. In hierarchical optimization we gave precedence to the optimization of the main objective, that is δ, and we subjected the optimization of s and d not to worsen the optimal value of δ. Weights w were set to 1.0 for energy, protein, fat and carbohydrates and to 0.1 for sodium, potassium, calcium, magnesium, phosphorus and iron. Table 1 shows the results obtained with ten instances, fixing $M_{\max} = 5$ and $C_{\max} = 5$, with objective δ.

The observed optimal values δ^* were close to zero, which means that the resulting mix well corresponded to the desidered proportion of nutrients.

We analyzed the dependence of the main objective δ from parameters $M_{\max}$ and $C_{\max}$. The results are shown in Table 2: the optimal value δ^* is very robust to variations of both parameters.

Table 1 Results and computing time for ten instances, when fixing $M_{max} = 5$ and $C_{max} = 5$ and optimizing δ

	Patients				Results					
	Weight (kg)	Height (cm)	Age (years)	Gender	Formulas used	Solutions (s)	δ^*	s	d	Time
1	87	186	38	M	2	2	0.0876	0.79	1777.5	0.10
2	56	164	41	F	3	0	0.0874	0.89	2070.6	0.10
3	73	178	55	M	3	2	0.0869	0.88	1822.8	0.11
4	75	168	64	F	5	2	0.0876	0.91	1684.9	0.22
5	90	184	69	M	2	2	0.0879	0.39	1719.2	0.10
6	66	163	73	F	3	2	0.0875	0.98	1891.5	0.10
7	58	170	77	M	3	0	0.0876	0.88	2008.4	0.11
8	78	172	82	F	4	2	0.0879	0.85	1644.2	0.09
9	65	187	85	M	3	2	0.0871	0.95	1889.3	0.13
10	46	154	94	F	2	2	0.0893	0.46	1131.4	0.09

Table 2 Optimal values of δ for different combinations of M_{max} (rows) and C_{max} (columns). Results have been obtained for a female patient with weight 75 Kg, height 168 cm, age 64 years

M_{max}	C_{max} 0 (%)	1 (%)	2 (%)	3 (%)
0		17.42	14.55	14.42
1	13.68	9.24	8.88	8.88
2	9.53	8.91	8.82	8.82
3	9.35	8.85	8.80	8.80
4	9.08	8.82	8.76	8.76
5	9.08	8.80	8.76	8.76

Finally, we observed that fixing δ to its optimal value, no room was left for any significant change in s and d.

Two-objectives optimization. We carried out a parametric analysis to explore the trade-off between δ and s, disregarding d, and between δ and d, disregarding s. Figures 1 and 2 show examples of the Pareto-optimal frontiers given by the parametric analysis of δ^* for different values of s and d, respectively, for a problem instance with $M_{max} = 5$ and $C_{max} = 5$ (on the left) and $C_{max} = 0$ (on the right).

The final software tool we have developed allows the user visualizing both Pareto-optimal regions obtained by the two parametric analyses.

Weighted optimization. We also defined a single weighted objective function $z = \delta + \alpha s + \beta d$ to be minimized. Results are not reported here, because—as expected—they turned out to be heavily dependent on the chosen values of α and β. However, the software tool allowed the user exploring the Pareto-optimal region in three dimensions to suitably tune α and β.

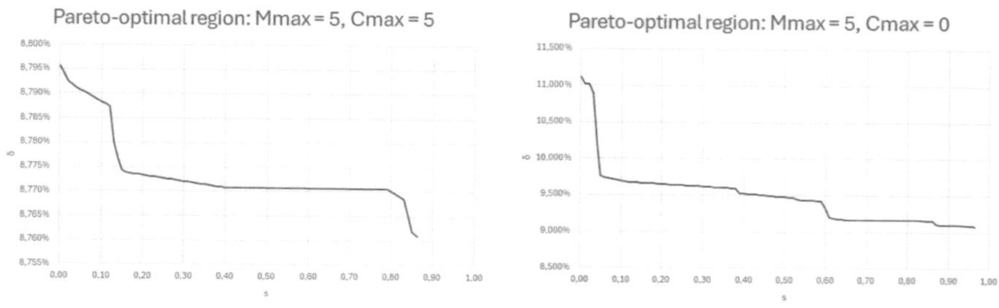

Fig. 1 Pareto-optimal region describing the trade-off between the two objectives δ (vertical axis) and s (horizontal axis)

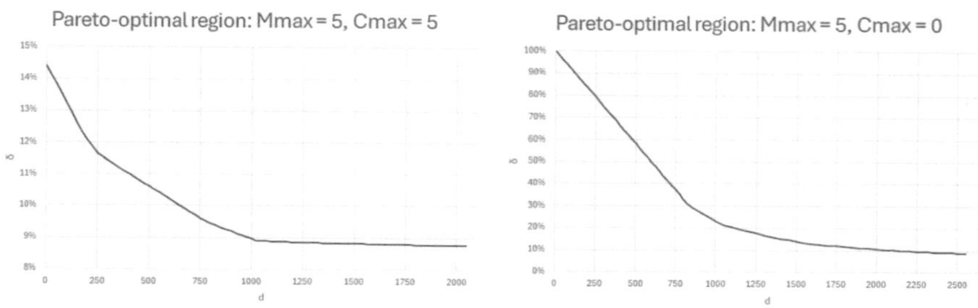

Fig. 2 Pareto-optimal region describing the trade-off between the two objectives δ (vertical axis) and d (horizontal axis)

4 Conclusions

This study originated from a real clinical need to personalize and optimize the nutrition of ICU patients. The variety and complexity of commercial formulas does not allow carrying out such a personalization without the support of a mathematical optimization model helping the clinicians to select and mix the right formulas in the right quantities to obtain the desired proportion of nutrients.

It should be remarked that the ease of use of the modelling language and the solver interface were instrumental for the success of the study, because they allowed O.R. experts and problem experts to efficiently interact through a common language. During the development of the model, the problem expert, equipped with a basic understanding of the modelling language, was able to test the model, evaluate the results and propose improvements. This allowed to fine-tune the model, by comparing several choices of some parameters, such as the maximum number of formulas and oral solutions to be used in the mix, the target value of each nutrient and the importance of each nutrient. The problem expert could also involve several colleagues, whose feedback was used to improve the model. This allowed to initially skip the implementation of an interface to make the model testable by physicians, thus saving time and effort; the mathematical model itself was used as the interface.

Such a virtuous cycle along which knowledge generated by the model is used to improve the model itself is inherent in model-driven approaches that are typical of decision science, where a mathematical model acts as a generator and an accumulator of knowledge.

The final purpose of this study was to produce a decision support system; for this reason, a software tool was eventually built, for a large-scale testing phase at ICUs in hospitals, which is foreseen in the next future. A further planned development is the extension of the database to additional commercial formulas for enteral and parenteral nutrition.

References

1. Singer, P., Blaser, A.R., Berger, M.M., Calder, P.C., Casaer, M., Hiesmayr, M., Mayer, K., Montejo-Gonzalez, J.C., Pichard, C., Preiser, J.-C., Szczeklik, W., van Zanten, A.R.H., Bischoff, S.C.: ESPEN practical and partially revised guideline: clinical nutrition in the intensive care unit. Clin. Nutr. **42**, 1671–1689 (2023)
2. Irving, S.Y., Albert, B.D., Mehta, N.M., Srinivasan, V.: Strategies to optimize enteral feeding and nutrition in the critically ill child: a narrative review. Pediatr. Med. **5** (2021)
3. Yuan, F., Yang, F., Zhang, W., et al.: Optimizing early enteral nutrition in severe stroke (OPENS): protocol for a multicentre randomized controlled trial. BMC Neurol. **19**(24) (2019)
4. ABBOTT Nutrition Catalog. https://nutrition.abbott/uk/adult
5. Kagan, I., Hellerman-Itzhaki, M., Bendavid, I., Statlender, L., Fishman, G., Wischmeyer, P.E., de Waele, E.: Pierre singer controlled enteral nutrition in critical care patients—a randomized clinical trial of a novel management system. Clin. Nutr. **42**(9), P1602–1609 (2023). September
6. George, J.: Stigler, the cost of subsistence. J. Farm Econ. **27**(2), 303–314 (1945)
7. Hillier, F.: Gerald Lieberman. McGraw-Hill, Introduction to Operations Research (2021)
8. Gnu Linear Programming Kit. https://www.gnu.org/software/glpk/
9. Fourer, R., Gay, D.M., Kernighan, B.W.: AMPL A Modelling Language For Mathematical Programming, 2nd edn. Duxbury Thomson (2003)
10. Python-MIP package. https://www.python-mip.com
11. Roza, A.M., Shizgal, H.M.: The Harris Benedict equation reevaluated: resting energy requirements and the body cell mass. Am. J. Clin. Nutr. **40**(1), 168–82 (1984)
12. Società Italiana Nutrizione Artificiale e metabolismo. https://www.sinpe.org/linee-guida-sinpe-espen.html

A Branch & Bound Algorithm for the Rainbow Spanning Forest Problem

Carmine Cerrone, Raffaele Dragone, and Anna Sciomachen

Abstract In this paper, we present a Branch & Bound algorithm for solving an NP-hard combinatorial optimisation problem denoted as Rainbow Spanning Forest Problem (RSFP). The RSFP consists of finding a spanning forest of an undirected, edge-coloured graph G with the minimum number of rainbow components, where a rainbow component of G is a subgraph of G that has all the edges with different colours. The proposed method is capable of determining the optimal solution within 0.4 s for small instances, thus proving to be computationally competitive with respect to the methods used by the most recent optimisation libraries and commercial SW environments available.

Keywords Branch & Bound · Rainbow Spanning Forest · Edge-coloured graph

1 Introduction

The rapid advancement of computer technology and the consequent development of powerful commercially available libraries and software environments make it possible to solve complex problems, especially combinatorial optimisation ones, for which, until very few years ago it was not possible to find a solution, except for instances with a limited number of decision variables. At the same time, however, the current considerable computational capacity has increased the challenge of researchers to develop new methods, either exact or heuristic, or variants of those already proposed in the literature, for which to compare the performance in terms of computational time and goodness of the solution.

C. Cerrone (✉) · R. Dragone · A. Sciomachen
Department of Economics and Business Studies, University of Genoa, Genoa, Italy
e-mail: carmine.cerrone@unige.it

R. Dragone
e-mail: raffaele.dragone@edu.unige.it

A. Sciomachen
e-mail: sciomach@economia.unige.it

M. Di Francesco et al. (eds.), *Operations Research: Closing the Gap Between Research and Practice*, AIRO Springer Series 15, https://doi.org/10.1007/978-3-031-90095-2_26

In this direction, in our work, we present a new exact Branch & Bound method for facing the Rainbow Spanning Forest Problem (RSFP), which was first proposed in the literature by Li and Zhang [1]. The RSFP consists of finding a spanning forest of an undirected, edge-coloured graph G with the minimum number of rainbow components, where a *rainbow* component of G is a subgraph of G that has all the edges with different colours. In [1], it is shown that RSFP is NP-Hard even for edge-coloured graphs with two colours. Carrabs et al. [2] show that the RSFP is NP-Hard on general graphs and trees. They also show that the problem can be solved in polynomial time if the optimal solution value is equal to 1 or if the graph is a path or a cycle. Different methods to deal with RSFP have been proposed in the literature in the last decade. In particular, an integer linear programming model for the RSFP was presented by Carrabs et al. [3], followed by Moreno et al. [4], who introduced a new and more efficient formulation. Ghoshal and Sundar [5] introduced two heuristics to deal with RSFP and, in a later work [6], proposed a new approach using a steady-state grouping genetic algorithm (SSGGA). In addition, Ferone et al. [7] have presented a variant of the RSFP problem denoted as the Rainbow Steiner Tree Problem (RSTP), together with a Lagrangian relaxation approach and a multi-start algorithm to solve it.

In this paper, we introduce a Branch & Bound (B&B) algorithm capable of identifying the optimal solution to the Rainbow Spanning Forest Problem. This algorithm uses various problem characteristics to minimise the number of nodes visited. An innovative aspect of the Branch & Bound method proposed here is that the reference mathematical programming model allows us to deal with graphs in which it is possible to have more than one arc between a pair of nodes i and j. Specifically, in the case of the RSFP, multiple arcs between a pair of nodes, if any, will have different colours. This extension was suggested by an application context of the problem, still under development, related to the need to determine a fleet of ships of different nationalities in the context of international cooperation for patrolling areas at risk of terrorist attacks. This application will generalise the case addressed in [8], where a mathematical programming model is proposed to favour international cooperation in a defence mission against piracy attacks on merchant ships in the Gulf of Aden. Note that in this case, each nation corresponds to a colour of the arc of the graph representing a route travelled in a risky sea area.

The remainder of this paper is organised as follows. Section 2 formally defines the problem and introduces a new application scenario. Section 3 describes the designed B&B algorithm. Section 4 presents the computational results obtained on benchmark instances from the literature. Finally, Sect. 5 offers our observations on this work and future developments of this technique.

2 Problem Definition

Consider $G = (V, E, L)$ as a connected undirected graph where V represents the set of n vertices, E the set of m edges, and L is a collection of l colours (also called labels). Further, define $l : E \rightarrow L$ as a function that maps each edge $e \in E$ to a

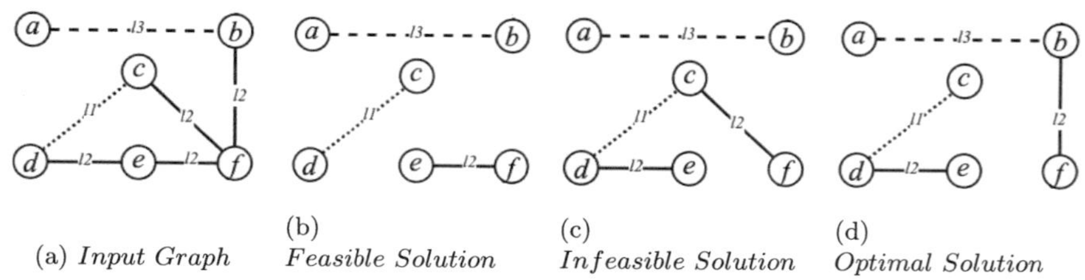

Fig. 1 **a** The input edge coloured graph. **b** A rainbow feasible spanning forest composed by three rainbow trees. **c** An infeasible rainbow spanning forest. **d** The optimal rainbow spanning forest composed by two trees

specific colour $l(e)$ within L. A rainbow spanning forest in G is defined as a spanning forest $F = (V_F, E_F, L_F)$, with $V_F = V$, $E_F \subseteq E$, and $L_F \subseteq L$, ensuring that each component is rainbow. In this context, a forest component is a connected, cycle-free graph. Thus, a rainbow component of F is a tree $T = (V_T, E_T, L_T)$, where $V_T \subseteq V$ and $E_T \subseteq E_F$, in which all edges have different colours, i.e. $|L_T| = |l(E_T)| = |E_T|$. If F comprises $\bar{c}$ rainbow components $T_1, \ldots, T_{\bar{c}}$, then it follows that $V_F = V = \bigcup_{i=1}^{\bar{c}} V_{T_i}$, $E_F = \bigcup_{i=1}^{\bar{c}} E_{T_i}$, and $L_F = \bigcup_{i=1}^{\bar{c}} L_{T_i}$. The goal of the Rainbow Spanning Forest Problem ($RSFP$) is to identify a rainbow spanning forest that minimises the number of rainbow trees.

Figure 1 provides a graphical representation of RSFP. Figure 1a shows an instance of the RSFP with a connected, undirected, edge-coloured graph with 6 vertices, 6 arcs, and 3 edge colours. Figure 1b shows a feasible solution for the RSFP with an objective function value of 3. In this solution, 3 rainbow components are identified. Figure 1c shows an infeasible solution for RSFP, as there is a component with more than one occurrence of the colour l2. Finally, Fig. 1d shows the optimal solution for RSFP. In this solution, 2 rainbow components were identified, with an objective function value of 2.

3 Branch & Bound

Figure 2 illustrates the flowchart related to the implemented B&B algorithm. The fundamental concept of this algorithm is to construct a B&B tree in which the choice associated with each node determines whether to include or exclude an edge in the solution. Specifically, for each edge in the input graph, a node in the B&B tree is created with two children: the first child explores solutions where the edge is included, while the second child investigates solutions excluding the edge. This procedure is recursively depicted. The input consists of a list containing the edges for which a decision has yet to be made and the current solution. Naturally, the graph G is also an input, but like other parameters, it is omitted for brevity's sake. The main steps of the B&B algorithm depicted in Fig. 2 are briefly explained as follows. Step A1 marks the initiation of the procedure. Step A2 checks if there are still edges on which to branch;

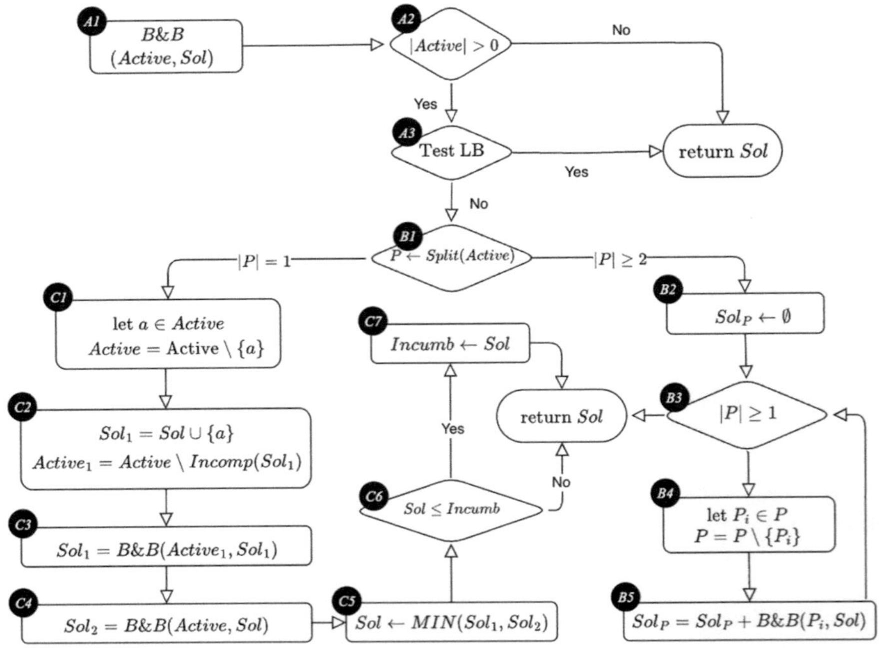

Fig. 2 Branch & Bound Flowchart

if not, it returns the current solution. Step A3 performs the lower bound check. This process uses the edges selected in *Sol* and those still available in *Active*. The details of the technique used for calculating the lower bound are presented in Sect. 3.1.

Step B1 plays a crucial role in this B&B, aiming to ascertain whether it is feasible to partition the input into isolated components. The function *Split*, considering both the edges used in *Sol* and those active in *Active*, examines if the residual graph induced by these edges forms isolated connected components. If at least two of these components exist, Steps B2–B5 are executed to solve the subproblems separately.

Steps C1–C7 are associated with the branching phase. Step C1 is tasked with selecting the active edge on which to branch. In Step C2, the solution Sol_1 is created by adding the selected edge to *Sol*. Then, $Active_1$ is formed by removing from the active edges not only the selected edge but also all other edges that would create subcycles in solution Sol_1, as well as edges that are no longer selectable due to repetitions of colours in Sol_1. Steps C3 and C4 are executed recursively on the two child nodes of the B&B. Step C5 is responsible for choosing the best solution produced by the two children. Finally, Steps C6 and C7 check for updates to the incumbent before returning the solution.

To elucidate the operation of the proposed algorithm, Fig. 3b illustrates the B&B tree obtained by solving the instance shown in Fig. 3a. In this execution of the B&B, the lower bound is not considered to better represent the scheme with which solutions are explored. The root of the B&B tree is represented by node N_1, which generates child N_2 by including the edge (a, b) in the solution, and generates child N_4 by

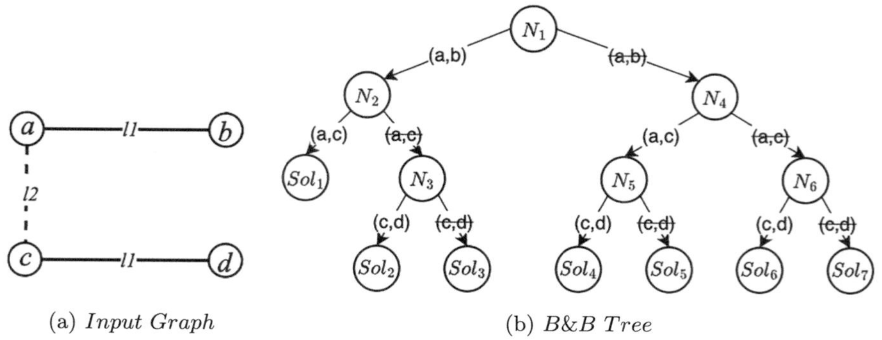

Fig. 3 Graphical representation of the Branch & Bound tree

excluding the edge (a, b) from the solution. Subsequently, node N_2 is executed. For node N_2, the choice becomes whether to include or exclude the edge (a, c) in the solution. By including the edge (a, c), $Sol_1 = \{(a, b), (a, c)\}$ is generated as the first solution. Sol_1 is generated because, after the addition of the edge (a, c) to the solution, the edge (c, d) is excluded from the set of active edges due to the $Incomp$ function executed in Step $C2$ of the B&B. The edge (c, d) is deemed incompatible because it is connected to the component a–b–c, which already contains an edge with the same label l_1 (solid black). At this point, node N_3 is reached, for which edge (a, c) is excluded from the solution. Node N_3 leads to the creation of the solution Sol_2, in which the edge (c, d) is included, and Sol_3, in which the edge (c, d) is not included. The next node to be processed is node N_4. All internal nodes in the tree are sequentially numbered from N_1 to N_6 to indicate the order in which the algorithm processes them. The same rule has also been used for the naming of the seven solutions from Sol_1 to Sol_7. Applying the lower bound using the calculation illustrated in Sect. 3.1 would yield $lb = 2$, leading to pruning of nodes N_3 and N_4.

This example clearly illustrates that, in the absence of bounds, the nodes explored by the proposed procedure would amount to $2^{|E|}$. For example, considering the 39 edges of the smallest graph used to test the proposed B&B, it became necessary to explore up to approximately 550 billion nodes. Centring the branching technique on whether to include an arc in the solution also enables the algorithm to solve the problem on multigraphs, where multiple arcs exist between pairs of nodes. In these multigraph scenarios, when branching on an edge (a, b), in the up-branch (i.e., when (a, b) is selected), all other edges $(a, b)'$ between the same pair of vertices cannot be selected and are thus fixed to 0.

3.1 Lower Bound

The number of nodes in the B&B tree to be visited is, of course, exponential with respect to the size of the problem. However, due to a robust lower bound (lb), it is possible to exclude from the search tree portions where it is definitely impossible to improve the incumbent solution. For a minimisation problem, obtaining a high lb is crucial; however, the calculation of lb requires additional computational time to analyse each node visited by the B&B. If lb is not sufficient to avoid exploring a branch of the tree, the time required for its calculation will negatively affect the overall performance of the algorithm. In this section, we present the lower bound used in this work. Specifically, we have chosen to use three different lbs, calculated in sequence. The first is very quick to compute, the second is also fast, and the third is more computationally demanding, but more effective. The first lower bound is always calculated at every iteration, while the second and third ones are only computed if the previous one is ineffective.

The first lower bound lb_1 considers that, for each edge added to the solution, the number of connected components decreases exactly by one unit $lb_1 = \max(1, |V| - |S| - |Active|)$. This bound assumes that, in the best-case scenario, all still active edges will be included in the solution. The advantage of such a simple bound is due to the computation speed.

The second lower bound takes into account the colours present in the set of active edges and the connected components that have already been created by the solution Sol we are building. Specifically, if the edges chosen in Sol create k connected components and if the distinct colours of the edges present in $Active$ are c, then $lb_2 = \lceil \frac{k}{c+1} \rceil$. This occurs because, to create heterochromatic components, it is possible to assemble the components into groups that consist of at most c edges and, consequently, $c + 1$ nodes. The example in Fig. 4 shows a graph with its optimal solution. For the graph in Fig. 4a, we will have $lb_1 = \max(1, |V| - |S| - |Active|) = \max(1, 10 - 13) = 1$ while $lb_2 = \lceil \frac{k}{c+1} \rceil = \lceil \frac{10}{5+1} \rceil = 2$.

The third lower bound considers the occurrences of colours present in the set of active edges. These colours are grouped into heterochromatic sets $E_1, E_2, \ldots$ of decreasing cardinality such that $E_1 \supset E_2 \supset \ldots$. By inserting an arc for each colour in E_1, it is possible to achieve a maximum reduction in the number of components from k to $k - |E_1|$. Subsequently, an arc is inserted for each colour in E_2, achieving a maximum reduction in the number of components from $k - |E_1|$ to $k - (|E_1| + |E_2|)$. For each set of colours E_i used to reduce the number of components, the lower bound can be increased by one unit because each set E_i contains all the colours present in the preceding sets E_{i-1}. The number of heterochromatic sets i needed to group all the k components of Sol so that $k - (|E_1| + |E_2| + \cdots + |E_i|) >= i$ gives us the lower bound i. Using the graph in Fig. 4a again, we obtain $|l_1| = 9$, $|l_2| = |l_3| = |l_4| = |l_5| = 1$. Consequently, we will have a heterochromatic set $A =$

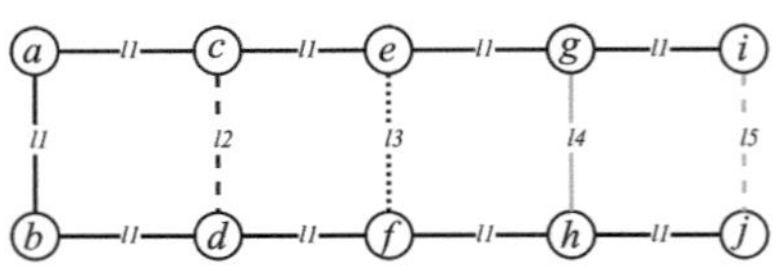

(a) *Input Graph*

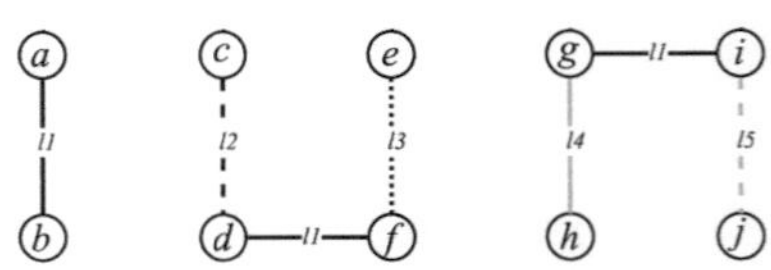

(b) *Optimal Solution*

Fig. 4 **a** Example edge coloured graph. **b** Optimal rainbow spanning forest

$\{l_1, l_2, l_3, l_4, l_5\}$ and then other eight sets composed only of the colour l_1. Therefore, to calculate lb_3, we use the set A, which reduces k from 10 to 5. Subsequently, we use in sequence two sets composed only of the label l_1 to reduce k to 3. Having now $k = 3$ and having used three sets, the procedure will end, returning $lb_3 = 3$.

4 Computational Experiments

In this section, we present the computational results produced by the Branch & Bound approach discussed in this work. The benchmark instances used were selected from those available in the literature and proposed by Carrabs et al. [3]. Among these test instances, we selected all those with a maximum of 30 nodes. The tests were carried out on an Apple Silicon M2 8-Core 3500 MHz CPU. The code was implemented in Java and executed in a single-thread mode. For all tests, a time limit of $1,000$ s was used. Whenever the algorithm required more computational time, the incumbent solution identified within this time limit was returned. The test instances are grouped into homogeneous scenarios based on the number of nodes n, the number of edges m, and the number of labels l. Each scenario contains exactly 5 instances. Table 1 shows the computational results and provides crucial information on the

Table 1 *B&B* Computational tests. Time in seconds

				Branch & Bound			
ID	n	m	l	Sol.	Time	#Nodes	#Split
1	20	39	3	6.6	0.01	10,596	139
2			6	4.2	0.02	27,697	391
3			11	2.2	0.04	117,188	554
4	20	58	3	5.6	0.07	223,125	3,137
5			6	3.2	0.38	1,364,752	13,049
6			12	2	0.01	12,916	52
7	20	77	4	4	0.29	717,461	20,437
8			7	3	0.01	32,758	113
9			13	2	0.01	7,204	61
10	30	74	4	8.4	7.30	16,597,208	288,589
11			7	5	60.10	147,868,252	1,299,359
12			13	3	15.47	47,729,646	149,954
13	30	117	4	6.8$^{(1)}$	566.00	611,037,383	11,344,981
14			7	4.2$^{(1)}$	275.88	593,021,441	5,935,315
15			14	2.8$^{(1)}$	202.73	795,038,199	323,249
16	30	161	4	6.6$^{(3)}$	923.35	524,380,115	15,552,338
17			8	4.4$^{(2)}$	406.52	145,077,502	108,021
18			15	2.2$^{(1)}$	201.08	48,538,951	7,503

execution time of the algorithm and the number of nodes visited in the B&B tree. The table presents data grouped into 18 scenarios, with the ID column indicating the scenario index. The data for each row are calculated as the values of the 5 instances within the scenario. The column *Sol* represents the average value of the objective function. Furthermore, if present, the superscript enclosed in parentheses indicates the number of instances for which the time limit was reached. The column $Time$ reports the average execution time in seconds. The column $\#Nodes$ represents the average number of nodes visited by the B&B algorithm. Finally, the $\#Split$ column indicates how many times the B&B's split function caused the problem to be divided into subproblems.

It should be noted that, for instances with 20 nodes, the computational times are extremely low, and the time limit is never reached. For instances with 30 nodes, the time limit is reached when considering the denser instances. This obviously occurs because as the number of edges increases, so does the average height of the B&B tree. Analysing the explored nodes ($\#Nodes$), it is easy to realise the exponential growth of the search tree. For instances with 30 nodes and 117 edges, it reaches almost 800 million nodes. As the number of edges increases to 161, the number of visited nodes appears to decrease almost. This is due to the higher computational complexity of each individual node, especially for the calculation of the lower bound. This leads to an average time per node of the B&B tree ranging from 0.38 s per million nodes to 1.7 s per million nodes. Finally, the column $\#Split$ shows that, for the larger instances, the problem is subdivided into sub-problems up to 15 million times.

Before concluding the commentary on the computational results, it is necessary to clarify that the B&B presented in this article should be compared with the mathematical models proposed in the literature [3, 4], to which we will refer as ILP and $ILPM$, respectively. From this comparison, it is evident that for small instances containing 20 nodes, the B&B is systematically comparable or faster than the mathematical models, with an average computational time in all 20 node instances of 0.09 s compared to 2.27 s for IMP and 0.37 s for $ILPM$, respectively. In this scenario, the B&B is also faster than the $ILPB$ model proposed in [4], which takes 0.14 s. However, the $ILPB$ model was excluded from the comparison because it benefits from an excellent heuristic. Considering that the same heuristic could also benefit the B&B, it was decided to implement this approach in a more elaborate future version of the B&B. Regarding instances with 30 nodes, while remaining comparable with ILP, for most instances, the computational times required by $ILPM$ are lower than those needed by the B&B. This gap becomes even more evident when considering instances with more than 30 nodes. These instances have not been reported because the B&B rarely manages to finish within its time limit. However, in these instances, the mathematical models often do not finish either, even using time limits of $10,800$ s. Comparison with other exact approaches proposed in the literature has suggested numerous insights for improving the B&B. These insights are an integral part of the conclusions of this work.

5 Conclusions

In this paper, we presented a new B&B algorithm for solving an NP-hard combinatorial optimisation problem denoted as Rainbow Spanning Forest Problem. The experimental results we have reported show that for problem instances of up to 20 nodes, the proposed method is able to determine the optimal solution within 0.4 s, thus proving to be computationally more efficient than the methods used by the most recent optimisation libraries and commercial SW environments available. To highlight this positive outcome, it is worth mentioning that this good computational time was achieved by taking into account instances with different ratios between the number of arcs and the number of colours. However, as the number of nodes increases, although the computational time required for the proposed B&B algorithm to find the optimal solution remains quite good, this performance is inferior to that which can be obtained using one of the above-mentioned SW. Additionally, note that only in a few instances the time limit of 1,000 s set by us has been reached. In any case, we aim to further improve the performance of the algorithm, especially in the case of large-size instances with a relatively low number of colours on the arcs. Drawing inspiration from the literature, particularly from the work of Moreno et al. [4], we find it intriguing to integrate a robust heuristic algorithm into our B&B framework. As another further research direction, we are working on the definition of the graph representing possible routes of merchant ships to define international cooperation for patrolling areas at risk of terrorist attacks. To accurately model this application scenario, the proposed B&B can solve instances represented by multigraphs in which multiple edges of different colours can exist between pairs of nodes.

References

1. Li, X., Zhang, X.: On the minimum monochromatic or multicolored subgraph partition problems. Theor. Comput. Sci. **385**(1–3), 1–10 (2007)
2. Carrabs, F., Cerrone, C., Cerulli, R., Silvestri, S.: On the complexity of rainbow spanning forest problem. Optim. Lett. **12**, 443–454 (2018)
3. Carrabs, F., Cerrone, C., Cerulli, R., Silvestri, S.: The rainbow spanning forest problem. Soft Comput. **22**, 2765–2776 (2018)
4. Moreno, J., Martins, S., Frota, Y.: A new approach for the rainbow spanning forest problem. Soft Comput. **24**, 3771–3780 (2020)
5. Ghoshal, S., Sundar, S.: Two heuristics for the rainbow spanning forest problem. Eur. J. Oper. Res. **285**(3), 853–864 (2020)
6. Ghoshal, S., Sundar, S.: A steady-state grouping genetic algorithm for the rainbow spanning forest problem. SN Comput. Sci. **4**(4), 321 (2023)
7. Ferone, D., Festa, P., Guerriero, F.: The rainbow Steiner tree problem. Comput. Oper. Res. **139**, 105621 (2022)
8. Sciomachen, A., Brignardello, M.: A mathematical programming model to promote international cooperation in defence missions against pirate attacks on merchant ship routes: the gulf of Aden case study. Int. J. Transp. Econ. 237–254 (2020)

Application of Artificial Intelligence Techniques in Battery Energy Storage Systems Optimization

Behzad Pirouz and Francesca Guerriero

Abstract This study aims to optimize battery energy storage systems by using artificial intelligence techniques to predict energy consumption and production from renewable sources. We propose a strategy for identifying appropriate intervals for battery storage. Then, we employed two artificial intelligence techniques, namely Extreme Gradient Boosting (XGBoost) and Long Short-Term Memory (LSTM), to make predictions on the data set. The XGBoost technique demonstrated better performance. In addition, by identifying the best and worst days for predicting energy consumption and energy generation from renewable sources, the risk associated with battery storage can be mitigated.

Keywords Optimization · AI techniques · Battery energy storage systems

1 Introduction

Artificial Intelligence (AI) techniques have been widely used for data prediction, especially data related to energy consumption and production (see [1, 2]). These techniques can gradually increase accuracy and reliability by continuously integrating feedback (see [3, 4]).

Despite the significant opportunities that renewable energies provide, they pose unique challenges for forecasting due to their variable nature and intensity (see [5, 6]). Due to this uncertainty in the output power of these energy systems, a specific storage capacity is required in the power system to ensure the reliability of the service (see [7, 8]).

B. Pirouz (✉) · F. Guerriero
Department of Mechanical, Energy and Management Engineering, University of Calabria, Rende, Italy
e-mail: behzad.pirouz@unical.it

F. Guerriero
e-mail: francesca.guerriero@unical.it

M. Di Francesco et al. (eds.), *Operations Research: Closing the Gap Between Research and Practice*, AIRO Springer Series 15, https://doi.org/10.1007/978-3-031-90095-2_27

AI techniques increase the accuracy of energy production predictions from renewable sources by processing and interpreting large amounts of data, including weather patterns (see [9, 10]).

Using AI to analyze and predict data enables battery energy storage systems (BESSs) management operators to comprehensively understand various factors, ranging from the resting state of charge of the BESS to the depth of its discharge, and increases the lifespan of BESS (see [11–13]).

Various AI techniques have been proposed for prediction, among them the most widely used approaches, especially for time series data, are the Extreme Gradient Boosting (XGBoost) and the Long Short-Term Memory (LSTM) (see [14, 15]).

Considering the importance of optimal storage in BESSs, the main goal of this research is to identify suitable time intervals for battery energy storage charging, by considering the production from renewable sources and the energy consumption. More specifically, we consider the specific situation, where photovoltaic panels are used and the aim is to identify the time slots in which the energy production is maximum and the energy consumption in the building is decreasing. Since the energy production from photovoltaic panels and the energy consumption are variable, an accurate prediction of their values is mandatory.

Data related to energy consumption and energy production from photovoltaic panels are available as time series. Among the various AI prediction techniques, XGBoost and LSTM are widely used in publications for time series data, so we use these techniques for prediction in our data set and compare the results. The comparison results showed that the performance of the XGBoost technique was better in prediction. On the other hand, by identifying the worst and best days of the model predictions in data sets, we try to reduce the risk of planning the charging of BESS.

The rest of the paper is organized as follows: Sect. 2 is devoted to materials and methods, performance metrics and selecting AI prediction techniques are introduced in Sect. 3, and Sect. 4 is devoted to computational experiments and results. The conclusion is discussed in Sect. 5.

2 Materials and Methods

Accurate prediction of energy consumption and photovoltaic (PV) energy production is crucial for devising an optimal plan for energy storage in a BESS. We considered the scenario in Fig. 1, where a building received energy from the power grid and PV system. It is assumed that the building is equipped with BESS for energy storage.

AI techniques can potentially revolutionize building energy storage management by integrating intelligent energy solutions that optimize battery performance. A critical aspect in this domain is the precise prediction of energy consumption and production from renewable sources like solar and wind, which can be influenced by environmental conditions.

To reduce energy consumption from the power grid, we assume that a portion of the energy generated by PV systems during the day is stored in the BESS. This

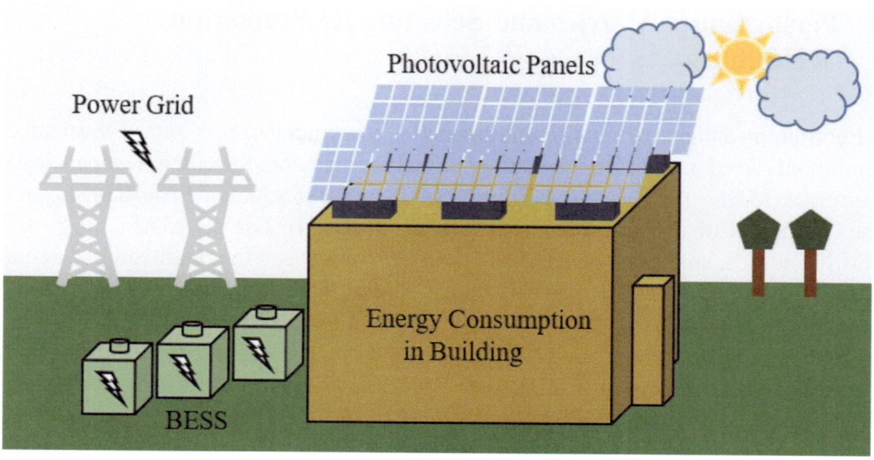

Fig. 1 The building receives energy from the power grid and PV systems and accesses the BESS for energy storage

Fig. 2 Energy consumption and energy generation from PV system in kWh and proposed appropriate intervals for BESS charging

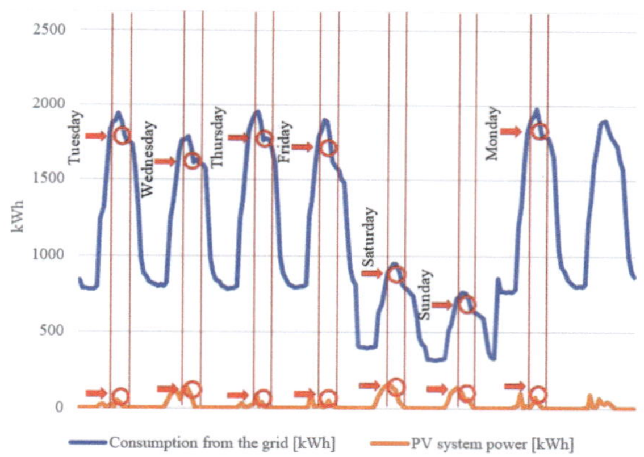

stored energy is then utilized to meet part of the building's energy consumption needs during the night.

In Fig. 2, data related to energy consumption from the power grid and energy production from PV systems in kWh during a week are displayed. In this figure, seven appropriate time intervals that can be considered for charging BESS each day are shown. These intervals correspond to periods when PV system production peaks and building energy consumption decreases.

We emphasize that determining the appropriate intervals for BESS charging and discharging depends on various parameters, while our primary goal in this study is to predict the energy consumption and production in the building as accurately as possible to determine the appropriate intervals as optimally as possible.

3 Performance Metrics and Selecting AI Prediction Techniques

After training and preparing a prediction model, it is crucial to assess its performance. Commonly used metrics for evaluating AI prediction techniques include mean absolute error (MAE), mean squared error (MSE), root mean squared error (RMSE), and the coefficient of determination (R-squared) (see [15]). For the MAE, MSE, and RMSE metrics, smaller values indicate a closer fit between the predicted and actual data. Conversely, for the R-squared metric, higher values signify a more accurate alignment between the prediction model and the actual data.

Various AI techniques have been proposed for prediction purposes. Among them, we cite Decision Tree, Artificial Neural Networks, Support Vector Regressor, Random Forest, Logistic Regression (see [16, 18]). In several scientific contributions, extensive comparisons among different prediction methods on various data sets have conducted. It has been consistently demonstrated that XGB and LSTM exhibit superior performance, especially when dealing with time series data sets (see [17]). Since the data set considered in this article is a time series type, we have used these two methods to make predictions (see [16]).

XGBoost represents an efficient and scalable iteration of the gradient boosting machine algorithm, demonstrating effectiveness in various machine learning applications. This method is an ensemble method, effectively handles classification and regression tasks and it is particularly suited to manage data with nonlinear features. The XGBoost methodology involves developing decision trees sequentially. Each new tree is constructed to fit the residual errors predicted by the previous trees (see [14]).

LSTM is a type of deep learning model, specifically a variant of Recurrent Neural Networks (RNN), designed to allow information persistence. It effectively addresses the vanishing gradient problem. LSTM is renowned for its ability to predict time series data (see [14]).

4 Computational Experiment and Results

In our considered case study, two data sets are available. The first set contains the energy production data from the PV panels, which are reported hourly. A total of 200 m2 of PV panels are available for energy production. The output power is reported in kilo Watts (kW) from the PV system, which reflects its performance. We have considered data related to energy production from PV panels for March 2016. Figure 3 illustrates these data presented as a time series for the mentioned month.

The second data set includes information on energy consumption, also presented as a time series for March 2016. Figure 4 visually represents the reported data regarding energy consumption during the specified month.

As depicted in Fig. 3, the energy production of the PV system exhibits noticeable variations on different days, likely attributed to environmental factors such as weather conditions. In contrast, the data shown in Fig. 4 highlight the weekly fluctuations in

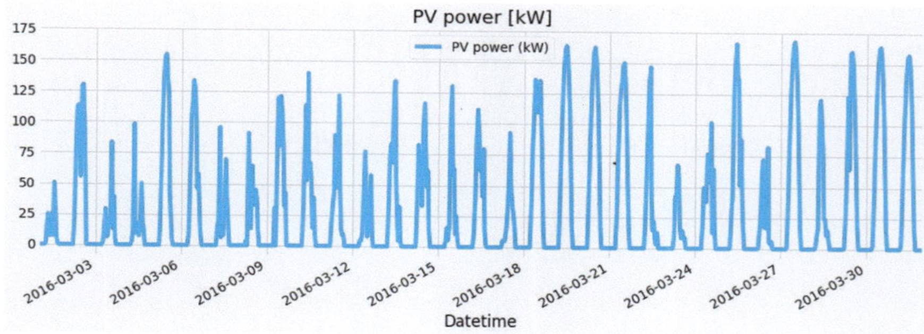

Fig. 3 Energy production reported data for March 2016

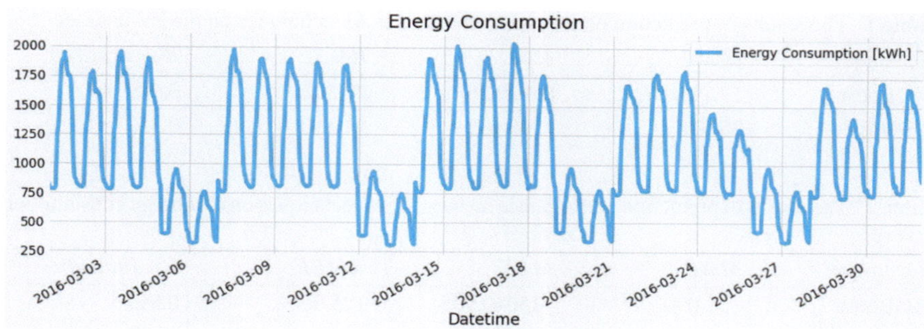

Fig. 4 Energy consumption reported data for March 2016

energy consumption. Especially since the building under study is an office building, energy consumption decreases significantly during the weekend holidays.

We assigned the first 75% of each data sets for training and the remaining 25% for testing. Figures 5 and 6 show the division between training and test data sets in PV and energy consumption.

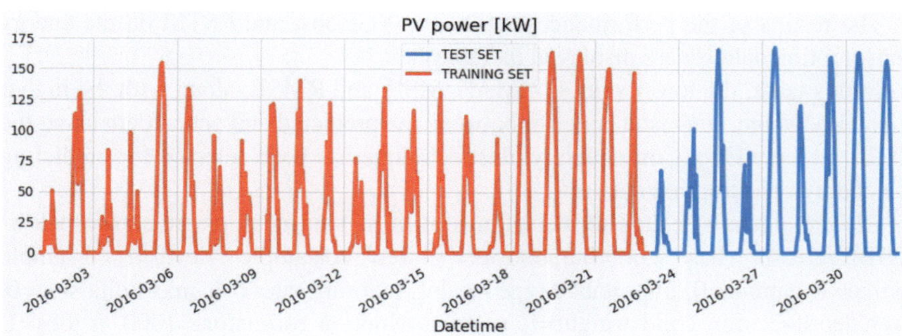

Fig. 5 Training and test PV data

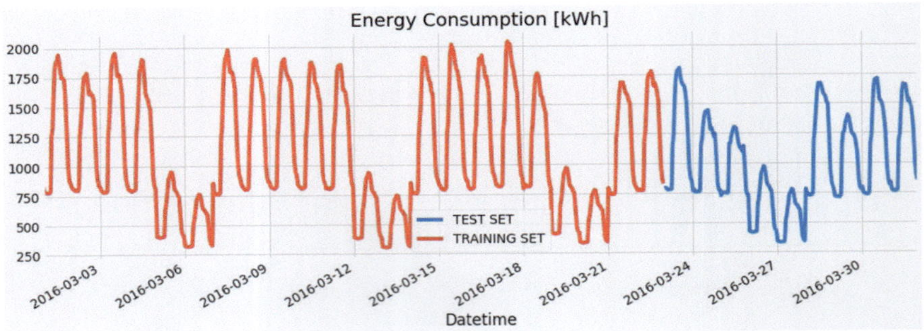

Fig. 6 Training and test Energy consumption data

Table 1 The results of prediction performance metrics for AI techniques on the PV data set

Technique	MAE	MSE	$RMSE$	$R - squared$
XGBoost	17.719	858.569	29.302	0.679
LSTM	25.078	1708.659	41.336	0.360

Table 2 The results of prediction performance metrics for AI techniques on the energy consumption data set

Technique	MAE	MSE	$RMSE$	R-$squared$
XGBoost	92.090	15640.905	125.064	0.897
LSTM	254.616	113195.470	336.445	0.252

The prediction performance metrics for XGBoost and LSTM AI techniques on the PV data set are presented in Table 1. This table demonstrates that the values of MAE, MSE, and RMSE, which are lower when there is a better fit between predicted and actual data, are smaller for the XGBoost technique. Furthermore, the R-squared value for XGBoost is higher, indicating a more accurate alignment between the prediction model and the actual data. Consequently, based on these findings, we have chosen to use XGBoost for predictions in the PV systems data set.

The results of the performance metrics for XGBoost and LSTM on the energy consumption data set are displayed in Table 2.

In this table, the lower values of MAE, MSE, and RMSE, along with the higher R-squared value, suggest a better fit between the predicted and actual data using the XGBoost technique. Consequently, this technique has been selected for predicting the energy consumption data set.

Also, the details related to the settings of these two methods are shown below: XGBRegressor: (base score=0.5, booster='gbtree', colsample bylevel=1, colsample bytree=1, gamma=0, importance type='gain', learning rate=0.1, max delta step=0, max depth=3, min child weight=1, missing=None, n estimators=1000, n jobs=1, nthread=None, objective='reg:linear', random state=0, reg alpha=0, reg lambda=1, scale pos weight=1, seed=None, silent=True, subsample=1).

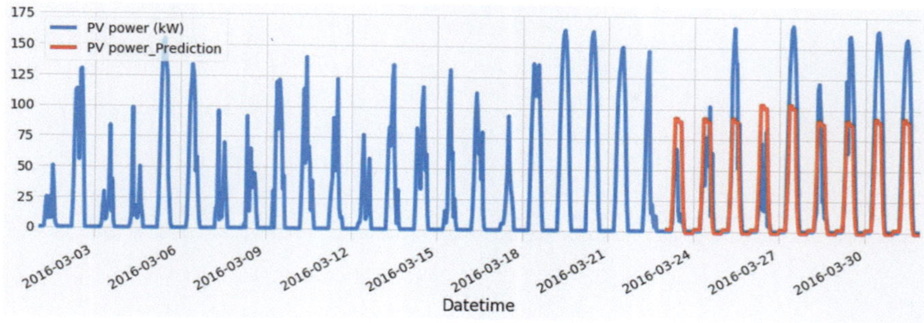

Fig. 7 Actual (blue part) and predicted (red part) PV data

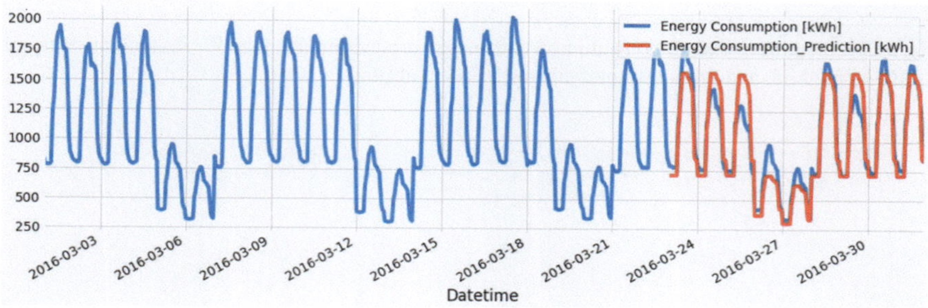

Fig. 8 Actual (blue part) and predicted (red part) energy consumption data

LSTM model: model = Sequential(), model.add (LSTM (128, return sequences=True, input shape = (X train.shape[1], 1))), model.add (LSTM (64, return sequences=False)), model.add(Dense(25)), model.add (Dense(10)), model.fit(X train, y train, batch size=2, epochs=10).

Figures 7 and 8 show the prediction results XGBoost on the testing data sets. The blue part represents the actual data, while the red represents the predicted data.

According to the results of XGBoost model, the predicted values for energy consumption and energy production from the PV system are illustrated in Fig. 9. Moreover, proposed intervals for BESS charging have been determined. As previously discussed, we suggest intervals when the predicted energy consumption is expected to decrease while the predicted energy production from the PV system remains at its peak. Ideally, multiple parameters should be simultaneously considered to ascertain these optimal intervals.

Ranking all days based on prediction error in both the PV production and energy consumption data sets enables us to evaluate the risk associated with optimal scheduling, particularly in determining suitable intervals for BESS charging.

Figures 10 and 11 depict the days with the worst and best prediction accuracy for each data set, respectively. As shown in Fig. 10a, on March 30th, there was a notably

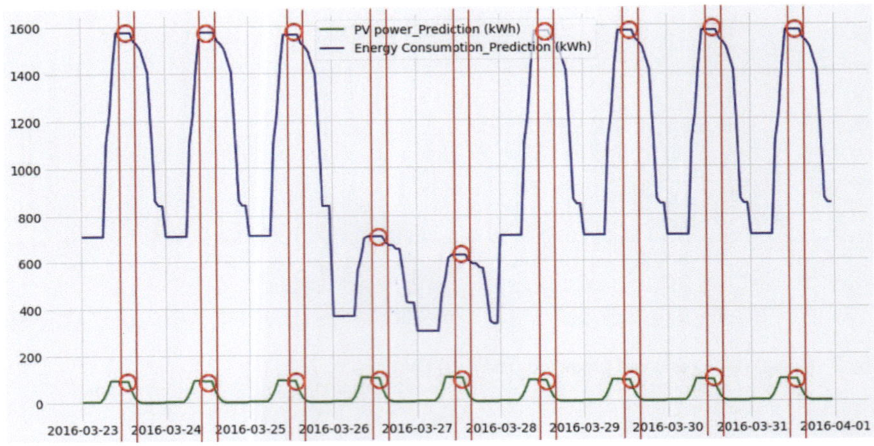

Fig. 9 Predictions of energy consumption and energy production from the PV system based on the XGBoost method, and proposed intervals for BESS charging

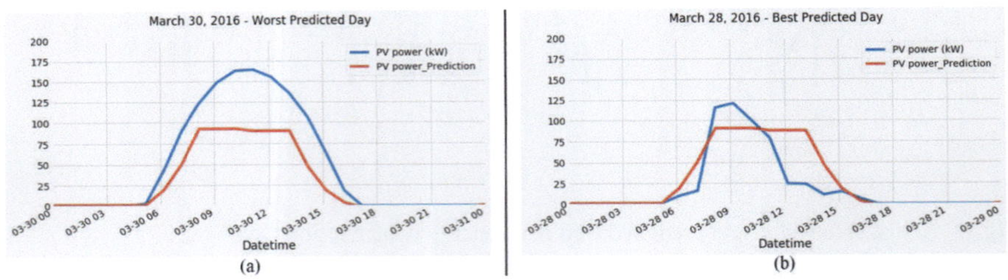

Fig. 10 **a** The worst prediction day and **b** the best prediction day for the PV production data set

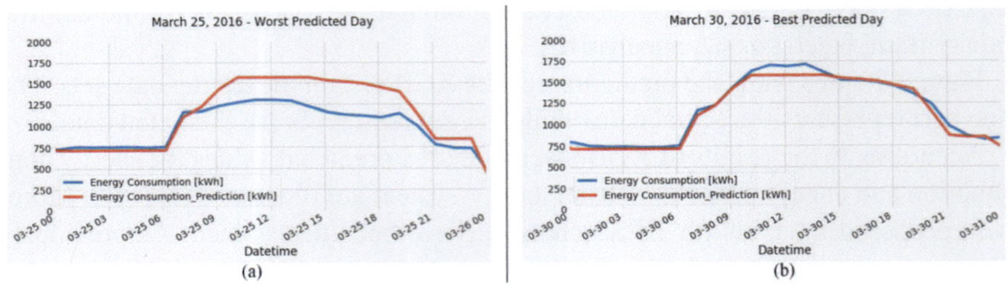

Fig. 11 **a** The worst prediction day and **b** the best prediction day for the energy consumption data set

large discrepancy between the predicted and actual amounts of energy production. Similarly, Fig. 11a reveals that on March 25th, during certain hours, the predicted energy consumption values were significantly higher than the actual consumption levels. Regarding Fig. 10b, it is evident that on March 28th, the predicted and actual values of energy production closely align. Specifically, Fig. 11b demonstrates a relatively accurate match between the predicted and actual values of energy consumption during most hours of March 30.

5 Conclusion

The variable and uncertain nature of energy production from renewable sources presents unique challenges. Artificial intelligence (AI) techniques are widely used for prediction in data sets of energy production from renewable sources. Also, due to the variable nature of energy consumption, AI techniques for prediction in data related to energy consumption are widely used. Predicting energy consumption and production from renewable sources as accurately as possible is essential for optimizing battery energy storage systems (BESS), mainly when renewable sources produce excess energy and planning energy consumption from BESS when the output from renewable sources is lower. In this study, we explore the effectiveness of two widely used AI techniques: Extreme Gradient Boosting (XGBoost) and Long Short-Term Memory (LSTM) on data sets related to energy consumption and renewable energy production. The primary focus is on identifying optimal intervals for BESS charging, a critical aspect dependent on accurate predictions of both energy production from PV systems and energy consumption. Utilizing the XGBoost and LSTM methods, our analysis revealed that the XGBoost technique outperformed in both data sets. Furthermore, we identified the best and worst prediction days based on error values, contributing to a less risky approach in BESS management planning. It is important to note that a portion of these error values may result from significant variations in weather conditions. These variations can lead to abrupt changes in both energy consumption and, notably, energy production from the PV system. Given the significant influence of weather parameters, the authors intend to explore their impact in future research endeavors. Moreover, to refine the model training in future research, the authors will incorporate data sets spanning several years.

Acknowledgements We acknowledge financial support from: PNRR MUR project PE0000013-FAIR.

References

1. Rojek, I., Mikołajewski, D., Mroziński, A., Macko, M.: Machine learning-and artificial intelligence-derived prediction for home smart energy systems with pv installation and battery energy storage. Energies **16**(18), 6613 (2023). https://doi.org/1996-1073/16/18/6613#
2. Mediavilla, M.A., Dietrich, F., Palm, D.: Review and analysis of artificial intelligence methods for demand forecasting in supply chain management. Procedia CIRP **107**, 1126–1131 (2022)

3. Rani, P., Dutta, K., Kumar, V.: Artificial intelligence techniques for prediction of drug synergy in malignant diseases: Past, present, and future. Comput. Biol. Med. **144**, 105334 (2022). https://doi.org/10.1016/j.compbiomed.2022.105334
4. Merembayev, T., Kurmangaliyev, D., Bekbauov, B., Amanbek, Y.: A Comparison of machine learning algorithms in predicting lithofacies: case studies from norway and Kazakhstan. Energies **14**(7), 1896 (2021). https://doi.org/10.3390/en14071896
5. Shrestha, A., Rajbhandari, Y., Gonzalez-Longatt, F.: Day-ahead energy-mix proportion for the secure operation of renewable energy-dominated power system. Int. J. Electr. Power Energy Syst. **155**, 109560 (2024). https://doi.org/10.1016/j.ijepes.2023.109560
6. McMaster, R., Noble, B., Poelzer, G.: Assessing local capacity for community appropriate sustainable energy transitions in northern and remote Indigenous communities. Renew. Sustain. Energy Rev. **191**, 114232 (2024). https://doi.org/10.1016/j.rser.2023.114232
7. Nie, Y., Li, X., Paletta, Q., Aragon, M., Scott, A., Brandt, A.: Open-source sky image datasets for solar forecasting with deep learning: a comprehensive survey. Renew. Sustain. Energy Rev. **189**, 113977 (2024). https://doi.org/10.1016/j.rser.2023.113977
8. Shi, X., Wang, J., Zhang, B.: A fuzzy time series forecasting model with both accuracy and interpretability is used to forecast wind power. Appl. Energy **353**, 122015 (2024). https://doi.org/10.1016/j.apenergy.2023.122015
9. Pop, T., Ungureanu, C., Pentiuc, R.D., Afanasov, C., Ifrim, V.C., Atănăsoae, P., Milici, L.D.: Off-grid hybrid renewable energy system operation in different scenarios for household consumers. Energies **16**(7), 2992 (2023). https://doi.org/10.3390/en16072992
10. Kallio, S., Siroux, M.: Exergy and exergy-economic approach to evaluate hybrid renewable energy systems in buildings. Energies **16**(3), 1029 (2023). https://doi.org/10.3390/en16031029
11. Rocha, H.R., Fiorotti, R., Louzada, D.M., Silvestre, L.J., Celeste, W.C., Silva, J.A.: Net zero energy cost building system design based on artificial intelligence. Appl. Energy **355**, 122348 (2024). https://doi.org/10.1016/j.apenergy.2023.122348
12. Al Smadi, T., Handam, A., Gaeid, K.S., Al-Smadi, A., Al-Husban, Y.: Artificial intelligent control of energy management PV system. Results Control Opt. **14**, 100343 (2024). https://doi.org/10.1016/j.rico.2023.100343
13. Khawaja, Y., Shankar, N., Qiqieh, I., Alzubi, J., Alzubi, O., Nallakaruppan, M.K., Padmanaban, S.: Battery management solutions for li-ion batteries based on artificial intelligence. Ain Shams Eng. J. 102213 (2023). https://doi.org/10.1016/j.asej.2023.102213
14. Runge, J., Saloux, E.: A comparison of prediction and forecasting artificial intelligence models to estimate the future energy demand in a district heating system. Energy **269**, 126661 (2023). https://doi.org/10.1016/j.energy.2023.126661
15. Bamisile, O., Cai, D., Oluwasanmi, A., Ejiyi, C., Ukwuoma, C.C., Ojo, O., Mukhtar, M., Huang, Q.: Comprehensive assessment, review, and comparison of AI models for solar irradiance prediction based on different time/estimation intervals. Sci. Rep. **12**(1), 9644 (2022). https://doi.org/10.1038/s41598-022-13652-w
16. Mirjalili, M.A., Aslani, A., Zahedi, R. and Soleimani, M.: A comparative study of machine learning and deep learning methods for energy balance prediction in a hybrid building-renewable energy system. Sustain. Energy Res. **10**(1), 8 (2022). https://doi.org/10.1186/s40807-023-00078-9
17. Alkahtani, H., Aldhyani, T.H. and Alsubari, S.N : Application of artificial intelligence model solar radiation prediction for renewable energy systems. Sustainability **15**(8), 6973 (2023). https://doi.org/10.3390/su15086973
18. Kaidi, R., Afoudi, Y., Omara, H., Achhab, M.A., Lazaar, M.: Enhanced aircraft time delay prediction using weighted hybrid ML and dimensionality reduction. In: International Conference on Artificial Intelligence and Green Computing, pp. 106–119. Springer Nature Switzerland, Cham (2023). https://doi.org/10.1007/978-3-031-46584-0_9

On Multiple Agents Dealing with Multiple Criteria via the Hypervolume Approach

Lorenzo Lampariello, Simone Sagratella, and Valerio Giuseppe Sasso

Abstract To deal with multi-agent problems where each agent wishes to optimize selfishly multiple criteria, given the choices of the other decision makers, we resort to the hypervolume approach and identify properties making it appealing also in this context.

Keywords Multi-objective optimization · Nash equilibrium problems · Variational inequalities

1 Introduction

The hypervolume approach belongs to the class of scalarization techniques that are relied upon to address multi-objective optimization problems. It essentially consists in the constrained maximization of the product of the (nonnegative) differences between some references values and the objective functions (to be minimized). Its foundations can be traced back to [3, 10], where, in a bi-objective context, some properties have been identified that make it attractive from many respects. We generalize such distinctive features whenever more than two objectives are involved and, taking into account the recent developments in [7], we obtain some more ones. We show that, under mild conditions, the approach is guaranteed to provide Pareto optima for the original multi-objective optimization problem. Possibly more interestingly, a decision maker who prefers a choice improving on the solution given by the hypervolume approach with respect to some objectives must be willing to

L. Lampariello (✉)
Roma Tre University, Rome, Italy
e-mail: lorenzo.lampariello@uniroma3.it

S. Sagratella · V. G. Sasso
Sapienza University of Rome, Rome, Italy
e-mail: sagratella@diag.uniroma1.it

V. G. Sasso
e-mail: sasso@diag.uniroma1.it

M. Di Francesco et al. (eds.), *Operations Research: Closing the Gap Between Research and Practice*, AIRO Springer Series 15, https://doi.org/10.1007/978-3-031-90095-2_28

accept a deterioration in terms of the product of the remaining objectives at least as large as the product of the achieved improvements. Unfortunately, these and other features come at the price of having to solve a nonconvex optimization problem: we show that the latter can actually be equivalently formulated as a convex optimization program, thus proving the viability of the approach. As a result, the hypervolume approach turns out to be also well-suited for contexts where different agents wish to optimize selfishly multiple criteria, given the choices of the other decision makers. This is the case of, e.g., multi-portfolio selection (see [2, 5, 6, 8, 9]), where Nash equilibrium problems arise involving multiple, agent-related criteria to be optimized simultaneously when trades from different clients' accounts are pooled for common execution. To the best of our knowledge, the very few results in the literature for such frameworks have so far been obtained relying on the classical weighted-sum method.

As for the paper organization, the work is divided in two main parts. In Sect. 2, the hypervolume approach is studied when dealing with (multi-objective, single-agent) optimization problems: some key properties are described in Sect. 2.1, while the distinctive features of the method are presented in Sect. 2.2. The concluding Sect. 3 is devoted to the extension of the hypervolume approach to multi-agent frameworks.

2 Relying on the Hypervolume Approach to Address Multi-objective Optimization Problems

We consider the multi-objective optimization problem

$$\begin{array}{ll} \underset{x}{\text{minimize}} & f(x) \triangleq \left(f_1(x), \ldots, f_m(x)\right)^{\mathsf{T}} \\ \text{s.t.} & x \in K, \end{array} \tag{1}$$

where $K \subseteq \mathbb{R}^n$ is a closed convex set and $f : \mathbb{R}^n \to \mathbb{R}^m$ is the vector of objective functions that are assumed to be convex.

In the following subsection, we recall and obtain preliminary results that are instrumental to lay down the main features of the approach in Sect. 2.2.

2.1 Preliminaries

We first report standard concepts of Pareto optimality.

Definition 1 A point $\widehat{x} \in K$ is Pareto optimal for (1) if $\forall x \in K$:

$$\exists j_x \in \{1, \ldots, m\} : f_{j_x}(\widehat{x}) < f_{j_x}(x), \text{ or } \forall j \in \{1, \ldots, m\} : f_j(\widehat{x}) \leq f_j(x).$$

The set of Pareto optimal points of (1) is denoted by P.

Definition 2 A point $\widehat{x} \in K$ is weak Pareto optimal for (1) if $\forall x \in K$:

$$\exists j_x \in \{1, \ldots, m\} \; : \; f_{j_x}(\widehat{x}) \leq f_{j_x}(x).$$

The set of weak Pareto optimal points of (1) is denoted by W.

We focus on (weak) Pareto optimal points $x \in K$ that satisfy some reference level (in terms of f_js) constraints, i.e.

$$f_j(x) \leq a_j, \; j = 1, \ldots, m,$$

for some a_js belonging to $\mathbb{R}$. This is particularly sensible e.g. in the context of portfolio selection, whenever one wishes to outperform some suitable benchmark index, in an enhanced index tracking spirit, see e.g. [3].

We assume the Slater condition to hold for the (constrained) system of inequalities above:

$$\exists \; \overline{x} \in K \text{ such that } f_j(\overline{x}) < a_j, \; j = 1, \ldots, m.$$

From now on, points satisfying the Slater condition are referred to as Slater points. It is easy to show that if the Slater condition is not satisfied, all points in $S_a \triangleq K \cap \{z \in \mathbb{R}^n : f(z) \leq a\}$ are weak Pareto optimal.

In order to address (1), we rely on a scalarization technique based on the use of suitable utility functions (see [7] for some recent developments and also the related achievement function approach [11]). More in detail, we aim at identifying Pareto optima of (1) by solving the following single-objective optimization problem:

$$\begin{array}{ll} \underset{x}{\text{maximize}} & h(x) \triangleq u(a_1 - f_1(x), \ldots, a_m - f_m(x)) \\ \text{s.t.} & x \in K, \end{array} \tag{2}$$

where

$$u(y) \triangleq \begin{cases} \prod_{j=1}^{m} y_j \text{ if } y \geq 0 \\ -\infty \;\; \text{otherwise.} \end{cases} \tag{3}$$

Hence,

$$h(x) = \begin{cases} \prod_{j=1}^{m} (a_j - f_j(x)) \text{ if } f(x) \leq a \\ \quad -\infty \qquad \text{otherwise.} \end{cases}$$

Note that for every $x \in S_a$, $h(x)$ can be interpreted as the hypervolume (to be maximized) of an m-dimensional hypercube $H_a(x)$ whose edge j length is $a_j - f_j(x) \geq 0$.

In the following proposition, whose proof can be tracked down in [7], we collect some properties of u.

Proposition 1 *The following statements hold:*

(i) *dom* $u = \mathbb{R}^m_+$;
(ii) u *is continuous on its domain;*
(iii) u *is continuously differentiable on the interior of its domain;*
(iv) u *is monotone on its domain, i.e.*

$$u(y+\delta) \geq u(y), \quad \forall y \in \mathbb{R}^m_+,\ \delta \in \mathbb{R}^m_+ : y+\delta \in \mathbb{R}^m_+;$$

(v) u *is strictly monotone on the interior of its domain, i.e.*

$$u(y+\delta) > u(y) \quad \forall y \in \mathbb{R}^m_{++},\ \delta \in \mathbb{R}^m_+ \setminus \{0\} : y+\delta \in \mathbb{R}^m_{++};$$

(vi) u *is pseudoconcave on the interior of its domain;*
(vii) u *is a* 0*-barrier, i.e.*

$$u(y) = 0, \quad \forall y \in bd\ dom\ u = \mathbb{R}^m_+ \setminus \mathbb{R}^m_{++}$$

$$u(y) > 0, \quad \forall y \in int\ dom\ u = \mathbb{R}^m_{++}.$$

While properties (i)–(vi) are rather standard, some comments are in order about (vii): the barrier property guarantees that u is constant on the border of the domain and increases just inside it. This condition plays a crucial role in some of the theoretical results below and is key for algorithmic developments (see [7]).

In Proposition 2, we focus on problem (2): leveraging the properties listed above, one can prove that Pareto optima of the original multi-objective problem (1) can be recovered by solving the single-objective problem (2).

Proposition 2 *The following statements hold:*

(i) *problem* (2) *is solvable;*
(ii) *solutions of* (2) *are Slater points and, in turn, belong to* P;
(iii) h *is pseudoconcave on the set of Slater points.*

Proof The claim in point (i) follows from closedness of K and continuity of u on its closed domain. As for point (ii), under the assumed Slater condition, the barrier property of u ensures that solutions of (2) are Slater points; in turn, the claim follows resorting to the strict monotonicity of u on the interior of its domain.

Finally, (iii) is due to the pseudoconcavity of u together with its monotonicity. For the detailed developments in this proof, we refer the reader to [7]. □

2.2 Main Features of the Approach

Addressing (1) via (2) has the following distinctive features:

- any solution x^* of (2) is Pareto optimal for (1) and such that $f_j(x^*) < a_j$, $j = 1, \ldots, m$ (see Proposition 2);

- let a decision maker opt for a choice $\widetilde{x} \in S_a$ improving on a solution x^* of (2) with respect to at least an objective. The product of the corresponding improvements obtained in terms of such objectives turns out to be smaller than the resulting deterioration in terms of the product of the remaining objectives, as detailed in Proposition 3;
- any solution x^* of (2), maximizing the hypervolume of hypercube $H_a(x)$, dominates the largest amount of solutions that dominate the reference a. Roughly speaking, x^* is a Pareto optimum that dominates the most the benchmark;
- finding a solution of (2) does not depend on the scales of possibly non homogeneous objectives.

Proposition 3 *Let x^* be a solution of (2). Assume $\widetilde{x} \in S_a$ is such that for some $\alpha_j > 1, a_j - f_j(\widetilde{x}) = \alpha_j(a_j - f_j(x^*))$, for every $j = 1, \ldots, p$, where $p < m$. Then,*

$$\prod_{j=p+1}^{m} (a_j - f_j(\widetilde{x})) \leq \frac{1}{\prod_{j=1}^{p} \alpha_j} \prod_{j=p+1}^{m} (a_j - f_j(x^*)).$$

Proof By optimality,

$$\begin{aligned}\prod_{j=1}^{p}(a_j - f_j(x^*)) \prod_{j=p+1}^{m} (a_j - f_j(x^*)) &\geq \prod_{j=1}^{p}(a_j - f_j(\widetilde{x})) \prod_{j=p+1}^{m} (a_j - f_j(\widetilde{x})) \\ &= \prod_{j=1}^{p}\alpha_j(a_j - f_j(x^*)) \prod_{j=p+1}^{m} (a_j - f_j(\widetilde{x})).\end{aligned}$$

In turn,

$$\prod_{j=p+1}^{m} (a_j - f_j(x^*)) \geq \prod_{j=1}^{p}\alpha_j \prod_{j=p+1}^{m} (a_j - f_j(\widetilde{x})),$$

and the claim follows. □

The advantages highlighted above come at the price of having to solve (to global optimality) a nonconvex optimization problem: in fact, the objective function in (2) is only pseudoconcave (on the set of Slater points). But, on the one hand, letting $\overline{x}$ be a Slater point, the latter property, together with the barrier condition, is sufficient for the following implication to hold:

$$\nabla h(\overline{x})^{\mathsf{T}}(x - \overline{x}) \leq 0 \; \forall x \in K \implies h(\overline{x}) \geq h(x) \; \forall x \in K,$$

meaning that, to identify global optima of (2), one can just target *stationary* solutions that satisfy the Slater condition, which is a reasonable task, in practice.

On the other hand, (2) can actually be equivalently formulated as a convex optimization problem. To this end, we introduce the modified problem

$$\begin{aligned} &\underset{x}{\text{maximize}}\ \widetilde{h}(x) \\ &\text{s.t.} \qquad x \in K, \end{aligned} \tag{4}$$

where

$$\widetilde{h}(x) = \begin{cases} \prod_{j=1}^{m} (a_j - f_j(x))^{\frac{1}{m}} & \text{if } f(x) \le a \\ -\infty & \text{otherwise.} \end{cases}$$

Similarly as done for (2), one can prove the properties below concerning problem (4).

Proposition 4 *The following statements hold:*

(i) problem (4) *is solvable;*
(ii) solutions of (4) *are Slater points and, in turn, belong to* P*;*
(iii) $\widetilde{h}$ *is concave.*

Proof The proof follows in the same steps as Proposition 2. Regarding (iii), the key ingredients are the monotonicity of $\prod_{j=1}^{m} y^{\frac{1}{m}}$ on $\mathbb{R}^m_+$ and its concavity, see [7]. □

In the light of Proposition 4, (4) is a convex, solvable problem.

We show that any solution of (4) is a solution of (2): as a consequence, one can recover Pareto optima for (1) that enjoy the distinctive features described above by addressing, instead of the nonconvex problem (2), the convex one (4).

Proposition 5 *Any solution* $\widehat{x}$ *of problem* (4) *is also solution to problem* (2).

Proof The claim is a consequence of the strict monotonicity of $(\bullet)^{\frac{1}{m}}$ on $\mathbb{R}_+$. We show the contrapositive: leaving aside the trivial case of $\widehat{x}$ not satisfying the Slater condition, assume $\widetilde{x} \in K$ exists such that

$$\prod_{j=1}^{m} (a_j - f_j(\widetilde{x})) > \prod_{j=1}^{m} (a_j - f_j(\widehat{x})),$$

and hence, $f_j(\widetilde{x}) < a_j$ for every j. Then, we have

$$\left[\prod_{j=1}^{m} (a_j - f_j(\widetilde{x}))\right]^{\frac{1}{m}} > \left[\prod_{j=1}^{m} (a_j - f_j(\widehat{x}))\right]^{\frac{1}{m}},$$

so that $\widehat{x}$ is not a solution of (4) and the claim is proven. □

Also the converse of the claim in Proposition 5 holds. Thus, the solution sets of the two problems coincide and no solutions are "lost" when dealing with (2) via (4).

Proposition 6 *Any solution x^* of problem* (2) *is also solution to problem* (4).

Proof For every $x \in K$ such that $f_j(x) < a_j$ for every j, we have

$$\prod_{j=1}^{m}(a_j - f_j(x^*)) \geq \prod_{j=1}^{m}(a_j - f_j(x)),$$

and

$$\left[\prod_{j=1}^{m}(a_j - f_j(x^*))\right]^{\frac{1}{m}} \geq \left[\prod_{j=1}^{m}(a_j - f_j(x))\right]^{\frac{1}{m}},$$

due to the monotonicity of $(\bullet)^{\frac{1}{m}}$. The claim follows considering also the trivial case of those $x \in K$ such that $a - f(x) \notin \mathbb{R}^m_{++}$. □

3 The Game-Theoretical Context

The results in Sect. 2 open the door to the generalization of the hypervolume approach to multi-agent frameworks.

We consider N agents, each one aiming at optimizing multiple objectives that possibly depend on the choices of the other agents. More in detail, for each $\nu \in \{1, \ldots, N\}$, m_ν criteria $(\theta_\nu)_j : \mathbb{R}^{n_\nu N} \to \mathbb{R}$, $j = 1, \ldots, m_\nu$ are addressed that depend on proprietary variables $x^\nu \in \mathbb{R}^{n_\nu}$, as well as on decisions from all the other agents, which are collected in vector $x^{-\nu}$:

$$x^{-\nu} \triangleq \begin{pmatrix} x^1 \\ \vdots \\ x^{\nu-1} \\ x^{\nu+1} \\ \vdots \\ x^N \end{pmatrix} \in \mathbb{R}^{(N-1)n_\nu}. \tag{5}$$

As is customary in the relevant literature, we denote by $x \in \mathbb{R}^{Nn_\nu}$ the vector formed by all the decision variables and, to emphasize the agent ν's ones within x, we write $(x^\nu, x^{-\nu})$ instead of x, still indicating the vector $x = (x^1, \ldots, x^\nu, \ldots, x^N)$.

Overall, for each ν, the following multi-objective, parametric (with respect to other agents related decisions $x^{-\nu}$) problem is tackled:

$$\begin{aligned} \underset{x^\nu}{\text{minimize}}\ & \theta_\nu(x^\nu, x^{-\nu}) \triangleq \left((\theta_\nu)_1(x^\nu, x^{-\nu}), \ldots, (\theta_\nu)_{m_\nu}(x^\nu, x^{-\nu})\right)^\mathsf{T} \\ \text{s.t.}\ \ & x \in K_\nu, \end{aligned} \tag{6}$$

where $K_\nu \subseteq \mathbb{R}^{n_\nu}$ is a closed convex set and $(\theta_\nu)_j$s are continuously differentiable and such that $(\theta_\nu)_j(\bullet, x^{-\nu})$, $j = 1, \ldots, m$ is convex for every $x^{-\nu} \in \prod_{\mu \neq \nu} K_\mu$.

Relying on the hypervolume approach, one is naturally led to consider the single-objective parametric problem

$$\begin{aligned} \underset{x}{\text{maximize}}\ & \widetilde{h}_\nu(x^\nu, x^{-\nu}) \\ \text{s.t.}\quad & x^\nu \in K_\nu, \end{aligned} \tag{7}$$

where

$$\widetilde{h}_\nu(x^\nu, x^{-\nu}) = \begin{cases} \prod_{j=1}^{m_\nu} (a_j^\nu - (\theta_\nu)_j(x^\nu, x^{-\nu}))^{\frac{1}{m_\nu}} & \text{if } \theta_\nu(x^\nu, x^{-\nu}) \leq a^\nu \\ -\infty & \text{otherwise.} \end{cases}$$

Note that $\widetilde{h}(\bullet, x^{-\nu})$ is concave for every $x^{-\nu} \in \prod_{\mu \neq \nu} K_\mu$, due to the assumed convexity of $(\theta_\nu)_j(\bullet, x^{-\nu})$ and by (iii) in Proposition 4, so that (7) turns out to be a convex problem with respect to proprietary variables x^ν.

Evidently, also the N agents related single-objective mathematical programs above are coupled to each other: their collection forms a Nash Equilibrium Problem (NEP), which is the problem to

$$\text{find } x \in K : \widetilde{h}_\nu(x^\nu, x^{-\nu}) \leq \widetilde{h}_\nu(v^\nu, x^{-\nu}) \quad \forall v^\nu \in K_\nu, \quad \nu = 1, \ldots, N, \tag{8}$$

where $K \triangleq K_1 \times \ldots \times K_N$. Any such point is termed equilibrium or solution of the NEP and

$$E \triangleq \left\{x \in K : \widetilde{h}_\nu(x^\nu, x^{-\nu}) \leq \widetilde{h}_\nu(v^\nu, x^{-\nu})\ \forall v^\nu \in K_\nu,\ \nu = 1, \ldots, N\right\}$$

is the set of equilibria of the NEP.

Let $\widehat{x}$ be an equilibrium of (8) and assume for every ν the (parametric) Slater condition

$$\exists\, \overline{x}^\nu \in K_\nu \text{ such that } (\theta_\nu)_j(\overline{x}^\nu, \widehat{x}^{-\nu}) < a_j^\nu,\ j = 1, \ldots, m_\nu$$

to hold. Then, since $\widehat{x}^\nu$ is an optimal solution of the parametric problem (7) corresponding to other agents' decisions $\widehat{x}^{-\nu}$, leveraging (ii) in Proposition 4, we have, for every ν, $(\theta_\nu)_j(\widehat{x}^\nu, \widehat{x}^{-\nu}) < a_j^\nu$, $j = 1, \ldots, m_\nu$ and $\widehat{x}^\nu$ turns out to be (parametric, with respect to $\widehat{x}^{-\nu}$) Pareto optimal for (7), i.e., $\forall x^\nu \in K_\nu$,

$$\exists j_{x^\nu} \in \{1, \ldots, m_\nu\} : (\theta_\nu)_{j_{x^\nu}}(\widehat{x}^\nu, \widehat{x}^{-\nu}) < (\theta_\nu)_{j_{x^\nu}}(x^\nu, \widehat{x}^{-\nu}),$$

or

$$\forall j \in \{1, \ldots, m_\nu\} : (\theta_\nu)_j(\widehat{x}^\nu, \widehat{x}^{-\nu}) \leq (\theta_\nu)_j(x^\nu, \widehat{x}^{-\nu}).$$

Since $(\theta_\nu)_j$s are assumed to be continuously differentiable, $\widetilde{h}_\nu(\bullet, x^{-\nu})$ turns out to be continuously differentiable on the open set of Slater points $\overline{x}^\nu \in K_\nu$ such that $(\theta_\nu)_j(\overline{x}^\nu, x^{-\nu}) < a_j^\nu$, $j = 1, \dots, m_\nu$. Hence, by means of the following proposition, we are able to characterize E as the solution set $\mathrm{SOL}(K, F)$ of the variational inequality $\mathrm{VI}(K, F)$ (see the fundamental [1, 4] for some recent theoretical developments concerning variational problems). The latter is the problem to

$$\text{find } x \in K : \ F(x)^\mathsf{T}(v - x) \geq 0 \quad \forall v \in K, \tag{9}$$

with

$$F(x) \triangleq - \begin{pmatrix} \nabla_1 \widetilde{h}_1(x) \\ \vdots \\ \nabla_N \widetilde{h}_N(x) \end{pmatrix},$$

and $\nabla_\nu \widetilde{h}_\nu(x)$ is the gradient of $\widetilde{h}_\nu(\bullet, x^{-\nu})$ at x^ν, where $x^\nu \in K_\nu$ is such that $(\theta_\nu)_j(x^\nu, x^{-\nu}) < a_j^\nu$ for every j.

Proposition 7 *The following statements hold:*

(i) let $\widehat{x} \in E$ and, for every ν, $\overline{x}^\nu \in K_\nu$ exist such that $(\theta_\nu)_j(\overline{x}^\nu, \widehat{x}^{-\nu}) < a_j^\nu$, $j = 1, \dots, m_\nu$. Then, $\widehat{x} \in \mathrm{SOL}(K, F)$;
(ii) let $\widehat{x}$ be such that, for every ν, $(\theta_\nu)_j(\widehat{x}) < a_j^\nu$, $j = 1, \dots, m_\nu$, and belong to $\mathrm{SOL}(K, F)$. Then, $\widehat{x} \in E$.

Proof (i) Due to (ii) in Proposition 4, we have, for every ν, $(\theta_\nu)_j(\widehat{x}^\nu, \widehat{x}^{-\nu}) < a_j^\nu$, $j = 1, \dots, m_\nu$ and $\widehat{x} \in E$ if and only if the first order condition

$$\nabla_\nu \widetilde{h}_\nu(\widehat{x}^\nu, \widehat{x}^{-\nu})^\mathsf{T}(x^\nu - \widehat{x}^\nu) \leq 0, \ \forall x^\nu \in K_\nu, \tag{10}$$

holds, because of the concavity of $\widetilde{h}_\nu(\bullet, x^{-\nu})$. Concatenating these relations, we obtain the claim.

(ii) For the converse, for every ν, suffice it to choose in

$$F(\widehat{x})^\mathsf{T}(v - \widehat{x}) \geq 0 \quad \forall v \in K,$$

$v = (x^\nu, \widehat{x}^{-\nu})$, with an arbitrary $x^\nu \in K_\nu$, to obtain again (10) and, thus, the claim. □

We remark that the mathematical model (6) naturally fits within the multi-portfolio context, where financial service providers manage multiple accounts from different clients. Since such accounts are usually accommodated simultaneously and the related trades are pooled for common execution, the transaction costs for each portfolio depend on the trades from all the accounts and the multiple portfolios turn out to be coupled to each other by the market impact of their transactions.

More in detail, for each account $\nu \in \{1, \dots, N\}$, one has to choose the fractions $y^\nu \in \mathbb{R}^K$ of a given account-dependent budget b^ν to be invested in K assets with the aim of optimizing multiple criteria such as portfolio's income $I_\nu(y^\nu) \triangleq b^\nu(\mu^\nu)^\top y^\nu$, where $\mu^\nu \in \mathbb{R}^K$ are expectations of the assets' returns r_ks, portfolio's risk $R_\nu(y^\nu) \triangleq \frac{1}{2}(b^\nu)^2(y^\nu)^\top \Sigma^\nu y^\nu$, where Σ^ν is the symmetric and positive semidefinite assets' returns covariance matrix, and transaction costs $TC_\nu(y^1, \ldots, y^N) \triangleq b^\nu(y^\nu)^\top \Omega \sum_{\lambda=1}^N b^\lambda y^\lambda$ where $\Omega \in \mathbb{R}^{K\times K}$ is the symmetric and positive semidefinite market impact matrix (see [5] for further information), which are responsible for the coupling between the accounts related problems. Clearly, other objectives can enter the picture such as, for example, sustainability-oriented criteria.

Resorting to the hypervolume approach, each account's single-objective scalarizing function $\widetilde{h}_\nu(x^\nu, x^{-\nu})$, to be maximized over the set of feasible portfolios K_ν, is given by

$$(I_\nu(y^\nu) - a_1^\nu)^{\frac{1}{3}}(a_2^\nu - R_\nu(y^\nu))^{\frac{1}{3}}(a_3^\nu - TC(y^\nu, y^{-\nu}))^{\frac{1}{3}}$$

if $(I_\nu(y^\nu), R_\nu(y^\nu), TC(y^\nu, y^{-\nu})) \leq a^\nu$, and $-\infty$ otherwise, where a_1^ν, a_2^ν and a_3^ν are some reference values.

We remark that questions arise naturally on identifying a set a suitable conditions guaranteeing, for the problem at hand, existence of solutions and the possibility of devising provably convergent solution procedures. We leave these topics to future research.

References

1. Bigi, G., Lampariello, L., Sagratella, S., Sasso, V.G.: Approximate variational inequalities and equilibria. Comput. Manag. Sci. **20**(1), 1–16 (2023)
2. Cesarone, F., Lampariello, L., Merolla, D., Ricci, J.M., Sagratella, S., Sasso, V.G.: A bilevel approach to ESG multi-portfolio selection. Comput. Manag. Sci. **20**(1), 1–24 (2023)
3. Cesarone, F., Lampariello, L., Sagratella, S.: A risk-gain dominance maximization approach to enhanced index tracking. Finance Res. Lett. **29**, 231–238 (2018)
4. Facchinei, F., Pang, J.-S.: Finite-Dimensional variational inequalities and complementarity problems (2003)
5. Lampariello, L., Neumann, C., Ricci, J.M., Sagratella, S., Stein, O.: Equilibrium selection for multi-portfolio optimization. Eur. J. Oper. Res. **295**(1), 363–373 (2021)
6. Lampariello, L., Sagratella, S., Sasso, V.G.: Nonsmooth Hierarchical multi portfolio selection. In: Cappanera, P., Lapucci, M., Schoen, F., Sciandrone, M., Tardella, F., Visintin, F. (eds) Optimization and Decision Science: Operations Research, Inclusion and Equity. ODS 2022. AIRO Springer Series, 9, pp. 37–46 (2022)
7. Lampariello, L., Sagratella, S., Sasso, V.G., Shikhman, V.: Scalarization via utility functions in multi-objective optimization, pp. 1–28 (2024). arXiv:2401.13831
8. Lampariello, L., Sagratella, S., Sasso, V.G.: Dealing with inexactness in hierarchical multi-portfolio selection. In: Bruglieri, M., Festa, P., Macrina, G., Pisacane, O. (eds) Optimization in Green Sustainability and Ecological Transition. ODS 2023. AIRO Springer Series, pp. 1–8 (2023)
9. Lampariello, L., Sagratella, S., Sasso, V.G., Stein, O.: Solving Multi-Follower Games. Optimization Online, pp. 1–26 (2023). https://optimization-online.org

10. Lootsma, F.A., Athan, T.W., Papalambros, P.Y.: Controlling the search for a compromise solution in multi-objective optimization. Eng. Optim.+ A35 **25**(1), 65–81 (1995)
11. Miettinen, K.: Nonlinear Multiobjective Optimization. Springer Science & Business Media (2012)

Existence and Approximation Results for Quasi Variational Inequalities with Linear Coupling Constraints

Lorenzo Lampariello, Simone Sagratella, and Valerio Giuseppe Sasso

Abstract We consider quasi variational inequality problems in which the feasible point-to-set mapping is defined by linear coupling constraints. We provide existence conditions easy to verify in practice. We analyze inexact solutions for different reformulations based on gap functions to obtain approximate solutions for the quasi variational inequality.

Keywords Quasi variational inequality · Existence results · Approximate solutions

1 Introduction

Quasi Variational Inequalities (QVI) are a widely used mathematical tool employed in the solution of Nash equilibrium problems and general variational problems, see, e.g., [4, 6]. Among the multitude of applications that can be handled by means of these mathematical tools, we want to recall one of the most considered in the recent literature on multi-agent optimization, that is, multi-portfolio selection [3, 8–11].

Motivation. The case in which the feasible set mapping of the QVI is defined by linear coupling constraints is not enough studied in the literature. See e.g. [5] and the references therein for some justifying examples in the context of Coulomb friction and transportation problems. Specifically, there is a lack of existence conditions easy to verify in practice. Moreover, reformulating the QVI by means of gap functions

---

L. Lampariello
Department of Business Studies, Roma Tre University, Rome, Italy
e-mail: lorenzo.lampariello@uniroma3.it

S. Sagratella (✉) · V. G. Sasso
Department of Computer, Control and Management Engineering Antonio Ruberti, Sapienza University of Rome, Rome, Italy
e-mail: sagratella@diag.uniroma1.it

V. G. Sasso
e-mail: sasso@diag.uniroma1.it

M. Di Francesco et al. (eds.), *Operations Research: Closing the Gap Between Research and Practice*, AIRO Springer Series 15, https://doi.org/10.1007/978-3-031-90095-2_29

and/or natural maps is not studied yet in the literature if one considers approximate solutions. We analyze QVIs with linear coupling constraints with the aim of filling these gaps. Notice that our results about approximate solutions of QVIs are also intended to generalize the results for variational inequalities given in [1].

2 Preliminaries

We recall some definitions for point-to-set mappings and fundamental results that are used in the forthcoming analysis.

Definition 1 A point-to-set mapping $S\colon \mathbb{R}^n \rightrightarrows \mathbb{R}^n$ is said to be outer-semicontinuous relative to a set $\Omega \subseteq \mathbb{R}^n$ at a point $\overline{x} \in \Omega$ if for every sequence $\{x^k\} \subseteq \Omega$ such that $x^k \to \overline{x}$, it holds that every accumulation point of $s^k \in S(x^k)$ belongs to $S(\overline{x})$.

Definition 2 A point-to-set mapping $S\colon \mathbb{R}^n \rightrightarrows \mathbb{R}^n$ is said to be outer-semicontinuous relative to a set $\Omega \subseteq \mathbb{R}^n$ if it is outer-semicontinuous relative to Ω at every point $\overline{x} \in \Omega$.

Definition 3 A point-to-set mapping $S\colon \mathbb{R}^n \rightrightarrows \mathbb{R}^n$ is said to be inner-semicontinuous relative to a set $\Omega \subseteq \mathbb{R}^n$ at a point $\overline{x} \in \Omega$ if for every point $\overline{s} \in S(\overline{x})$, a sequence $\{x^k\} \subseteq \Omega$ exists such that $x^k \to \overline{x}$, and $s^k \to \overline{s}$ for some $s^k \in S(x^k)$.

Definition 4 A point-to-set mapping $S\colon \mathbb{R}^n \rightrightarrows \mathbb{R}^n$ is said to be inner-semicontinuous relative to a set $\Omega \subseteq \mathbb{R}^n$ if it is inner-semicontinuous relative to Ω at every point $\overline{x} \in \Omega$.

Definition 5 A point-to-set mapping $S\colon \mathbb{R}^n \rightrightarrows \mathbb{R}^n$ is said to be continuous relative to a set $\Omega \subseteq \mathbb{R}^n$ if it is both inner- and outer-semicontinuous relative to Ω at every point $\overline{x} \in \Omega$, i.e.,

$$\forall \{x^k\} \subseteq \Omega\colon x^k \to \overline{x} \quad \text{it holds that} \quad \lim_{k\to\infty} S(x^k) = S(\overline{x}).$$

Definition 6 A point-to-set mapping $S\colon\mathbb{R}^n \rightrightarrows \mathbb{R}^n$ is said to be bounded on $\Omega \subseteq \mathbb{R}^n$ bounded if

$$\forall x \in \Omega \quad \text{it holds that} \quad S(x) \subseteq \Omega.$$

Lemma 1 *(From [2, Theorem 4.3.3]) Let $f : \mathbb{R}^{n+n} \to \mathbb{R}$ be continuous. Let $S : \mathbb{R}^n \rightrightarrows \mathbb{R}^n$ be closed-valued and continuous relative to $\Omega \subseteq \mathbb{R}^n$. Let the problem*

$$\text{minimize}_y \; f(y, x) \quad \text{s.t.} \quad y \in S(x),$$

be convex and solvable for every $x \in \Omega$. The value-function

$$\phi(x) \triangleq \min_y \{f(y, x) \mid y \in S(x)\}$$

is continuous relative to Ω.

Lemma 2 *(From [2, Theorem 3.1.1]) Let* $g : \mathbb{R}^{n+n} \to \mathbb{R}^m$ *be continuous. Let* $Y : \mathbb{R}^n \rightrightarrows \mathbb{R}^n$ *be outer-semicontinuous relative to* $\Omega \subseteq \mathbb{R}^n$*. The point-to-set mapping*

$$S(x) \triangleq \{y \in \mathbb{R}^n \ : \ g(y, x) \leq 0\} \cap Y(x)$$

is outer-semicontinuous relative to Ω*.*

Lemma 3 *(From [13, Corollary 5.20]) Let* $S : \mathbb{R}^n \rightrightarrows \mathbb{R}^n$ *be single-valued, bounded on* $\Omega \subseteq \mathbb{R}^n$*, and outer-semicontinuous relative to* Ω*. The mapping* S *is continuous relative to* Ω*.*

Lemma 4 *(From [7, Theorem 2.1.18]) Let* $S : \mathbb{R}^n \rightrightarrows \mathbb{R}^n$ *be continuous relative to* $\Omega \subseteq \mathbb{R}^n$ *convex and compact. Assume that*

$$\forall x \in \Omega \quad \text{it holds that} \quad \emptyset \neq S(x) \subseteq \Omega.$$

A point $\widehat{x} \in \Omega$ *exists such that*

$$\widehat{x} \in S(\widehat{x}).$$

Given a nonempty, closed, and convex set $C \subseteq \mathbb{R}^n$. We denote with P_C the projection operator over C: for every $x \in \mathbb{R}^n$

$$P_C(x) \triangleq \arg\min\{\|y - x\|_2^2 \mid y \in C\}.$$

Lemma 5 *Let* $C \subseteq \mathbb{R}^n$ *be nonempty, closed, and convex. The projection mapping* P_C *is single-valued and continuous with respect to* $\mathbb{R}^n$*.*

For every $x \in \mathbb{R}^n$*, it holds that* $P_C(x)$ *is the unique vector such that*

$$(P_C(x) - x)^\top (y - P_C(x)) \geq 0, \quad \forall y \in C.$$

For every $x, y \in \mathbb{R}^n$*, it holds that*

$$\|P_C(x) - P_C(y)\|_2 \leq \|x - y\|_2.$$

3 Problem Definition

The finite-dimensional Quasi Variational Inequality (QVI) is the problem of finding $\widehat{x}$ such that

$$\widehat{x} \in K(\widehat{x}) \tag{1}$$

$$F(\widehat{x})^\top (y - \widehat{x}) \geq 0, \ \ \forall y \in K(\widehat{x}), \tag{2}$$

where $F : \mathbb{R}^n \to \mathbb{R}^n$ and $K : \mathbb{R}^n \rightrightarrows \mathbb{R}^n$. We denote by $\text{SOL}(F, K)$ the solution set of $\text{QVI}(F, K)$. Notice that QVI is actually "a double problem": a fixed-point problem, see (1), and a (parametric) variational inequality, see (2).

The following is an existence result that can be found, e.g., in [12], and for which we provide a demonstrative proof. In the following sections we consider some specific structures for K that allow one to simplify the conditions given in the theorem below.

Theorem 1 *Let $\Omega \subseteq \mathbb{R}^n$ be a compact convex set such that*

$$\emptyset \neq K(x) \subseteq \Omega, \ \ \forall x \in \Omega. \tag{3}$$

If K is continuous relative to Ω, convex-valued and closed-valued, and F is continuous, then $SOL(F, K) \neq \emptyset$.

Proof Let us define the operator $G(x) \triangleq P_{K(x)}(x - F(x))$.

Since $K(x)$ is nonempty, closed, and convex for every $x \in \Omega$, by Lemma 5, G is a single-valued mapping from Ω to Ω. Thus, it is also bounded on Ω.

Let $\phi(x) \triangleq \min\{\|y - [x - F(x)]\|_2^2 \mid y \in K(x)\}$. Since K is continuous relative to Ω, convex-valued and compact-valued, and F is continuous, we can resort to Lemma 1 to say that ϕ is continuous relative to Ω. Therefore, since $G(x) = \{y \in \mathbb{R}^n : \|y - [x - F(x)]\|_2^2 \leq \phi(x)\} \cap K(x)$, G turns out to be outer-semicontinuous relative to Ω, in view of Lemma 2, thanks to the continuity of F and the outer-semicontinuity of K. A fortiori, G is continuous relative to Ω, due to Lemma 3.

In the light of Lemma 4, $\widehat{x} \in \Omega$ exists such that $\widehat{x} = G(\widehat{x})$.

Therefore, (1) holds, and, by Lemma 5, for every $y \in K(\widehat{x})$ it holds that

$$F(\widehat{x})^\top (y - \widehat{x}) = (\widehat{x} - \widehat{x} + F(\widehat{x}))^\top (y - \widehat{x}) \geq 0,$$

i.e., (2) holds. □

Notice that actually most of the conditions given in Theorem 1 for K and Ω are fundamental for obtaining the existence of a point satisfying the fixed-point problem (1), see Lemma 4. The only additional assumptions to get a point satisfying also condition (2) are the continuity of F and the convex- and closed-valuedness of K.

4 Reformulations

We consider some reformulations for the QVI that could be useful from a practical solvability perspective. To equivalently reformulate QVIs, consider the following solvable convex parametric (with respect to x) optimization problem:

$$\text{minimize}_v \ F(x)^\top (v - x) \quad \text{s.t.} \quad v \in K(x). \tag{4}$$

Accordingly, let

$$\varphi(x) \triangleq \min_v \{F(x)^\top (v - x) \mid v \in K(x)\}$$

be the value-function for (4). We also define the natural map associated to the QVI:

$$f(x) \triangleq \|x - P_{K(x)}(x - F(x))\|_2.$$

The following result provides a thoretical basis for the reformulations of the QVI we consider.

Proposition 1 *The following relations are equivalent:*

(i) $\widehat{x} \in SOL(F, K)$
(ii) $f(\widehat{x}) = 0$
(iii) $\widehat{x} \in \arg\min_v \{F(\widehat{x})^\top (v - \widehat{x}) \mid v \in K(\widehat{x})\}$
(iv) $\widehat{x} \in K(\widehat{x})$, $\varphi(\widehat{x}) \geq 0$.

Proof (i) $\Longleftrightarrow$ (ii) Relations (1) and (2) are equivalent to

$$\widehat{x} \in K(\widehat{x}), \quad [\widehat{x} - (\widehat{x} - F(\widehat{x}))]^\top (y - \widehat{x}) \geq 0, \quad \forall y \in K(\widehat{x}),$$

which is equivalent to $\widehat{x} = P_{K(\widehat{x})}(\widehat{x} - F(\widehat{x}))$, thanks to Lemma 5.

(i) $\Longrightarrow$ (iii) Relations (1) and (2) entail (iii), observing that $F(\widehat{x})^\top (\widehat{x} - \widehat{x}) = 0$.

(iii) $\Longrightarrow$ (iv) Inclusion (iii), since $F(\widehat{x})^\top (\widehat{x} - \widehat{x}) = 0$, implies (iv).

(iv) $\Longrightarrow$ (i) the claim holds observing that (2) follows from $\varphi(\widehat{x}) \geq 0$. □

Under the assumptions of Theorem 1, we define the following finite quantities:

$$D \triangleq \max_{u,v \in \Omega} \|u - v\|_2, \quad F^{\max} \triangleq \max_{u \in \Omega} \|F(u)\|_2.$$

With the following result, we are able to provide some relations between approximate solutions for the QVI and for its natural map reformulation. This can be very useful in practical solving QVIs, mainly whenever a stopping criterion for a numerical procedure needs to be defined.

Proposition 2 *Assume all the assumptions in Theorem 1 to hold. Let $x \in \Omega$ and consider two cases:*

(i) $\|P_{K(x)}(x) - x\|_2 \leq \varepsilon_1$, $F(x)^\top (y - x) \geq -\varepsilon_2$, $\forall y \in K(x)$;
(ii) $f(x) \leq \varepsilon_3$.

If (i) holds then (ii) is verified with $\varepsilon_3 = \sqrt{(D + F^{\max})\varepsilon_1 + \varepsilon_2}$.

If (ii) holds, then (i) is verified with $\varepsilon_1 = 2\varepsilon_3$ and $\varepsilon_2 = (D + F^{\max})\varepsilon_3$.

Proof Let $z = P_{K(x)}(x - F(x))$. If (i) holds, then

$$\begin{aligned}
-\varepsilon_2 \leq F(x)^\top (z - x) &= [x - z - (x - z - F(x))]^\top (z - x) \\
&= -\|x - z\|_2^2 - (x - F(x) - z)^\top (z - x) \\
&= -\|x - z\|_2^2 - (x - F(x) - z)^\top (z - P_{K(x)}(x)) \\
&\quad -(x - F(x) - z)^\top (P_{K(x)}(x) - x)
\end{aligned}$$

$$\begin{aligned}
&\le -\|x - z\|_2^2 - (x - F(x) - z)^\top (P_{K(x)}(x) - x) \\
&\le -\|x - z\|_2^2 + (\|x - z\|_2 + \|F(x)\|_2)\|P_{K(x)}(x) - x\|_2 \\
&\le -\|x - z\|^2 + (D + F^{\max})\varepsilon_1,
\end{aligned}$$

where the second inequality holds thanks to Lemma 5. Thus,

$$f(x) \le \sqrt{(D + F^{\max})\varepsilon_1 + \varepsilon_2}.$$

If (ii) holds, then, as for feasibility,

$$\begin{aligned}
\|P_{K(x)}(x) - x\|_2 &= \|P_{K(x)}(x) - P_{K(x)}(z) + z - x\|_2 \\
&\le \|P_{K(x)}(x) - P_{K(x)}(z)\|_2 + \|z - x\|_2 \\
&\le 2\|z - x\|_2 \le 2\varepsilon_3,
\end{aligned}$$

where the second inequality is due to Lemma 5.

Moreover, thanks again to Lemma 5, we have

$$[z - (x - F(x))]^\top (y - z) \ge 0, \quad \forall y \in K(x),$$

and, thus, for all $y \in K(x)$,

$$F(x)^\top (y - x) \ge (x - z)^\top (y - z) + F(x)^\top (z - x) \ge -(D + F^{\max})\varepsilon_3. \qquad \square$$

5 Problems with Linear Coupling Constraints

We consider the case of a set $K(x)$ that is described by means of functional coupling constraints that are given by linear equalities. Specifically, from now on, we assume the following blanket conditions to hold:

(A1) $K(x) = \{y \in Y : c(y, x) = 0\}$, where $Y \subseteq \mathbb{R}^n$ is nonempty, closed and convex, and $c : \mathbb{R}^n \times \mathbb{R}^n \to \mathbb{R}^m$ is an affine mapping, that is,

$$c(y, x) \triangleq Ay + Bx + d,$$

with A, B $m \times n$ matrices and $d \in \mathbb{R}^m$;

(A2) F is continuous and monotone on Y.

In Theorem 2, which follows from Theorem 1, we give some sufficient conditions for the existence of a solution of QVI(F, K) under the following assumption.
Assumption (B) Let Y be bounded and any of the following conditions hold:

(i) Y is polyhedral and for every $x \in Y$, $c(y_x, x) = 0$ for some $y_x \in Y$;
(ii) for every $x \in Y$, $c(y_x, x) = 0$ for some $y_x \in \operatorname{relint} Y$.

We remark that the claim in the following result holds for any F continuous, not necessarily monotone.

Theorem 2 *If Assumption (B) holds, then K is continuous relative to Y and, in turn, $SOL(F, K) \neq \emptyset$.*

Proof First, we show that, under (A1), K is inner semicontinuous relative to Y.

As for (i) in Assumption (B), K is continuous, see e.g. [13, Example 9.35] and, a fortiori, inner semicontinuous relative to Y.

Concerning (ii) in Assumption (B), one can write $Y = \{y \in \mathbb{R}^n : \overline{A}y + \overline{d} = 0\} \cap \overline{Y}$, for some $p \times n$ matrix $\overline{A}$, $\overline{d} \in \mathbb{R}^p$ and $\overline{Y} \subseteq \mathbb{R}^n$ with $\operatorname{int} \overline{Y} \neq \emptyset$, and $\overline{Y}$ compact and convex. Thus, since $K(x) = \{y \in \overline{Y} : \overline{c}(y, x) = 0\}$, where

$$\overline{c}(y, x) \triangleq \begin{pmatrix} A \\ \overline{A} \end{pmatrix} y + \begin{pmatrix} B \\ 0 \end{pmatrix} x + \begin{pmatrix} d \\ \overline{d} \end{pmatrix},$$

for every $x \in Y$, there exists $y_x \in \operatorname{int} \overline{Y}$ such that $\overline{c}(y_x, x) = 0$.

We show that K is inner semicontinuous relative to Y, that is for any $\widetilde{x} \in Y$ and $\widetilde{y} \in K(\widetilde{x})$, and any sequence $x^k \underset{Y}{\rightarrow} \widetilde{x}$, a sequence $\{y^k\}$ exists such that $y^k \in K(x^k)$, and $y^k \rightarrow \widetilde{y}$. Let $\overline{C}(x) \triangleq \{y \in \mathbb{R}^n : \overline{c}(y, x) = 0\}$. The mapping $\overline{C}$ is ([13, Example 9.35]) continuous relative to Y: thus, $\overline{c}(\widetilde{y}, \widetilde{x}) = 0$, and v^k exists such that $\overline{c}(v^k, x^k) = 0$, $v^k \rightarrow \widetilde{y}$. By the continuity of $\overline{C}$, and due to the existence of $y_{\widetilde{x}} \in \operatorname{int} \overline{Y}$ with $\overline{c}(y_{\widetilde{x}}, \widetilde{x}) = 0$, we can define $y_{x^k} \in \operatorname{int} \overline{Y}$ such that $\overline{c}(y_{x^k}, x^k) = 0$, $y_{x^k} \rightarrow y_{\widetilde{x}}$. Since $y_{x^k} \in \operatorname{int} \overline{Y}$ for every k, due to the convexity of $\overline{Y}$, we can consider $y^k \triangleq (1 - \tau^k)v^k + \tau^k y_{x^k}$, where $\tau^k \in [0, 1)$ is the smallest scalar such that $y^k \in \overline{Y}$. Moreover, we also have $\overline{c}(y^k, x^k) = 0$ and, thus, $y^k \in K(x^k)$ for every k. Recalling that $v^k \rightarrow \widetilde{y} \in \overline{Y}$, we get $y^k \rightarrow \widetilde{y}$.

The outer semicontinuity of K relative to Y (see Lemma 2) implies the continuity of K relative to Y, and $\text{SOL}(F, K) \neq \emptyset$ follows taking in Theorem 1 $\Omega = Y$. □

Remark 1 We consider (see assumption (A1)) a feasible set K defined by functional coupling constraints that are only equalities. This is done without loss of generality: consider a solvable QVI(G, M) with G Lipschitz continuous and monotone on V, $M(z) = \{v \in V : Dv + Ez + t \leq 0\}$, where $V \subseteq \mathbb{R}^n$ is nonempty, compact and convex, and with D, E $m \times p$ matrices and $t \in \mathbb{R}^m$. Take

$$n = p + m, \quad x = \begin{pmatrix} z \\ s \end{pmatrix}, \quad F(x) = \begin{pmatrix} G(z) \\ 0 \end{pmatrix}$$

$$Y = V \times \prod_{i=1}^{m} [0, u_i], \quad u_i = - \min_{(v,z) \in V \times V} D_{i*}v + E_{i*}z + t_i \in [0, +\infty)$$

moreover, K and c are defined as in assumption (A1), with

$$A = \begin{pmatrix} D & I \end{pmatrix}, \quad B = \begin{pmatrix} E & 0 \end{pmatrix}, \quad d = t.$$

It is standard to show that $\widehat{z} \in \text{SOL}(G, M)$ and $\widehat{s} = -(D + E)\widehat{z} - t$ if and only if $\widehat{x} = (\widehat{z}^\top\ \widehat{s}^\top)^\top \in \text{SOL}(F, K)$. As for the resulting QVI(F, K), K falls within the class of constraints defined by assumption (A1), and $Y \subseteq \mathbb{R}^n$ is also bounded, as well as F verifies assumption (A2). Note that, however, whenever G is strongly monotone on V, F is only monotone. Therefore, the framework we consider encompasses all QVIs with linear functional coupling constraints.

The following result better specifies Proposition 2 for the linear coupling constraint case we are considering.

Proposition 3 *Let*

(i) $x \in Y$, $\|c(x, x)\|_2 \le \varepsilon_1$, $F(x)^\top(y - x) \ge -\varepsilon_2$, $\forall y \in K(x)$;
(ii) $x \in Y$, $f(x) \le \varepsilon_3$.

If (i) holds and $\kappa \ge 0$ *exists such that*

$$\|P_{K(x)}(x) - x\|_2 \le \kappa\, \varepsilon_1, \tag{5}$$

then (ii) is verified with $\varepsilon_3 = \sqrt{(D + F^{\max})\kappa\, \varepsilon_1 + \varepsilon_2}$.

If (ii) holds, then (i) is verified with $\varepsilon_1 = \|A\|_2\, \varepsilon_3$ *and* $\varepsilon_2 = (D + F^{\max})\varepsilon_3$.

Proof Let $z = P_{K(x)}(x - F(x))$. If (i) holds, then the proof is similar to that of Proposition 2:

$$\begin{aligned}
-\varepsilon_2 &\le F(x)^\top(z - x) = [x - z - (x - z - F(x))]^\top(z - x)\\
&= -\|x - z\|_2^2 - (x - F(x) - z)^\top(z - x)\\
&= -\|x - z\|_2^2 - (x - F(x) - z)^\top(z - P_{K(x)}(x))\\
&\quad -(x - F(x) - z)^\top(P_{K(x)}(x) - x)\\
&\le -\|x - z\|_2^2 - (x - F(x) - z)^\top(P_{K(x)}(x) - x)\\
&\le -\|x - z\|_2^2 + (\|x - z\|_2 + \|F(x)\|)\|P_{K(x)}(x) - x\|\\
&\le -\|x - z\|^2 + (D + F^{\max})\kappa\, \varepsilon_1,
\end{aligned}$$

where the last inequality is due to (5). Finally, $f(x) \le \sqrt{(D + F^{\max})\kappa\, \varepsilon_1 + \varepsilon_2}$.

If (ii) holds, then, as for feasibility,

$$\|c(x, x)\|_2 = \|c(z, x) - c(x, x)\|_2 \le \|A\|_2 f(x) \le \|A\|_2\, \varepsilon_3.$$

While to obtain the bound ε_2 the proof is the same of that of Proposition 2. □

Condition (5) is verified e.g. whenever Y is polyhedral, as in case (i) of Assumption (B). This is due to the renowned Hoffman error bound result that we report for completeness.

Lemma 6 *(From [7, Lemma 3.2.3]) A constant $\kappa > 0$ exists such that*

$$\|P_{K(x)}(x) - x\|_2 \leq \kappa \|c(x, x)\|_2,$$

for every $x \in Y$ such that $K(x) \neq \emptyset$.

References

1. Bigi, G., Lampariello, L., Sagratella, S., Sasso, V.G.: Approximate variational inequalities and equilibria. Comput. Manag. Sci. **20**, 43 (2023)
2. Bank, B., Guddat, J., Klatte, D., Kummer, B., Tammer, K.: Non-Linear Parametric Optimization. Akademie-Verlag, Berlin (1982)
3. Cesarone, F., Lampariello, L., Merolla, D., Ricci, J.M., Sagratella, S., Sasso, V.G.: A bilevel approach to ESG multi-portfolio selection. Comput. Manag. Sci. **20**, 1–24 (2023)
4. Dreves, A., Sagratella, S.: Nonsingularity and stationarity results for quasi-variational inequalities. J. Optim. Theory Appl. **185**, 711–743 (2020)
5. Facchinei, F., Kanzow, C., Sagratella, S.: QVILIB: a library of quasi-variational inequality test problems. Pac. J. Optim. **9**, 225–250 (2013)
6. Facchinei, F., Kanzow, C., Sagratella, S.: Solving quasi-variational inequalities via their KKT conditions. Math. Prog. **144**, 369–412 (2014)
7. Facchinei, F., Pang, J.-S.: Finite-Dimensional Variational Inequalities and Complementarity Problems. Springer (2003)
8. Lampariello, L., Sagratella, S., Sasso, V.G.: Nonsmooth hierarchical multi portfolio selection. In: Cappanera, P., Lapucci, M., Schoen, F., Sciandrone, M., Tardella, F., Visintin, F. (eds.) Optimization and Decision Science: Operations Research, Inclusion and Equity. ODS 2022, vol. 9, pp. 37–46. AIRO Springer Series (2022)
9. Lampariello, L., Sagratella, S., Sasso, V.G., Shikhman, V.: Scalarization via utility functions in multi-objective optimization, pp. 1–28 (2024). arXiv:2401.13831
10. Lampariello, L., Sagratella, S., Sasso, V.G.: Dealing with inexactness in hierarchical multi-portfolio selection. In: Bruglieri, M., Festa, P., Macrina, G., Pisacane, O. (eds.) Optimization in Green Sustainability and Ecological Transition. ODS 2023, pp. 1–8. AIRO Springer Series (2023)
11. Lampariello, L., Sagratella, S., Sasso, V.G., Stein, O.: Solving multi-follower games. Optim. Online, 1–26 (2023). https://optimization-online.org
12. Pang, J.-S., Fukushima, M.: Smith, T.F., Waterman, M.S.: Quasi-variational inequalities, generalized Nash equilibria, and multi-leader-follower games. Comput. Manag. Sci. **2**, 21–56 (2009)
13. Rockafellar, R.T., Wets, J.B.: Variational Analysis. Springer (1998)

A Comparison of Solution Approaches for the Port Rail Shunting Optimization: A Case Study

Daniela Ambrosino, Veronica Asta, and Camilla Traversa

Abstract The focus of this study is to optimize the shunting operations in the port area for the transfer of trains from the rail network to maritime terminals and vice versa, with the aim of managing all activities with the available resources while minimizing train dwell times. Two methods are used to schedule and optimize these activities. The two methods are then compared, in terms of computational time and goodness of solution. Preliminary results are given.

Keywords Maritime container terminal · Port rail management · Heuristic approaches

1 Introduction and Problem Description

Ports are essential gateways to international markets and form an ecosystem composed of transportation, energy, and industry systems, helping to build strategic corridors with added value for the economy. In addition, due to their integration into a comprehensive global network, ports provide transportation links for the mobility, delivery and receipt of goods, which, using other transport mode links, reach their destination from where they are then delivered to consumers.

Seaports therefore play a very complex and strategic role not only in terms of domestic activities and the provision of exports and imports, but also because of their role as a crucial hub in connecting with other transport networks. The analysis of the modal split in freight transport [1] stresses this fact. Moreover, port-rail links add

D. Ambrosino (✉)
Department of Economics, University of Genoa, Genoa, Italy
e-mail: daniela.ambrosino@unige.it

D. Ambrosino · V. Asta · C. Traversa
OPTIMeasy - University of Genoa Spin Off, Genoa, Italy
e-mail: veronica.asta@optimeasyway.com

C. Traversa
e-mail: camilla.traversa@optimeasyway.com

M. Di Francesco et al. (eds.), *Operations Research: Closing the Gap Between Research and Practice*, AIRO Springer Series 15, https://doi.org/10.1007/978-3-031-90095-2_30

value to port users by reducing overall transportation costs and increasing business activities through extension with the hinterland. For this reason, it is essential to promote and invest in transport connections.

In the previous Multiannual Financial Framework, the European Commission focused on shifting traffic to sustainable transport modes (rail and inland waterways) and thus on financing rail infrastructure and connections with other modes of transport [2].

Planned interventions in northern Italian ports [3] include the installation of new equipment at Vado Ligure station, the construction of new tracks at the P.S.A Genoa Pra' Terminal, and the implementation of new tracks and technological equipment in Trieste.

Having explained the importance of linking ports and railways, it is necessary to understand how these two infrastructures are actually connected. Typically, the general layout of a rail-sea interchange includes three elements: a railway station, a shunting area, and a sea terminal area, as sketched in Fig. 1. The railway station is connected to both the entire rail network on one side and the shunting area on the other. The shunting area allows the transfer of trains from the rail network to the terminals and vice versa. For what concerns the import cycle, goods are transported by sea from the shipping company to the port terminal. All the activities related to the ship unloading, the goods storage in the terminal yard and the goods loading on trains are executed under the responsibility of the maritime terminal. When the train is loaded, all the operations required to transfer the train from the terminal to the tracks outside the port and other ancillary operations, such as unloading cars, transferring individual cars, and bringing in/extracting empty cars, are entrusted to the shunting operator (see [4] for more details); all these activities are defined as shunting operations in [5].

Thus, the first train handover is from the terminal operator to the shunting manager and occurs when the train is on a terminal track (straddling the blue area and the yellow area in Fig. 1). The second handover of the train is from the shunting manager to the railroad company and usually occurs when the train is on a terminal track outside the port area (straddling the yellow area and the red area in Fig. 1). From this point, the train will travel to the destination inland terminal on the national rail network, under the control of the infrastructure manager.

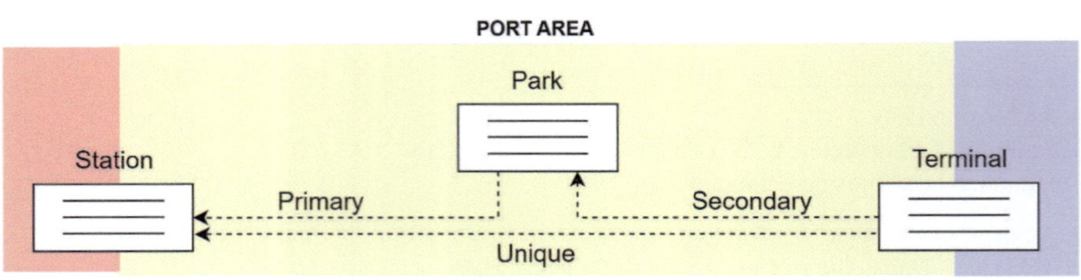

Fig. 1 The port area and maneuvers required by the import cycle

In order to transfer trains from the terminal to the station (and vice versa), specific manoeuvres must be performed. Note that in this paper the whole trains must be transferred within the port area, due to the fact that all the wagons composing a train have the same origin/destination maritime terminal.

According the availability of resources and the train requirements, the following manoeuvres can be executed:

- unique shunting manoeuvre: the train is moved directly from the terminal to the station, in the case of import cycle (vice versa in case of export cycle), when all the required resources are available and no other ancillary operations are required;
- the secondary shunting and then the primary shunting manoeuvre in the case of import cycle, and vice versa in case of export cycle: the import train is moved from the terminal to a shunting park and then from the park to the station (vice versa in case of export cycle), when either at least one of the required resources is not available or one ancillary operation is required; moreover, the train has to wait its correct time for either exiting the port area or entering the terminal.

These manoeuvres can be optimized through operational research models [6]. Activities must be scheduled to respect time constraints, handling procedure characteristics, and to achieve the best utilization of terminal resources that can be shared among different terminal activities [7].

In more detail, the problem under investigation concerns the schedule of the shunting operations that a set of export and import trains must perform within a given time horizon, having a specific port area, with parks and rail trucks, locomotives and human resources, time windows limitations and specific instant time to respect for entering and exiting the shunting area.

The paper is organized as follows: a brief literature review is reported in Sect. 2. The solution approaches are described in Sect. 3, while a case study and preliminary results are given in Sect. 4. Conclusions are reported in Sect. 5.

2 Literature Review

Operations research approaches can be useful as support for decision makers to improve the process and foster the development of rail transportation [8]. One of the first studies defining the port rail shunting problem was [4], that described the entire rail transportation process in the context of intermodal transport, and highlighted critical points that can be improved. In [9] an operation-time-space network has been proposed to model various logistical problems involving temporal and spatial decisions, and among the cited problems a rail shunting scheduling problem has been investigated: an application of the operation-time-space network and its model are described.

The major part of the literature is devoted to the two crucial and essential issues in railway operations that are scheduling and train routing. In [10] the authors develop a joint optimization model in a multiresolution time-space network by coupling conflict-free routing constraints to the train scheduling process. In [11] the authors implemented the simultaneous allocation of railcars and locomotives to plan train formation according to schedule. The authors developed a mixed integer linear programming model with the goal of maximizing the railway company's profits from satisfying customer demand from freight trains in the rail network. Railway station capacity, locomotive traction, train cancellation, and active and dead consistency of locomotives are also considered in this mathematical model.

Often the schedule regards either a portion of the train or a wagon/car, and the related more popular problems are known as train unit scheduling problem (TUSP) [12] or train unit assignment problem (TUAP) [13].

A critical problem encountered in a rail freight system is the determination of the best formation plan on a limited-capacity physical network. The formation plan is the foundation of rail operations, the basis of train scheduling, terminal management, and resource planning. In [14] a nonlinear binary programming model with the objective of minimizing the total costs of accumulation, reclassification, and travel distances while satisfying different practical conditions is proposed. To solve the mathematical model, an efficient heuristic solution approach based on simulated annealing is developed. They use a penalty function method to address capacity constraints and a customized methodology for operational requirements.

3 Methodology

The scheduling problem introduced in Sect. 1 has been solved by using different approaches. A flow model has been derived from [5] and used to solve small instances; a constructive heuristic has been used to obtain an initial feasible solution by assigning shunting operations following the initial sequence of trains. Then, the initial solution has been improved by applying two different algorithms: a Local Search and a Simulated Annealing. In the following sub-sections the heuristics approaches are briefly described while interested readers can obtain the MILP model from [5].

3.1 A Constructive Approach

A simple constructive heuristic has been proposed to obtain an initial feasible solution; given a list of trains to serve (i.e. a sequence of trains), it is necessary to define and initialize the resource timelines. To each resource (tracks, shunting teams, etc.), it is assigned a timeline in such a way to manage the resource's occupancy as a time function. A resource is busy when it manoeuvres and free when does not manoeuvre.

The constructive approach is based on the following steps; for each train in the given sequence, steps 1 and 2 are repeated until an initial solution x is obtained. Please note that, it is also possible to return a solution with some trains not served.

For each train z of the sequence:

Step1: Allocation of shunting operations of train z:
for each operation of train z
search for time interval greater than the operation's duration in which all necessary resources are available to execute it;
Step2: Updating timelines and costs:
update the resources' capacity;
compute the total cost for train z stopover.

3.2 The Local Search Algorithm

The local search algorithm tries to improve the initial solution obtained by the constructive approach described in Sect. 3.1. Given a current solution x and its cost c_x, the neighbourhood of that solution is defined thanks to two different moves, i.e. a *Swap* move and an *Insert* one, used in a given sequence of trains. The length of the sequence of trains and the maximum number of iterations to execute are defined.

The *Swap* move swaps the trains related to two random indexes in the sequence of trains under consideration; the *Insert* move takes two random indexes and inserts the element that is in the position of the first index into the position identified by the second one. To alternate the two moves, the first is used in even iterations while the second is used in odd ones.

The local search algorithm is based on the following steps that are repeated until the maximum number of iterations is reached. Then the best solution is returned.

For each iteration k:

Step1: *generate* randomly two variables a and b, such that:
$a \neq b$
$a, b <=$ number of trains in the considered sequence;
perform a move
if k is even
do swap move: trains with indices a and b are swapped;
else (if k is odd)
do insert move: the train with index a is inserted into the sequence at position $b - 1$ (immediately before the train with index b);
Step 2: *compute*
by using the constructive heuristic introduced in Sect. 3.1 the solution x_n on the new sequence and its cost c_n;
if $c_n < c_x$
$x_c = x_n$ (the new solution become the current solution).

3.3 The Simulated Annealing Algorithm

To avoid the risk of ending up in a local optimum, a simulated annealing algorithm is used. Transitions that correspond to worsening are accepted, in order to explore more comprehensively the set of feasible solutions. The probability of accepting worsening solutions gradually decreases over the course of the search procedure until it tends to zero. In the simulated annealing algorithm, the temperature, cooling value and acceptance probability are valued and the maximum number of iterations is defined.

Thus, the following Steps 1, 2 and 3 are repeated until the maximum number of iterations is reached. Then, the best solution is returned.

For each iteration k:

Step1: *generate* randomly two variables a and b, such that:
  $a \neq b$
  $a, b <=$ number of trains in the considered sequence;
perform a move
  if k is even
    do swap move: trains with indices a and b are swapped;
  else (k is odd)
    do insert move: the train with index a is inserted into the sequence at position $b - 1$ (immediately before the train with index b);

Step 2: *compute*
  by using the constructive heuristic introduced in Sect. 3.1 the solution x_n on the new sequence and its cost c_n;
  the temperature t and acceptance probability p (based on the new temperature);

Step 3: *generate* a random number r:
  if $c_n < c_x$
    $x_c = x_n$ (the new solution become the current solution);
  else
    if r is less than the acceptance probability p
      $x_c = x_n$ (the new solution become the current solution).

4 Case Study and Results

In this section, a case study is reported with the results of several experiments. All the approaches have been implemented in Python language (version 3.9.7) and solved with the commercial solver Gurobi Optimizer version 10.0.0, on an Intel(R) Core (TM) i3-10110U CPU @ 2.10 GHz 2.59 GHz, 8.00 GB RAM).

The case study concerns a system consisting of a railway station considered as a buffer, two marshalling yards and two maritime terminals, as depicted in Fig. 2 and each zone is described below.

Starting from the buffer areas (represented by rectangles):

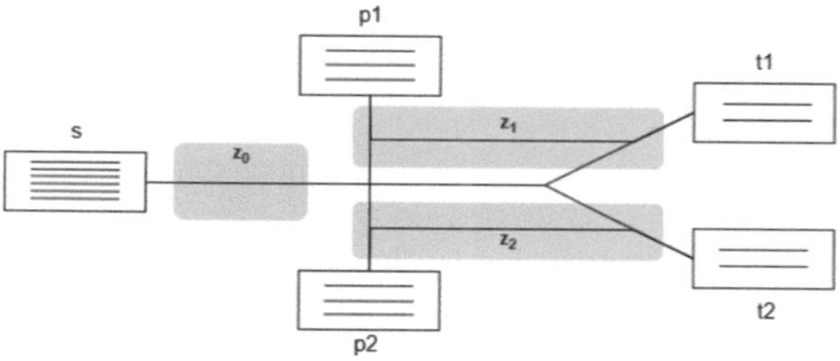

Fig. 2 Case study port area configuration

- s represents the railway station, where trains start and end shunting operations and can wait after their arrival or before their departure;
- p1 and p2 represent the marshalling yards, where the necessary checks can be carried out or the waiting operations for trains arriving from the railway station or terminals can be performed;
- t1 and t2 represent the terminals, inside which one train at a time can enter and perform loading/unloading operations; the terminals are the origin or destination of import or export trains passing through the port.

The manoeuvring areas z_0, z_1, z_2, represented by rounded grey rectangles, are the areas where manoeuvres are carried out. It is not permitted to perform different types of manoeuvres at the same time in the same manoeuvring area.

Both import and export trains must be handled in this system. The export cycle is here explained. An export train arrives at the buffer station according to its arrival time and must be inside the destination terminal after completing its shunting. Two cases may occur. In the first case, since no technical inspection is to be carried out, if the areas to be crossed are free and the terminal is empty, the train will be transported directly to the terminal through a single shunting. In the second case, if the train has to stop in the park because technical inspections have to be carried out or if the single shunting cannot be carried out because either the zones or the terminal are occupied; the team, performing a primary shunting, will bring the train into the park. After the stop, the train will be transported into the terminal by means of a secondary shunting. If the marshalling or staging areas are congested, the export train may also stop at the station before starting the process.

Figure 3 shows a part of the operation-time-space network used: (a) for the export cycle and (b) for the import cycle.

Specifically, the green arcs depicted in (a) represent the export cycle to terminal t1 and in (b) the import cycle from terminal t1. The red arcs depicted in (a) represent the export cycle to terminal t2 and in (b) the import cycle from terminal t2.

Taking export trains headed to terminal t1 as an example (green arcs (a)): starting from station s, they can either:

- perform unique shunting by crossing the z_0 and z_1 zones (vertical arrows s $\rightarrow z_0$ and $z_0 \rightarrow z_1$) arriving directly at the terminal (curved arrow from $z_1 \rightarrow$ t1);
- perform primary shunting first by crossing the z_0 and z_1 zones (vertical arrows s $\rightarrow z_0$ and $z_0 \rightarrow z_1$) arriving at the p1 park (curved arrow from $z_1 \rightarrow$ p1) and then

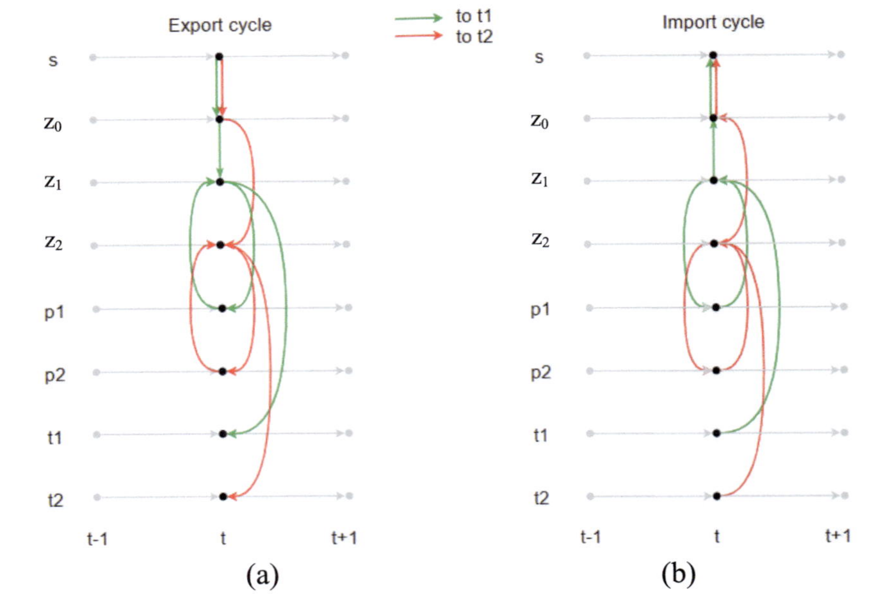

Fig. 3 Operation-time-space network an example of import and export flows

perform secondary shunting by crossing the z_1 zone again (curved arrow from p1 $\rightarrow z_1$) arriving at the terminal t1 (curved arrow from $z_1 \rightarrow$ t1).

Particular attention should be paid to the z_0 area, which is traversed by both green and red arcs. This is the crux of the case study layout: the sharing of this infrastructure makes the problem complex. Therefore, it is not possible to solve the problem by considering separate paths, but it is necessary constrain the crossing of this zone.

Primary and secondary operations for an export train that has to reach terminal 1 are here below depiched in Fig. 4.

Some experimental campaigns have been developed in order to validete the proposed solution approaches. In Table 1 the results obtained by the local search and the simulated annealing on different instances caractherized by a planning horizon of one day, 7 pairs of trains, 3 of which are headed to Terminal 1 and 4 to Terminal 2; no train has to perform the technical verifications, no time limitation is considered

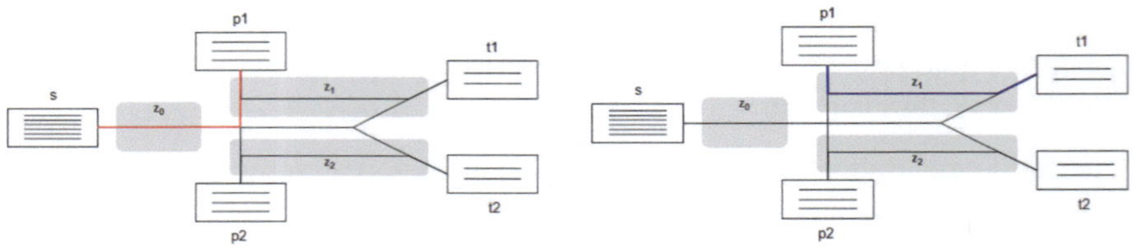

Fig. 4 Primary and secondary shunting representation for trains heading to t1

for entering/exiting terminals and a single shunting team is considered. Stopovers at the terminal are minimized, maintaining a minimum of 90 min to carry out loading/ unloading operations. The number of iterations for both two algorithms is 1000.

From Table 1 it is quiete evident that the local search approach performs better than the simulated annealing algorithm. Both approaches arrive at the same solutions in all performed tests but the required CPU time is lower when the local search is used.

A different experimental campaign to investigate the problem complexity has been performed to compare the solutions obtained with the mathematical formulation derived by [5] and the heuristic approach. Specifically, the execution time and the gap obtained in each solution were compared, starting with a simple case with only 1 pair of trains and reaching a complex case with 15 pairs of trains to be served in a single day.

Looking at the obtained results reported in Table 2, it can be seen that, by increasing the number of train pairs from 1 up to 9 (above the red dashed horizontal line in the table), with both solving methods the optimal solution was obtained (gap = 0%). However, the CPU time is quite different: with the heuristic approach, the CPU time ranges from 9 to 13,1 s, while the mathematical model requires from 7.3 to 355 s. Turning to the second part of the table (below the red dashed line) where the number of trains increases, the complexity of the problem increases too.

Thanks to the mathematical model it is possible to obtain the optimal solution, even if the CPU time grows up to 2881 s (more than 48 min). The proposed heuristics

Table 1 Comparison between LS and SA results

	Local search		Simulated annealing	
Test	F.ob	Run time	F.ob	Run time
0	630	5,02	630	1.09
1	630	0.55	630	0.47
2	630	4.59	630	4.43
3	630	0.45	630	4.59
4	630	0.54	630	6.42
5	630	0.52	630	6.61
…	…	…	…	…
95	630	4.52	630	44.35
96	630	4.45	630	4.44
97	630	4.49	630	44.66
98	630	0.05	630	4.46
99	630	4.61	630	4.43
Avg	630	3.57	633	8.47
Min	630	0.05	630	0.44
Max	630	5.02	720	50.3

Table 2 Results comparison

# Train pairs	Best bound	Mathematical model			Heuristic approach		
		Solution	Gap (%)	Run time	Solution	Gap (%)	Run time
1	90	90	0.0	7.3	90	0.0	10.1
2	180	180	0.0	14.2	180	0.0	10.3
3	270	270	0.0	43.4	270	0.0	9.0
4	360	360	0.0	59.8	360	0.0	9.9
5	450	450	0.0	107.0	450	0.0	9.6
6	540	540	0.0	160.9	540	0.0	10.7
7	630	630	0.0	185.0	630	0.0	11.7
8	720	720	0.0	239.0	720	0.0	12.2
9	810	810	0.0	355.0	810	0.0	13.1
10	900	900	0.0	438.0	903	0.3	14.0
11	990	990	0.0	611.0	998	0.8	15.5
12	1080	1080	0.0	798.0	1097	1.6	17.3
13	1170	1170	0.0	1170.0	1216	3.9	16.4
14	1260	1260	0.0	1950.0	1332	5.7	17.4
15	1350	1350	0.0	2881.0	1477	9.4	17.9

permits to obtain solutions in less than 20 s with a gap ranging from 0.3 to 9.4%. Note that the best bound, on average, is obtained by the MILP model in 78% of the time spent to return the optimal solution.

Figures 5 and 6 report the solutions obtained by using the MILP model and the heuristic approach for managing 10 couple of trains.

5 Conclusions

The purpose of this work is to find a solution approach that can be used in a DSS in port areas, for supporting rail shunting managers. By optimizing the schedule of the operations required to transfer trains in the port area, it is possible to realize an efficient usage of the resources and thus to increase the efficiency; finally, this permits to shuntng managers to increment the number of trains served every day.

Specifically, the application of a mathematical model was compared with a heuristic approach. Two algorithms were selected for the heuristic approach: a Local Search and a Simulated Annealing. Both algorithms seems to be promising, in particular the local search. Further experimental tests will be realized, on both mathematical model and euristic approaches, for including them in a decision tool for supporting shunting managers in ports characterized by different infrastructures.

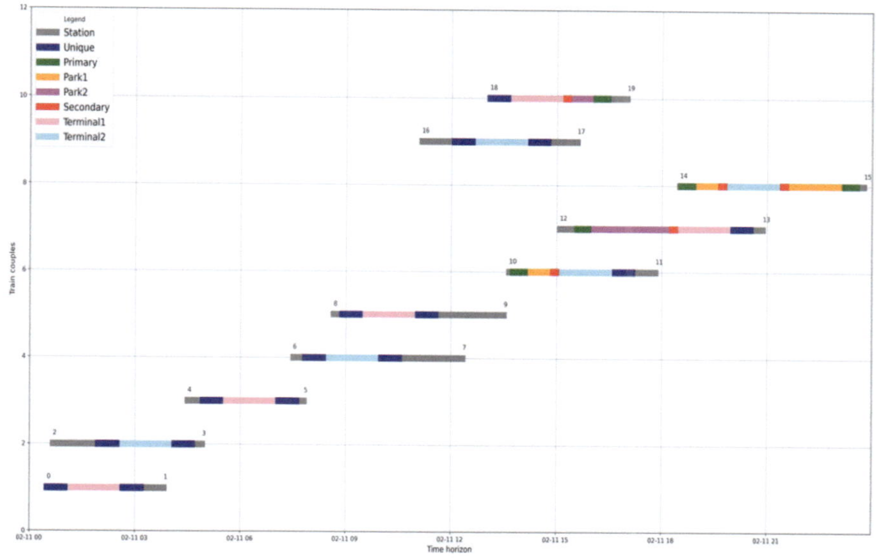

Fig. 5 Representation of the solution obtained with the MILP model

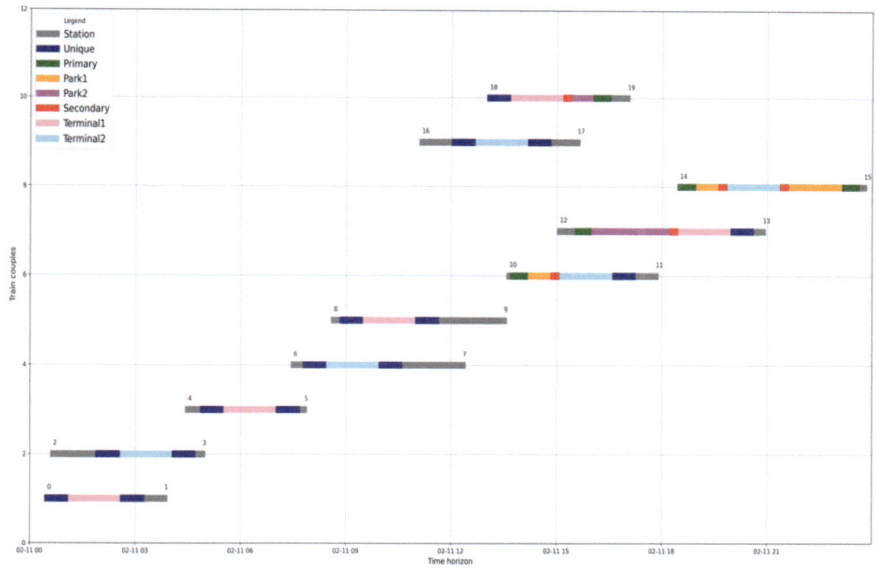

Fig. 6 Representation of the solution obtained with the heuristic approach

References

1. https://ec.europa.eu/eurostat/statistics-explained/index.php?title=Freight_transport_statistics_-_modal_split&oldid=563291
2. Luica, P.: Ports' strength is given by rail connections. https://www.railwaypro.com/wp/ports-strength-is-given-by-rail-connections/
3. Curti Sacchi, P.: Porti e ferrovie sempre più connessi nei progetti Rfi. https://www.trasportoeuropa.it/notizie/ferrovia/porti-e-ferroviesempre-piu-connessi-nei-progetti-rfi/
4. Ambrosino, D., Asta, V.: Intermodality and rail transport: focus on port rail shunting operations. In: Paolucci, M., Sciomachen, A., Uberti, P. (eds.) Advances in Optimization and Decision Science for Society, Services and Enterprises. AIRO Springer Series, vol 3. Springer, Cham (2019)
5. Ambrosino, D., Asta, V., Crainic, T.G.: Port rail shunting scheduling problem. Techincal report CIRRELT, 2022-02
6. Stahlbock, R., Vob, S.: Operations research at container terminals: a literature update. OR Spectr. **30**, 1–52 (2008)
7. Legato, P., Mazza, R.M.: A decision support system for integrated container handling in a transshipment hub. Decision Support Syst. (2018)
8. Ambrosino, D., Asta, V., Crainic, T.G.: Optimization challenges and literature overview in the intermodal rail-sea terminal. Transp. Res. Procedia **52**, 163–170 (2021)
9. Ambrosino, D., Asta, V.: An innovative operation-time-space network for solving different logistic problems with capacity and time constraints. Networks **78**, 350–367 (2021)
10. Wang, E., Yang, L., Li, P., Zhang, C., Gao, Z.: Joint optimization of train scheduling and routing in a coupled multi-resolution space-time railway network. Transport. Res. Part C: Emerg. Technol. **147**, 103994 (2023)
11. Allafeepour, A., Tavakoli, A.H., Arvin, A.: An optimization integrated approach for simultaneous allocation of railcars and locomotives for train formation based on a pre-designed time schedule. J. Rail Transport Plan. Manag. **25**, 100366 (2023). ISSN 2210-9706
12. Lin, Z., Kwan, R.S.: A 2 phase approach for real world train unit scheduling. Public Transp. **6**, 35–65 (2014)
13. Cacchiani, V., Caprara, A., Toth, P.: A fast heuristic algorithm for the train unit assignment problem. In: 12th Workshop on Algorithmic Approaches for Transportation Modelling, Optimization and Systems. Schloss Dagstuhl-Leibniz-Zentrum fuer Informatik (2012)
14. Lin, B., Zhao, Y., Lin, R., Liu, C.: Integrating traffic routing optimization and train formation plan using simulated annealing algorithm. Appl. Math. Modell. **93**, 811–830 (2021). ISSN 0307-904X

Two-Dimensional Irregular Knapsack Problem with Defects from a Dynamic Textile Environment

Layane Rodrigues de Souza Queiroz, Thiago Alves de Queiroz, Dario Vezzali, and Manuel Iori

Abstract The production of clothing, garments, footwear, and other related goods involves cutting fabric and leather rolls. High-quality goods must be free of defects, and the cutting process must be optimized to improve all stages. The problem emerges from a dynamic textile production environment in which a large object is first scanned for defective areas and then cut into smaller irregular items free of defects. The time to define the cutting pattern is short and depends on knowing the defects first. This paper proposes an iterated local search-based heuristic as a solution method to the two-dimensional irregular knapsack problem, which is an NP-hard problem. The algorithm starts by generating an initial random solution, which is then improved using perturbation and local search operators based on swap and insertion moves. Items are positioned in the large object with the bottom-left positioning rule. Experiments are conducted on instances from the literature, evaluating the solution quality for different time limits. The algorithm achieves satisfactory solutions from as early as a 30 s time limit, occupying an average of 75.63% of the object area. The occupied area increases as the time limit increases, reaching 77.78% on average for a time limit of 120 s.

Keywords Two-dimensional knapsack problem · Irregular items · Defective area · Iterated local search

L. R. de Souza Queiroz
Institute of Mathematics and Statistics, Federal University of Goiás, Goiânia-GO, Brazil
e-mail: layanequeiroz@ufg.br

T. Alves de Queiroz (✉)
Institute of Mathematics and Technology, Federal University of Catalão, Catalão-GO, Brazil
e-mail: taq@ufcat.edu.br

D. Vezzali · M. Iori
Department of Sciences and Methods for Engineering, University of Modena and Reggio Emilia, Reggio Emilia, Italy
e-mail: dario.vezzali@unimore.it

M. Iori
e-mail: manuel.iori@unimore.it

M. Di Francesco et al. (eds.), *Operations Research: Closing the Gap Between Research and Practice*, AIRO Springer Series 15, https://doi.org/10.1007/978-3-031-90095-2_31

1 Introduction

The sale of clothing and other related products has increased over the years. The Italian textile and apparel sector experienced a strong export recovery in 2021, growing about 18%. This sector is the second largest in the country and will undergo a profound transformation in the coming years, for example, to comply with sustainable production goals [11]. A common problem in this sector is the efficient use of raw materials in the cutting process. Cutting is one of the main factors that add value to products, and determining good cutting patterns is essential to reduce scraps, speed up the production process, and increase profits.

In this paper, we address the problem of cutting rolls of fabric or leather to produce clothing, footwear, and sports goods. These rolls are two-dimensional objects cut (e.g., by laser machines) to obtain small, irregularly shaped items. Items are assembled (e.g., by hand or sewing machines) to produce the final goods. The cutting process may involve some stages in which several machines are interconnected by a conveyor belt.[1] First, the large object is scanned and checked for defects (and quality areas) to confirm its dimensions. This process can take from a few seconds (e.g., in a fully automated process) to some minutes (e.g., when the intervention of an operator is needed). Therefore, the time to define the cutting pattern is short and dependent on this analysis. The large object is cut following the cutting pattern, which must respect the object dimensions, ensure no overlapping between items, and guarantee that the resulting items are free of defects. The objective is to maximize the area cut from the large object.

The problem in this work belongs to a class of NP-hard problems, namely the two-dimensional irregular knapsack problem (2IKP). The literature on the knapsack problem and variants is wide when items are regular, particularly rectangles and boxes. Recent surveys can be found in [3, 4, 10, 17], where the authors reviewed the main characteristics of the problem and its variants and discussed state-of-the-art solution methods, both exact and heuristic. When irregular items are considered, the number of papers in the literature decreases [13]. In particular, one of the most significant problems is handling the geometry of items to obtain a feasible cutting pattern. In this sense, the tutorial in [2] discussed using the raster method, phi-functions, no-fit polygons, or direct trigonometry. More recent approaches have combined the raster method with no-fit polygons [20, 24] and no-fit polygons with the direct trigonometry method [6].

The literature on the 2IKP focuses mainly on heuristic and metaheuristic algorithms. This is the case of [7], where the authors developed a constructive heuristic that positions the items in the large object according to the occupancy ratio of the convex hull. In [1], heuristic algorithms based on Lagrangian relaxation are used to solve an application in the leather industry. In [15], a general heuristic randomly chooses from six positioning rules the one to be applied to position the items of a sequence. The rules are bottom-left, left-bottom, vertical zig-zag, horizontal zig-zag, spiral, and anti-spiral. In [21], the authors proposed a biased random-key genetic

[1] One example in the cutting of leather: https://www.youtube.com/watch?v=wryrHu-a8nA.

algorithm and a variable neighborhood search for one of the problems addressed in [15], outperforming by a large amount the general heuristic. An evolutionary algorithm, which uses a mutation operator to add, remove, move, or rotate items, and a crossover operator to swap items from their positions, was proposed in [19].

Concerning the cutting of irregularly shaped items in the presence of defects, a quick location and movement heuristic that transforms irregular shapes into polygons, determines the referral point of each item, and discretizes the large object in a grid of points, is proposed in [14]. The items are then positioned following the large object's occupancy ratio and the best rotation to use. In [18], the large object may have quality areas besides defects. The author proposed heuristics based on the bottom-left positioning rule. In [5], the authors addressed the problem of cutting large slates with irregular shapes and defects. They proposed a horizontal-level heuristic embedded in a genetic algorithm to position the items. A two-stage stochastic programming model to the 2IKP with uncertain defects was proposed in [22]. The solutions of the model are evaluated in terms of the expected value of perfect information and the value of the stochastic solution, measuring the impact that uncertainties have on the problem.

Unlike the reviewed literature, we operate in a dynamic environment where not all information is known in advance. In particular, the defects on the large object become known during the scanning phase, limiting the time available to determine a (good) cutting pattern. To solve the problem, we propose an iterated local search (ILS) that uses the bottom-left positioning rule to quickly position the items. New solutions are generated using swap and insertion movements.

The remainder of the paper is organized as follows. Section 2 reports the formal definition of the problem and the method used to guarantee the feasibility of the solutions. Section 3 presents the proposed heuristic algorithm, providing details on how a problem solution is encoded, how the initial solution is generated, how the local search is performed, and how the solutions are perturbed. Section 4 gives the results of the computational experiments performed on instances from the literature, evaluating the impact of different time limits on the solution quality. Concluding remarks and directions for future work are discussed in Sect. 5.

2 Problem Definition

This work addresses a variant of the 2IKP in which the large object, although having known dimensions, has defective areas. Since we operate in a dynamic cutting environment, these areas are detected in the first stage. Therefore, in the next stage, the resulting items must be free of defects when cutting the large object. The period at which the large object exits the first stage and enters the second stage is denoted by T (in seconds and generally depends on the environment). The cutting pattern to be processed in the second stage must be defined at this period.

A cutting pattern is a feasible positioning of irregular items in a large object, such that no item exceeds the object dimensions, no item overlaps one another, and no item overlaps a defective area. The objective of the 2IKP is to obtain a cutting pattern with the maximum occupancy area; thus, the sum of the area of the items cut cannot extrapolate the large object area. For this problem, we assume the large object is rectangular with dimensions (L, H): L is associated with the x-axis, and H is associated with the y-axis. The large object is located in the first quadrant of the Cartesian plane with origin at (0, 0). We have n items in set I: each item i has area a_i, can be rotated in discrete angles in R_i, and has a reference vertex (with the smallest y-coordinate and breaking ties by the smallest x-coordinate). The defects are modeled as (irregular) items in set D.

To determine a cutting pattern, we assume that the large object is discretized into a grid of points, and the items are positioned according to their reference vertex at the points of this grid [24]. To guarantee that the items do not exceed the object dimensions, we compute the inner-fit raster: each item is translated internally, always touching one of the object sides; the path constructed by the reference vertex results in the inner-fit polygon; the polygon is discretized as in the raster method. Also, to guarantee that the items do not overlap with each other (and with defective areas), we compute the inner-fit polygon: for each pair of items, one item is fixed while the other translates around the fixed one, always remaining in touch but without overlapping; the path constructed by the reference vertex of the translating item results in the no-fit polygon; the polygon is then discretized following the raster method. More details on how to compute the inner-fit and no-fit rasters can be obtained in [16, 24].

3 Iterated Local Search-Based Heuristic

A cutting pattern for the 2IKP variant addressed in this work must be computed within T seconds. We propose an ILS-based heuristic algorithm to achieve this goal and produce good solutions in short computing times. The ILS has been successfully applied to solve many combinatorial optimization problems, including cutting and packing. In [9], the authors developed an ILS to solve an overlap minimization problem and obtain high-quality solutions for the two-dimensional irregular strip packing. In [12], the ILS was applied to the bin packing problem with weighted precedence constraints between pairs of items. Recently, a multi-start ILS for the corridor allocation problem was developed in [8]. The authors used a variable neighborhood descent in the local search phase and neighborhood structures based on swap, insert, inverse, cross, and block insert moves. A detailed overview of the ILS can be found in [23].

The proposed ILS is described in Algorithm 1. The time limit for a solution, T, is given as input. The best solution, S, is returned as output. A vector of integers represents a solution. Each position of this vector contains the index of an item in I. Moreover, following up on the good results by [21], we assume this vector also has k additional positions filled with value -1. Positions containing value -1 do not represent an item in I, but instead, they indicate a flag value later used by the positioning rule to evaluate the cutting pattern and obtain the objective function value (i.e., the occupied area of the large object). In this way, a solution vector has size $n + k$. The initial solution, S_0, is generated by randomly filling the solution vector with the indices of the items in I and the k additional -1 values. The solution is then passed to the local search procedure, which returns the best solution S. The while loop in lines 4–10 iterates until the remaining time T^* is reached. At each iteration, a perturbed solution S_1 is generated. Then, S_1 is passed to the local search procedure, resulting in solution S_2, which updates the best solution S if it shows a better area occupancy than S.

Algorithm 1 Iterated Local Search developed for the 2IKP with defects

1: $S_0 \leftarrow$ randomly generated initial solution
2: $S \leftarrow$ Local-Search(S_0)
3: $T^* \leftarrow$ difference between T and the time spent on lines 1 and 2
4: **while** time limit T^* is not reached **do**
5:     $S_1 \leftarrow$ perturbed solution (starting from S)
6:     $S_2 \leftarrow$ Local-Search(S_1)
7:     **if** $area(S_2) > area(S)$ **then**
8:         $S \leftarrow S_2$
9:     **end if**
10: **end while**

In the perturbation procedure, we randomly select three positions of the solution vector, i, j, and l; and we then swap the items associated with these positions ($i \rightarrow j$, $j \rightarrow l$, and $l \rightarrow i$). In the local search procedure, we perform the following swap and insertion moves. Swap consists of considering all pairs of positions i and j, swapping their contents, and continuing until the first improvement is found; insertion consists of considering all pairs of positions i and j, inserting the content of position i between positions $j-1$ and j, continuing until the first improvement is found. The solution resulting from (swap) is used as input to (insertion), while the solution resulting from (insertion) is allocated to S_2. Other moves were also tested (e.g., performing the moves on sequences of positions instead of single positions) but proved time-consuming, thus incompatible with the imposed time limit T.

Whenever a solution needs to be evaluated (e.g., in the local search procedure), the function $area()$ is used. This function transforms the solution vector into a cutting pattern and returns its occupied area. We apply the bottom-left positioning rule to position the items following the order of the solution vector (from left to right). If there are $m = 1, 2, \ldots, k$ positions with -1 values before the position of an item $j \in I$, the positioning rule must skip the next m feasible points of the grid, and thus item j will be positioned only at the $m+1$ feasible point. This strategy should avoid premature convergence, diversifying the cutting patterns the placement rule can generate [21]. Otherwise, when m is zero, item j is arranged in the first feasible bottom- and left-most point of the large object's grid. If, in any case, there is no feasible point to j, then this item is kept out of the cutting pattern. The angle in which item j is considered is obtained from the remainder of the division of its position by $|R_j|$. For example, supposing item j is associated with 3 different rotations (e.g., angles of 0°, 45°, and 90° in relation to the x-axis), and its index is in the 5th position of the solution vector. Then, item j will be positioned using its third rotation since the remainder of 5 divided by 3 is 2 (i.e., rotated by 90°).

To improve the algorithm's efficiency, we consult a hash table before obtaining the cutting pattern and its area to check whether that solution vector has already been evaluated. Similarly, after evaluating a solution vector, the solution vector and its area are added to the hash table. This strategy avoids evaluating the same solution vector multiple times, thus saving computing time.

4 Computational Results

The proposed ILS was coded in C++. The computational experiments were run on a computer with an Intel(R) Xeon(R) E3-1245 3.50 GHz processor, 32 GB of RAM, and Ubuntu 18.04 LTS as the operating system. The objective is to assess the performance of the proposed algorithm in solving different-size instances of the 2IKP with defects within the imposed time limit $T = \{30, 60, 90, 120\}$ s.

The instances used to evaluate the ILS were proposed by [22]. Information on the instances and the results obtained by the ILS can be found in Table 1. Each row reports the instance name, the number of items n, the large object dimensions H and L, and

Table 1 Results of the ILS for different time limits T in seconds

Name	n	H	L	$T = 30$		$T = 60$		$T = 90$		$T = 120$	
				Area	GAP	Area	GAP	Area	GAP	Area	GAP
blazewicz2	14	15	11	78.98	30.65	80.25	28.57	80.25	28.57	81.53	26.56
blazewicz3	21	15	17	78.95	24.62	82.59	19.12	82.59	19.12	82.59	19.12
blazewicz4	28	15	22	80.75	24.62	81.52	23.43	82.61	21.80	82.61	21.80
fu	12	38	29	84.55	17.08	85.56	15.71	86.84	14.00	87.20	13.52
fu7	7	38	19	75.77	30.13	75.77	30.13	75.77	30.13	75.77	30.13
fu8	8	38	20	76.33	30.49	79.79	24.83	79.79	24.83	79.79	24.83
poly1a	15	40	13	70.02	14.37	72.17	10.96	74.41	7.61	74.41	7.61
poly1b	15	40	13	73.93	30.65	75.39	28.11	76.95	25.51	76.17	26.79
poly1e	15	40	10	65.31	12.89	69.39	6.25	70.03	5.28	70.03	5.28
rco2	14	15	13	92.51	9.25	92.51	9.25	92.51	9.25	92.51	9.25
shapes4	16	40	16	58.86	72.04	59.49	70.21	59.49	70.21	59.49	70.21
shapes5	20	40	20	63.13	60.00	63.13	60.00	63.64	58.73	63.64	58.73
shapes7	28	40	28	69.42	45.08	70.14	43.59	71.22	41.41	71.94	40.00
shapes9	34	40	33	67.68	45.95	68.29	44.64	69.21	42.73	69.82	41.48
shapes15	43	40	40	65.33	53.46	67.59	48.33	68.34	46.69	69.10	45.09
shirts2-4	26	40	14	80.34	24.01	80.89	23.18	80.89	23.18	80.98	23.04
shirts3-6	39	40	21	86.06	15.22	86.06	15.22	86.66	14.42	86.60	14.50
shirts4-8	52	40	28	85.12	16.22	85.61	15.55	85.61	15.55	87.81	12.65
shirts5-10	65	40	35	83.84	17.82	84.81	16.48	85.56	15.45	85.81	15.11
Average	–	–	–	75.63	30.24	76.89	28.08	77.49	27.08	77.78	26.62
Std. Dev.	–	–	–	9.09	17.46	8.86	17.63	8.76	17.63	8.85	17.55

the results for each time limit T: the occupied area, which is computed as the total area of the arranged items divided by the large object area (i.e., without considering the defective areas); the percentage gap, which is computed as the percentage change between the total area of all items divided by the large object area (i.e., without considering the defective areas), and the occupied area in the final solution. We consider three different types of defective areas (modeled by a triangle, a square, and a rhombus) randomly positioned in the large object for all these instances. This corresponds to scenario #8 of the equiprobable case as proposed by [22]. Besides that, the inner-fit rasters, no-fit rasters, and grid of points were computed considering a scale of 1 point per unit of distance. For the number of feasible points that the bottom-left positioning rule can skip, we set $k = 0.25n$ as in [21].

Table 1 reports the results obtained with the ILS for different time limits. The last two rows give the average and standard deviation values for columns (Area) and (GAP). It is worth highlighting that for only two instances (i.e., "fu7" and "rco2"), no improvement was obtained by the ILS as the time limit T increased. For the others, at least one better cutting pattern was obtained if the time limit was increased. This is confirmed by instances "fu", "shapes7", "shapes9", "shapes15", and "shirts5-10", where the occupied area also increased as the time limit increased/changed. In the case of instances "blazewicz3", "fu8", "shapes4", and "shapes5", the ILS improved the cutting pattern only once and could not get improved solutions with $T = 90$ and $T = 120$. Also, for two instances (i.e., "poly1b" and "shirts3-6"), the best cutting pattern was obtained for $T = 90$, while for $T = 120$, the results were slightly worse. On average, the occupied area is 75.63%, 76.89%, 77.49%, and 77.78% for $T = 30$, $T = 60$, $T = 90$, and $T = 120$, respectively. When going from $T = 30$ to $T = 120$, the average GAP reduced from 30.24 to 26.62%. This represents an overall reduction of 3.62 percentage points.

Figures 1 and 2 show the best solution obtained by the ILS for instances "blazewicz3" and "fu", respectively. In the former, the cutting pattern for $T = 30$ improves when $T = 60$, but no other improvement is obtained as the time limit

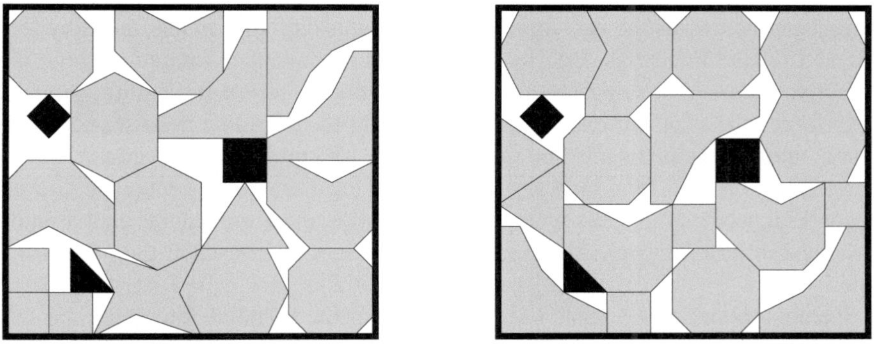

(a) $T = 30$ and $Area = 78.95\%$ (b) $T = 60$ and $Area = 82.59\%$

Fig. 1 Cutting patterns of instance "blazewicz3"

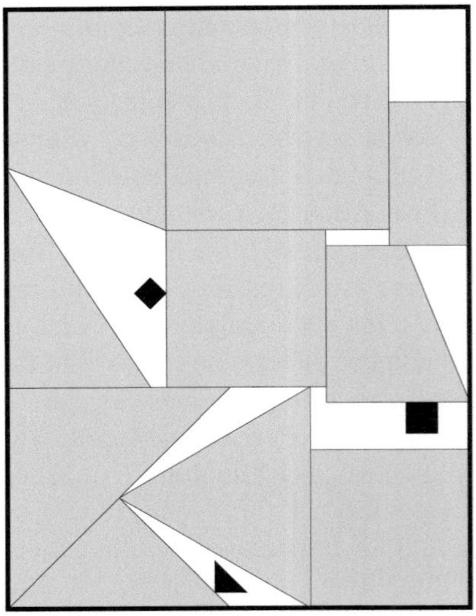

(a) $T = 30$ and $Area = 84.55\%$

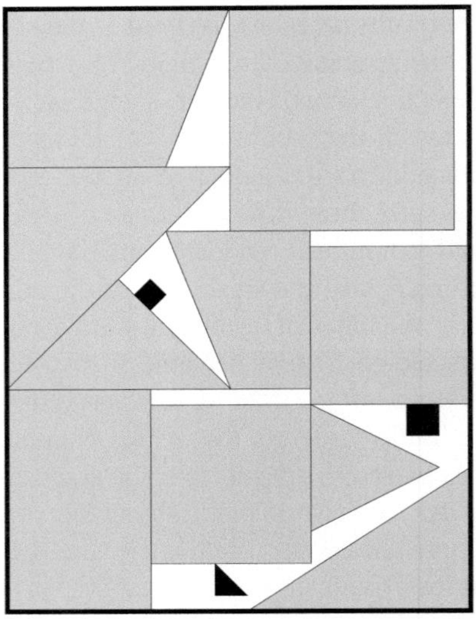

(b) $T = 120$ and $Area = 87.20\%$

Fig. 2 Cutting patterns of instance "fu"

increases. In the latter, the cutting pattern improves every time the time limit increases. The black-colored items represent the defective areas. It is important to mention that we cannot compare our results with those of [22] since these authors solved the problem with uncertain data using a stochastic integer linear programming model. Besides that, while their model solved each instance considering eight scenarios, we instead solved only one of the scenarios.

5 Concluding Remarks and Future Work

This paper addresses the two-dimensional irregular cutting problem faced by textile and related industries. Due to a dynamic environment, information about the defective areas of the large object is known shortly before the cutting process. This forces the decision-makers to react quickly to determine a satisfactory cutting pattern whose items must be free of defects. The problem is solved through an iterated local search for which perturbation and local search procedures are developed. As a vector of integers represents the solution, these procedures are based on swap and insertion moves. Also, the bottom-left heuristic is used as the positioning rule.

Computational experiments on instances from the literature show that the ILS obtains high-quality solutions with an occupied area between 58.86 and 92.51%, within a 30 s time limit. The solution quality improves as the time limit increases, reaching average values of occupied area of 77.78% within a 120 s time limit. Only a few instances did not achieve any improvement in the cutting pattern as the time limit increased.

After all, we observe there is room for improvement in different directions. New ways of representing the problem solution and determining the cutting pattern could be considered. For example, instead of using only the bottom-left positioning rule, other rules already proposed in the literature could be adapted and tested [15, 21]. Another direction could be testing the ILS on larger instances having more items and defects.

Acknowledgements The authors thank the financial support provided by the National Council for Scientific and Technological Development (CNPq) [grant numbers 405369/2021-2, 408722/2023-1, 315555/2023-8, 444679/2024-3, and 407005/2024-2], the State of Goiás Research Foundation (FAPEG), the National Recovery and Resilience Plan (NRRP), Mission 04 Component 2 Investment 1.5-NextGenerationEU, Call for tender n. 3277 dated 30/12/2021, Award Number: 0001052 dated 23/06/2022.

References

1. Baldacci, R., Boschetti, M.A., Ganovelli, M., Maniezzo, V.: Algorithms for nesting with defects. Discret. Appl. Math. **163**, Part 1, 17–33 (2014)
2. Bennell, J.A., Oliveira, J.F.: A tutorial in irregular shape packing problems. J. Oper. Res. Soc. **60**, 93–105 (2009)
3. Cacchiani, V., Iori, M., Locatelli, A., Martello, S.: Knapsack problems—An overview of recent advances. Part I: single knapsack problems. Comput. Oper. Res. **143**, 105692 (2022)
4. Cacchiani, V., Iori, M., Locatelli, A., Martello, S.: Knapsack problems—An overview of recent advances. Part II: multiple, multidimensional, and quadratic knapsack problems. Comput. Oper. Res. **143**, 105693 (2022)
5. Chen, K., Zhuang, J., Zhong, S., Zheng, S.: Optimization method for guillotine packing of rectangular items within an irregular and defective slate. Mathematics **8**(11), 1914 (2020)
6. Cherri, L.H., Mundim, L.R., Andretta, M., Toledo, F.M., Oliveira, J.F., Carravilla, M.A.: Robust mixed-integer linear programming models for the irregular strip packing problem. Euro. J. Oper. Res. **253**(3), 570–583 (2016)
7. Dalalah, D., Khrais, S., Bataineh, K.: Waste minimization in irregular stock cutting. J. Manuf. Syst. **33**(1), 27–40 (2014)
8. Durmaz, E.D., Şahin, R.: An efficient iterated local search algorithm for the corridor allocation problem. Expert Syst. Appl. **212**, 118804 (2023)
9. Imamichi, T., Yagiura, M., Nagamochi, H.: An iterated local search algorithm based on nonlinear programming for the irregular strip packing problem. Discret. Optim. **6**(4), 345–361 (2009)
10. Iori, M., de Lima, V.L., Martello, S., Miyazawa, F.K., Monaci, M.: Exact solution techniques for two-dimensional cutting and packing. Euro. J. Oper. Res. **289**(2), 399–415 (2021)
11. ITMA Services: the Italian textile industry: a key economic sector (2022). https://itma.com/admin/itma/media/itma/ITMA/The_Italian_Textile_Industry/The-Italian-Textile-Industry.pdf
12. Kramer, R., Dell'Amico, M., Iori, M.: A batching-move iterated local search algorithm for the bin packing problem with generalized precedence constraints. Int. J. Prod. Res. **55**(21), 6288–6304 (2017)
13. Leão, A.A.S., Toledo, F.M.B., Oliveira, J.F., Carravilla, M.A., Alvarez-Valdes, R.: Irregular packing problems: a review of mathematical models. Euro. J. Oper. Res. **282**(3), 803–822 (2020)

14. Lee, W.C., Ma, H., Cheng, B.W.: A heuristic for nesting problems of irregular shapes. Comput.-Aided Des. **40**(5), 625–633 (2008)
15. Mundim, L.R., Andretta, M., Carravilla, M.A., Oliveira, J.F.: A general heuristic for two-dimensional nesting problems with limited-size containers. Int. J. Prod. Res. **56**(1–2), 709–732 (2018)
16. Mundim, L.R., Andretta, M., Queiroz, T.A.: A biased random key genetic algorithm for open dimension nesting problems using no-fit raster. Expert Syst. Appl. **81**, 358–371 (2017)
17. Oliveira, O., Gamboa, D., Silva, E.: An introduction to the two-dimensional rectangular cutting and packing problem. Int. Trans. Oper. Res. **30**(6), 3238–3266 (2023)
18. Reijntjes, I.E.: Solving a nesting problem using the collision free region. In: Master in Mathematics and Computer Science, Eindhoven University of Technology, Eindhoven, Netherlands (2016)
19. Román, A.E.: An Evolutionary Algorithm for Solving the Two-Dimensional Irregular Shape Packing Problem Combined with the Knapsack Problem. Polytechnic University of Catalonia, Barcelona, Spain, Mestrado em artificial intelligence (2020)
20. Sato, A.K., Martins, T.C., Gomes, A.M., Tsuzuki, M.S.G.: Raster penetration map applied to the irregular packing problem. Euro. J. Oper. Res. **279**(2), 657–671 (2019)
21. Souza, Queiroz, L.R., Andretta, M.: Two effective methods for the irregular knapsack problem. Appl. Soft Comput. **95**, 106485 (2020)
22. Souza, Queiroz, L.R., Andretta, M.: A stochastic optimization model for the irregular knapsack problem with uncertainty in the plate defects. Pesquisa Operacional **42**, e259930 (2022)
23. Stützle, T., Ruiz, R.: Iterated Local Search. Handbook of Heuristics, pp. 579–605. Springer International Publishing, Cham (2018)
24. Toledo, F.M.B., Carravilla, M.A., Ribeiro, C., Oliveira, J.F., Gomes, A.M.: The dotted-board model: a new MIP model for nesting irregular shapes. Int. J. Prod. Econ. **145**(2), 478–487 (2013)

The Ellipsoidal Separation Machine

Antonio Frangioni, Enrico Gorgone, and Benedetto Manca

Abstract We build on the general proposal of Grzybowski et al. [16] that defines the concept of separation of two finite point sets by means of a convex set by choosing the latter, roughly speaking, as the minimum volume ellipsoid that intersects the convex combinations of all pairs of points of different class. The corresponding fitting problem is non-convex, hence we tackle it heuristically via an iterative algorithm of the block-Gauss-Seidel type solving alternatively an SDP program and a quadratically constrained (convex) program. The thusly computed separating ellipsoid is used to classify new points by means of a newly defined score based on the relative fraction of the original points that are properly separated from them. This necessarily leads to being unable to classify points—e.g, those inside the ellipsoid—making ours inherently a classifier with reject, as opposed with most proposals in the literature where a reject function is bolted upon a standard classifier. This feature can be relevant in cases where an incorrect classification may be more damaging than explicitly refusing to assign a label, indicating uncertainty. We will provide numerical experiments comparing the quality of the ellipsoidal classifier with that of standard approaches endowed with a rejection function.

Keywords Data classification · Convex analysis · Nonsmooth optimization

---

A. Frangioni
Dipartimento di informatica, Università di Pisa, Pisa, Italy
e-mail: frangio@di.unipi.it

E. Gorgone (✉) · B. Manca
Dipartimento di Matematica e Informatica, Università degli Studi di Cagliari, Palazzo delle Science, Cagliari, Italy
e-mail: egorgone@unica.it

B. Manca
e-mail: bmanca@unica.it

M. Di Francesco et al. (eds.), *Operations Research: Closing the Gap Between Research and Practice*, AIRO Springer Series 15, https://doi.org/10.1007/978-3-031-90095-2_32

1 Introduction

Pattern classification is a supervised learning technique that, on the basis of similarities in the data, assigns samples to one of the datasets using a given classification rule [10, 22, 25]. For binary classification, the most popular technique is the Support Vector Machine (SVM) approach [7, 9, 26], where the classification rule is a hyperplane which splits the space into two half-spaces, each of them containing (ideally) only points with the same label. If equipped with the kernel trick it also allows to build nonlinear separation surfaces by nonlinearly mapping the input space into a larger space where linear separation is easier to achieve.

As all classifiers, SVM makes non-obvious assumptions. For instance, it classifies a sample that belongs to the half-space of points with a given label even if it is "very far" from each of the original inputs, while common sense would dictate that such a point should rather be considered an outlier and consequently not classifiable. Such an indication can be useful in many real-world applications, such as finance, healthcare and product inspection [6, 11, 17, 19, 21, 24]. Indeed, Classification with Rejection (CwR) has been introduced [8] whereby, if the risk of misclassification is too high, classification is not performed and the sample is rejected. The rejection option can be implemented in a straightforward way in SVM by analyzing the margin area around the hyperplane [12, 15, 23, 27]. Nevertheless, outliers far away from both the labelled points and out of the margin region would still be confidently classified.

We approach CwR in a novel way by using the concept of separation of two (convex) sets by another (convex) set in the sense of Grzybowski et al. [16], whereby two points x^+ and x^- are separated by a (convex) set $\mathcal{S} \subset \mathbb{R}^n$ if the segment with extremes x^+ and x^- traverses $\mathcal{S}$, i.e.,

$$conv(\{x^+, x^-\}) \cap \mathcal{S} \neq \emptyset \,. \tag{1}$$

The arbitrary set $\mathcal{S}$ generalizes the hyperplane which linearly separates x^+ and x^- in SVM; in fact, any separating hyperplane ensures the satisfaction of (1). The definition can be easily extended to two arbitrary sets X^+ and X^- by requiring that (1) holds for all pairs $(x^+, x^-) \in X^+ \times X^-$. Furthermore, in the context of classification X^+ and X^- are usually given as a finite set of points, i.e., $X^+ = \{x_1^+, x_2^+, \dots, x_{n+}^+\}$ and $X^- = \{x_1^-, x_2^-, \dots, x_{n-}^-\}$. The use of a "small"—bounded—set $\mathcal{S}$ instead of a hyperplane may help to not classify points that lie "too far" from the original data, since $\mathcal{S}$ is not able to separate points far from both the convex hulls of X^+ and X^-, and from itself.

To put the mechanism in action is not an easy task. In literature, a few approaches have been proposed and only using piecewise-linear separators [1, 2, 13]. In this work we rather choose an ellipsoid as the set $\mathcal{S}$ in order to exploit the powerful SemiDefinite Programming (SDP) techniques that have already successfully used in the classification framework [3]. Ellipsoidal sets in $\mathbb{R}^n$ require only "a few ingredients" to be constructed: a positive semidefinite matrix $0 \preceq S \in \mathbb{R}^{n \times n}$ and a vector

$\gamma \in \mathbb{R}^n$. Furthermore, the volume of an ellipsoid is a convex function of its parameters, which means that finding a ellipsoidal sets with minimum volume, such as the Löwner-John ellipsoid [18], is a convex problem. Defining $I^+ = \{1, \ldots, n+\}$, $I^- = \{1, \ldots, n-\}$, and $P = \{(i, j) : i \in I^+, j \in I^-\}$ the set of all pairs of (indices of) points in the two sets to separate, we can write a first version of the separation problem as follows:

$$\min vol(S) \tag{2a}$$
$$(x_{ij} - \gamma)^\top S(x_{ij} - \gamma) \leq 1 \qquad (i, j) \in P \tag{2b}$$
$$(x_i^+ - \gamma)^\top S(x_i^+ - \gamma) > 1 \qquad i \in I^+ \tag{2c}$$
$$(x_j^- - \gamma)^\top S(x_j^- - \gamma) > 1 \qquad j \in I^- \tag{2d}$$
$$x_{ij} = \alpha_{ij} x_i^+ + (1 - \alpha_{ij}) x_j^- \qquad (i, j) \in P \tag{2e}$$
$$\alpha_{ij} \in [0, 1] \qquad (i, j) \in P \tag{2f}$$
$$S \succeq 0 \tag{2g}$$

where $x_{i,j}$ represent a point inside the segment between x_i^+ and x_j^- parametrized by the variable α_{ij} inside the interval [0, 1]. In (2a), $vol(S)$ is the volume of the ellipsoid; (2b) imposes that every the segment joining points with different labels touches $\mathcal{S}$, whereas (2c) and (2c) impose that the points of X^+ and X^- are left outside of it. We will refer to this problem as the Ellipsoidal Separation Machine (ESM). The formulation, as it is written, contains several cross-products between variables and therefore it is not an SDP program. In Sect. 2 we will introduce a reformulation of (2) which makes the problem easier to solve. Then, in Sect. 3 we will discuss a novel classification criterion with a rejection option which differentiates ESM to other binary classifiers. Finally, in Sect. 4 we will discuss some numerical experiments on synthetic datasets.

2 Reformulation of the ESM Problem

The formulation (2) of the ESM problem contains bilinear terms of the form $x^\top Sx$ where both x and S are variables; this makes the problem difficult to solve. Moreover, the quantity $vol(S)$ requires a significant number of constraints and auxiliary variables to be exactly represented, thus we will introduce a linear approximation for the volume of the separating ellipsoid.

2.1 Modelling the $vol(S)$ Term

Given an ellipsoid (S, γ), its i-th semi-axis is given by $1/\lambda_i$ where λ_i is i-th eigenvalue of the matrix S. Therefore, it is possible to express the volume of the ellipsoid as

$$vol(S) = \prod_{i=1}^{n} \frac{1}{\lambda_i}.$$

As shown in [5] the logarithm of the previous expression is SDP-representable by introducing several semi-definite constraints and auxiliary variables which result in an increase of the size of the program. In order to avoid this issue, we consider the box-size of the ellipsoid (S, γ) given by

$$B(S)^2 = \sum_{i=1}^{n} \frac{1}{\lambda_i} = \mathrm{tr}(\mathrm{S}^{-1}).$$

This choice is motivated by the fact that to minimize the box-size of S it is only necessary to introduce an auxiliary matrix variable $T \in \mathbb{R}^{n \times n}$ and the following constraint

$$\begin{pmatrix} S & I \\ I & T \end{pmatrix} \succeq 0 \Leftrightarrow \mathrm{tr(\,T\,)} \geq \mathrm{tr(\,S^{-1}\,)} \tag{3}$$

Therefore, by minimizing tr(T) under the constraint (3) we also minimize the box-size of the ellipsoid (S, γ).

2.2 Avoiding Bilinear Term in γ and S

In order to linearize the condition for a point to belong or not to an ellipsoid (S, γ) in $\mathbb{R}^n$ we will use the following fact (for more details we refer to [14]): any ellipsoid (S, γ) in $\mathbb{R}^n$ can be considered as the intersection of an ellipsoid $(\tilde{S}, \mathbf{0})$ in $\mathbb{R}^{n+1}$ with an appropriate hyperplane $\mathcal{H} \subseteq \mathbb{R}^{n+1}$.

Moreover, if we consider the following decomposition of the matrix $\tilde{S}$

$$\tilde{S} = \begin{pmatrix} s & t^\top \\ t & F \end{pmatrix}, \quad s \in \mathbb{R}^+, t \in \mathbb{R}^n \text{ and } F \succeq 0 \in \mathbb{R}^{n \times n}, \tag{4}$$

and we compute the quantities $d = -F^{-1}t$ and $\delta = 1 - d^\top F d$, the following holds for every $x \in \mathbb{R}^n$

$$(1, x)^\top \tilde{S}(1, x) \leq 1 \Leftrightarrow (x - d)^\top \frac{F}{1 - \delta}(x - d) \leq 1. \tag{5}$$

Thus, we can reformulate the constraints (2b)–(2d) using $\tilde{S}$ instead of (S, γ) as

$$\tilde{x}_{ij}^\top \; \tilde{S} \; \tilde{x}_{ij} \leq 1 \qquad (i\,,\, j) \in P \tag{6a}$$
$$(\tilde{x}_i^+)^\top \; \tilde{S} \; \tilde{x}_i^+ > 1 \qquad i \in I^+ \tag{6b}$$
$$(\tilde{x}_j^-)^\top \; \tilde{S} \; \tilde{x}_j^- > 1 \qquad j \in I^- \tag{6c}$$

with $\tilde{x}_{ij} = (1, x_{ij})$, $\tilde{x}_i^+ = (1, x_i^+)$, $\tilde{x}_j^- = (1, x_j^-)$ and then reconstruct the ellipsoid (S, γ), satisfying the corresponding constraints in $\mathbb{R}^n$, by setting $S = \frac{F}{1-\delta}$ and $\gamma = d$, according to (5).

2.3 Final Formulation

We can now state the final formulation for the ESM problem:

$$\min \mathrm{tr}(\mathrm{T}) + \mathrm{C}_1 \sum_{(\mathrm{i,j})\in\mathrm{P}} \beta_{\mathrm{ij}} + \mathrm{C}_2(\sum_{\mathrm{i}\in\mathrm{I}^+} \nu_{\mathrm{i}}^+ \sum_{\mathrm{j}\in\mathrm{I}^-} \nu_{\mathrm{j}}^-) + \mathrm{C}_3\|\mathrm{F}\|_{\mathrm{F}}^2 \tag{7a}$$
$$\tilde{x}_{ij}^\top \; \tilde{S} \; \tilde{x}_{ij} \leq 1 + \beta_{ij} \qquad (i\,,\, j) \in P \tag{7b}$$
$$(\tilde{x}_i^+)^\top \; \tilde{S} \; \tilde{x}_i^+ \geq 1 - \nu_i^+ \qquad i \in I^+ \tag{7c}$$
$$(\tilde{x}_j^-)^\top \; \tilde{S} \; \tilde{x}_j^- \geq 1 - \nu_j^- \qquad j \in I^- \tag{7d}$$
$$\tilde{S} = \begin{pmatrix} s & t^\top \\ t & F \end{pmatrix} \succeq 0 \tag{7e}$$
$$x_{ij} = \alpha_{ij} x_i^+ + (1 - \alpha_{ij}) x_j^- \qquad (i\,,\, j) \in P \tag{7f}$$
$$\beta_{ij} \geq 0 \qquad (i\,,\, j) \in P \tag{7g}$$
$$\nu_i^+ \geq 0 \qquad i \in I^+ \tag{7h}$$
$$\nu_j^- \geq 0 \qquad j \in I^- \tag{7i}$$
$$\alpha_{ij} \in [0\,,\, 1] \qquad (i\,,\, j) \in P \tag{7j}$$

where we have added the slack variables β and $\nu^\pm$ in order to obtain a feasible solution even when the input points are not perfectly separable by an ellipsoid. Moreover, we also minimize the Frobenius norm of the matrix F to prevent the separating ellipsoid to be degenerate, i.e., having too large eigenvalues. The hyper-parameters C_1, C_2 and C_3 will weight the different terms in the objective function.

We observe that problem (7) is still difficult to solve since there are nonlinear terms in the constraints (7b). However, it is possible to use a Block-Gauss-Seidel approach to obtain a solution. In fact, by fixing the variable α, and thus the quantities $\tilde{x}_{ij}$, the problem reduces to a linear SDP:

$$
\begin{array}{ll}
\min \mathrm{tr}(\mathrm{T}) + \mathrm{C}_1 \sum_{(\mathrm{i},\mathrm{j})\in \mathrm{P}} \beta_{\mathrm{ij}} + \mathrm{C}_2(\sum_{\mathrm{i}\in \mathrm{I}^+} \nu_{\mathrm{i}}^+ \sum_{\mathrm{j}\in \mathrm{I}^-} \nu_{\mathrm{j}}^-) + C_3\|F\|_F^2 & \\
\tilde{x}_{ij}^\top \, \tilde{S} \, \tilde{x}_{ij} \le 1 + \beta_{ij} & (i\,,\,j) \in P \\
(\tilde{x}_i^+)^\top \, \tilde{S} \, \tilde{x}_i^+ \ge 1 - \nu_i^+ & i \in I^+ \\
(\tilde{x}_j^-)^\top \, \tilde{S} \, \tilde{x}_j^- \ge 1 - \nu_j^- & j \in I^- \\
\tilde{S} = \begin{pmatrix} s & t^\top \\ t & F \end{pmatrix} \succeq 0 & \\
\beta_{ij} \ge 0 & (i\,,\,j) \in P \\
\nu_i^+ \ge 0 & i \in I^+ \\
\nu_j^- \ge 0 & j \in I^-
\end{array}
\tag{8}
$$

On the other hand, by fixing the variable $\tilde{S}$, the problem reduces to a SOCP in the variables α and β:

$$
\begin{array}{ll}
\min C_1 \sum_{(i,j)\in P} \beta_{ij} & \\
\tilde{x}_{ij}^\top \, \tilde{S} \, \tilde{x}_{ij} \le 1 + \beta_{ij} & (i\,,\,j) \in P \\
\beta_{ij} \ge 0 & (i\,,\,j) \in P \\
\alpha \in [0, 1] &
\end{array}
\tag{9}
$$

The algorithm we used to obtain a solution works as follows: we fix the values of the α_{ij} to $1/2$ for every i and j. We solve (8) to obtain a starting separating ellipsoid $\tilde{S}$ and the optimal value of the objective function $\phi^* = \mathsf{val}(8)$. Then, we set the estimate $\bar{\phi}$=ϕ^* and we alternatively solve (9) and (8) a number r of times. If the relative difference between $\bar{\phi}$ and the objective function value $\mathsf{val}(8)$ of the last SDP solved is less than a threshold ε the algorithm stops, otherwise we update $\bar{\phi}$ with $\mathsf{val}(8)$ and we repeat the process. A scheme of the algorithm is given in Algorithm 1.

Input: $\alpha = 1/2 \times \mathbf{1}_{kl}$
$\tilde{S}, \phi^* \leftarrow$ solve (8) **do**
  $\bar{\phi} = \phi^*$ **for** *it=1 to r* **do**
    $\alpha \leftarrow$ solve (9)
    $\tilde{S}, \phi^* \leftarrow solve(8)$
  end
while $|\phi^* - \bar{\phi}| > \varepsilon$;
return $\tilde{S}$

Algorithm 1: Ellipsoidal separation$(C_1, C_2, C_3, r, \varepsilon)$

3 The Classification Criterion with Rejection Option

Once we obtain a separating ellipsoid (S, γ) as a solution of problem (7), we need to define a classification criterion for any point $z \in \mathbb{R}^n$ with respect to (S, γ). According to the definition of separability of convex set we are considering, the natural criterion is that z is separated from X^+, and thus classified as a point of X^-, if the following hold:

– $S \cap [z, x^+] \neq \emptyset, \forall x^+ \in X^+$;
– $S \cap [z, x^-] = \emptyset, \forall x^- \in X^-$.

We will use the symmetric relationship to define the classification of z as a point of X^+. However, this criterion leaves many points as not classified; for example, all the points in the interior of the ellipsoid are neither separated from X^+ nor X^-. Also, points may be only partially separated from X^+ or X^-. This led us to defining the following classification criterion: let $\bar{X}^\pm = X^\pm \setminus S$, then for any $z \in \mathbb{R}^n$ we compute the number $n^\pm(z)$ of points in $X^\pm$ that are completely separated from z. Then we can define the following scores:

$$r^\pm(z) = \frac{n^\mp(z)}{|\bar{X}^\mp|} - \frac{n^\pm(z)}{|\bar{X}^\pm|} \in [-1, 1]. \tag{10}$$

Observe that if $r^\pm(z) = 1$ the point z is completely separated from $\bar{X}^\mp$ and thus we can classify z as a point of $X^\pm$. In order to consider also points that are almost separated from one of the two classes, we consider a threshold $\tau \in (0, 1]$ and we will classify a point z as a point of $X^\pm$ if $r^\pm(z) > \tau$. Whenever the absolute value of the two scores $r^\pm(z)$ are small, we will consider the point z as rejected or not classified. The presence of rejected points is a specific feature of ESM, and is not considered in other binary classifier based on ellipsoids [3, 4]. We believe that this feature can be useful in classification tasks where the presence of false positive is a significant problem, since ESM classifies a point as a member of one of the two classes only if there is a strong evidence that the point indeed belongs to the class. Figure 1 shows a simple example in $\mathbb{R}^2$ where the red points labeled 1 and 2 and the blue point labeled 1 are perfectly separated with respect to the ellipsoid, while the red point labeled 3 and the blue point labeled 2 are only partially separated. The green points inside of the ellipsoid are rejected since all the segments connecting them to red or blue points intersect the ellipsoid.

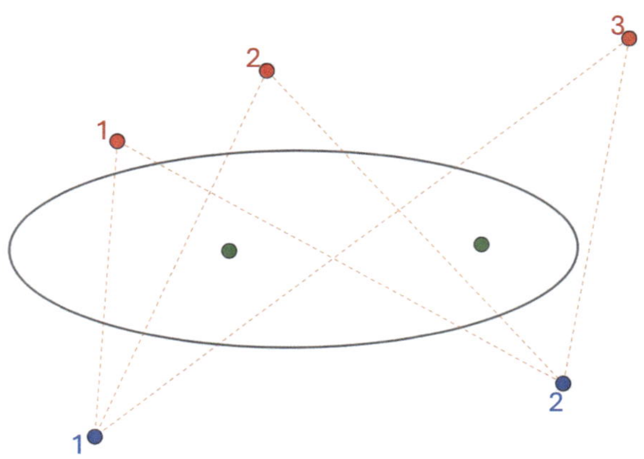

Fig. 1 Example in $\mathbb{R}^2$ of a separating ellipsoid with some points perfectly separated (points 1 and 2 red and point 1 blue), two points partially separated (points 3 red and 2 blue) and two rejected points (green points)

4 Numerical Experiments

We have implemented the CwR approach described in the previous sections using Python as programming language. As optimization solver we have used MOSEK under the Fusion API which allowed us to solve the SDP program (8) and the quadratic constraint program (9). We have decided to implement the classifier as a scikit-learn [20] estimator in order to be able to use its pipeline (i.e., grid-search for hyper-parameters, random train-test splitting, etc.) Since our algorithm depends on several hyper-parameters, we have first run some preliminary experiments in order to reduce the size of the grid on which perform the search for the best ones. The search space we have decided to use is the following: $C_1 \in [0.001, 0.01, 0.1]$, $C_2 \in [1.0, 10, 100, 1000]$, $C_3 \in [0.1, 1.0, 10, 100]$. The other parameters have been fixed since our preliminary experiments showed that they didn't influence significantly the results or the values chosen were the best among all the values we have considered: $\tau = 1e - 8, \varepsilon = 0.1, r = 2$.

For the assessment of the quality of the classifier we have considered the following scores:

- the accuracy with rejection (AR) given by the formula

$$\sum_{x \in X_{wc}} \frac{1}{n} - \sum_{x \in X_{rej}} \frac{\xi}{n}, \tag{11}$$

  where the first sum is taken over the elements of the set X_{wc} of well classified points and the second sum over the set X_{rej} of rejected points. While the quantity $\xi \in [0, 0.5]$ is the rejection cost parameter which weights each rejected point and n is the number of samples considered;
- the accuracy (A) on the classified points, i.e., not considering the rejected points;
- the percentage of well classified points (WC);
- the percentage of misclassified points (MC);
- the percentage of rejected points (R).

We believe that the more interesting characteristic of our classifier is the existence of the rejection zone, therefore we have used the accuracy with rejection as score to determine the best hyper-parameters. In order to be able to compare the ellipsoidal separation machine we have considered the Support Vector Machine (SVM) binary classifier with different kernel: Gaussian (G-SVM), Polynomial (P-SVM) and Sigmoid (S-SVM). By default SVM doesn't have a rejection zone, but it is very natural to consider all the points inside its margin as rejected, as it is regularly done in the CwR literature.

Due to the large amount of time required to the interior point method to solve the semi-definite program (8), we were able to run only preliminary experiments on synthetic datasets. All the numerical experiments have been executed on a Intel i5-4460 3.00 GHz 4-core with 132 GB of RAM under a i686 GNU/Linux operating

Table 1 Parameters used to generate the synthetic dataset via the scikit-learn Python package

Dataset	class_sep	n_cluster_per_class
1	0.5	2
2	0.5	4
3	0.5	6
4	1.0	2
5	1.0	4
6	1.0	6
7	1.5	2
8	1.5	4
9	1.5	6

system. We have considered synthetic dataset for binary classification generated randomly using a specific module of the Python package sklearn (https://scikit-learn.org/stable/modules/generated/sklearn.datasets.make_classification.html). This package allows to generate datasets with different characteristics, and we have considered four parameters:

- n_samples: the number of points in the dataset;
- n_features: the dimension of the feature space;
- class_sep: a parameter controlling how much the two classes of the dataset are separable. Higher values correspond to easier classification task;
- n_cluster_per_class: the number of clusters each class have.

We have generated all the dataset with 100 samples and 7 features in order to obtain the results in a reasonable amount of time. Table 1 summarizes the other characteristics of the synthetic datasets we have considered:

During the experiments we have split the dataset into training (70%) and testing (30%) and performed a 3-fold cross validation on the training data in order to obtain the best hyper-parameters. The results are illustrated in Figs. 2, 3, 4, 5, 6, 7, 8, 9 and 10 where the five scores are represented in a radar charts, which is a useful way to represent the results of the different classifier with respect to different metrics. Each axis of the radar charts corresponds to one of the metric we have considered to evaluate the classifiers: Accuracy with rejection (AR), Accuracy on the classified points (A), percentage of well classified points (WC), misclassified points (MC) and rejected points (R). For each classifier, the vertices along the axis of the plot represent the value of the corresponding metric.

The results show that ESM is competitive with SVM-based approaches; it performs similarly to S-SVM and tends to reject less points than G-SVM and P-SVM, having a slightly worse score in terms of accuracy with rejection. However, the crucial observation is that the radar charts for ESM are quite different from those of the SVM-based approaches: this indicates that the classification is done in a substantially different way, and therefore that the classification approach is indeed novel.

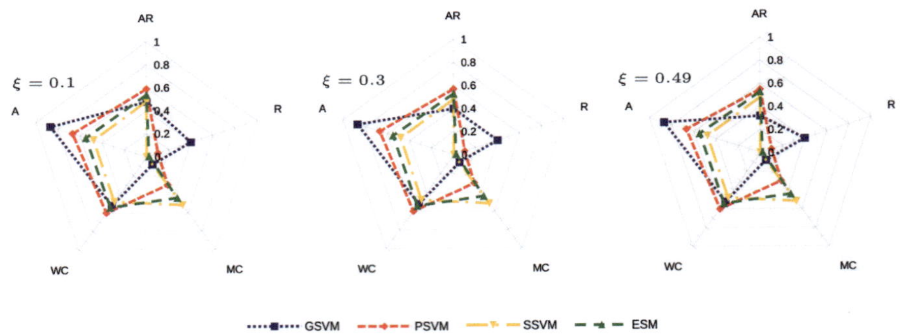

Fig. 2 Numerical results for the synthetic dataset with class separability equal to 0.5 and number of clusters equal to 2

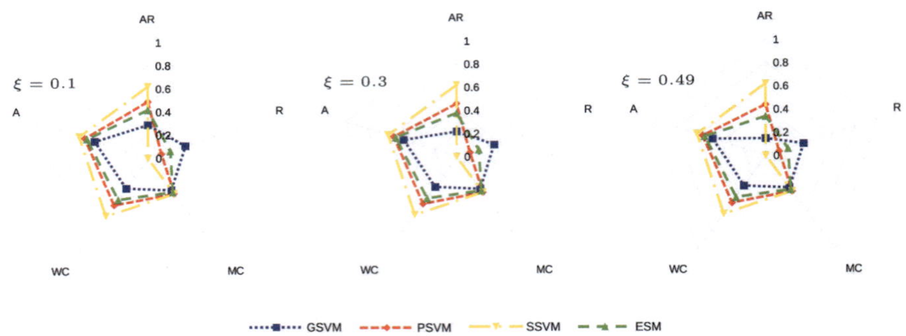

Fig. 3 Numerical results for the synthetic dataset with class separability equal to 0.5 and number of clusters equal to 4

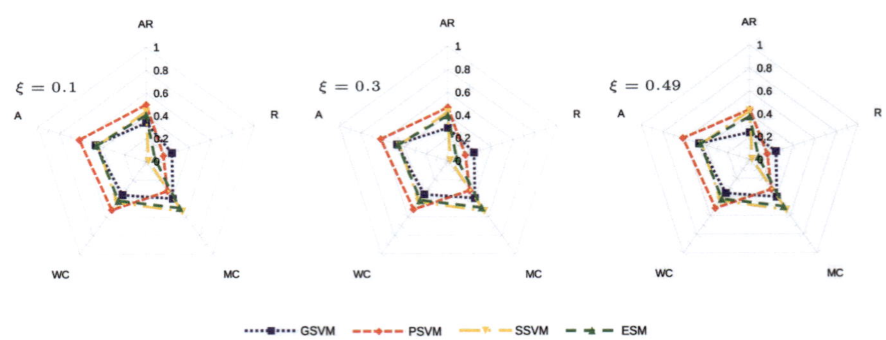

Fig. 4 Numerical results for the synthetic dataset with class separability equal to 0.5 and number of clusters equal to 6

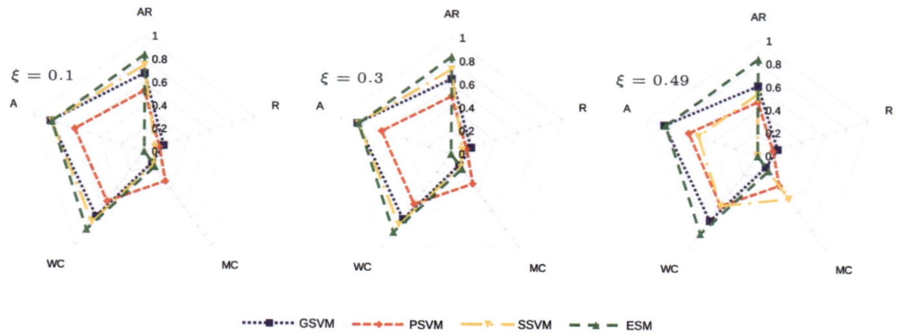

Fig. 5 Numerical results for the synthetic dataset with class separability equal to 1.0 and number of clusters equal to 2

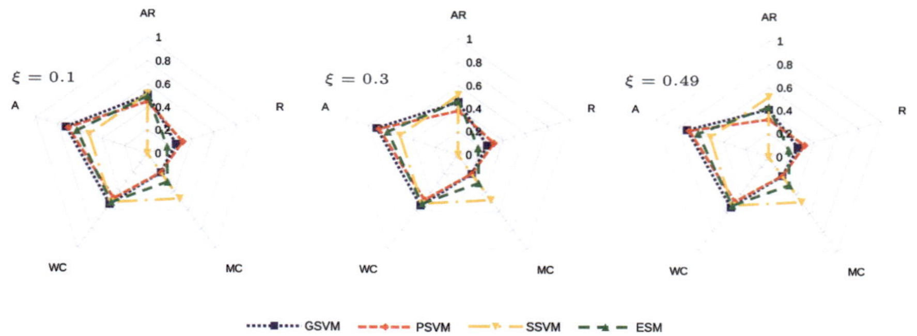

Fig. 6 Numerical results for the synthetic dataset with class separability equal to 1.0 and number of clusters equal to 4

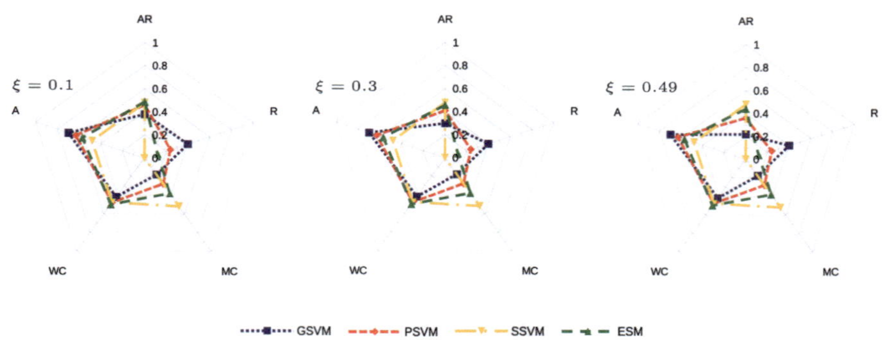

Fig. 7 Numerical results for the synthetic dataset with class separability equal to 1.0 and number of clusters equal to 6

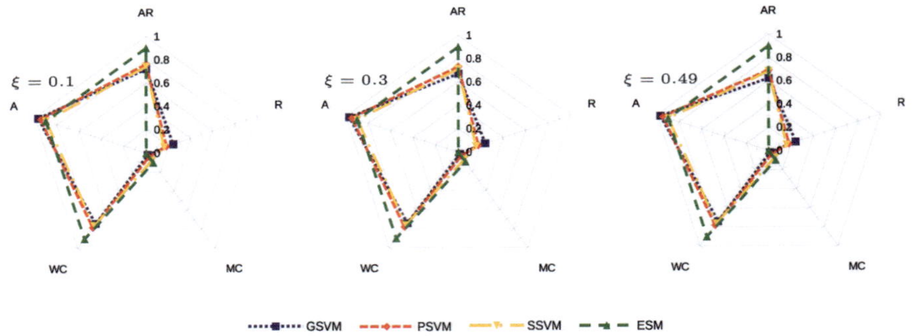

Fig. 8 Numerical results for the synthetic dataset with class separability equal to 1.5 and number of clusters equal to 2

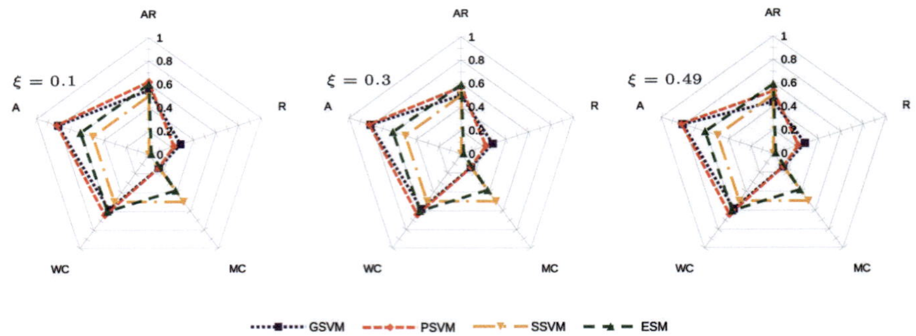

Fig. 9 Numerical results for the synthetic dataset with class separability equal to 1.5 and number of clusters equal to 4

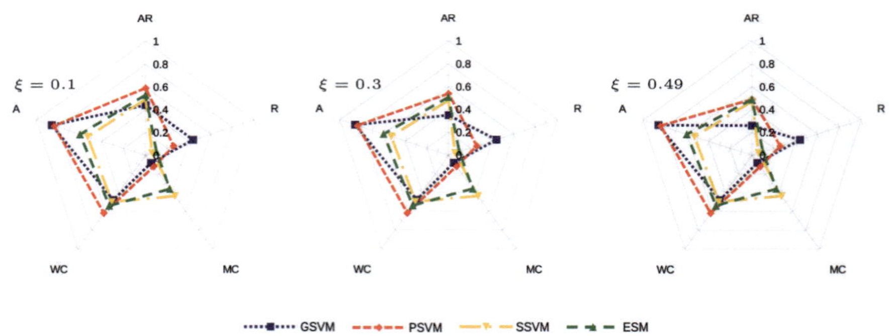

Fig. 10 Numerical results for the synthetic dataset with class separability equal to 1.5 and number of clusters equal to 6

5 Conclusions

We have developed ESM, a novel binary classifier, with a rejection option whose training phase involves a non-linear semi-definite program with a specific structure. This characteristic allowed us to solve the problem using a Block Gauss-Seidel approach which iteratively solve a linear SDP and a SOCP. The preliminary experiments on small synthetic datasets show that ESM can obtain good results against kernel SVM with an artificial rejection zone (the margin). Future works will revolve around algorithmic strategies to reduce the time necessary to solve the linear SDP, for example by considering a lagrangian relaxation which can give an approximate solution in faster times.

Acknowledgements The third author was supported by PON R&I 2014-2020 (CUP F25F21002210003).

References

1. Astorino, A., Gorgone, E., Gaudioso, M., Pallaschke, D.: Data preprocessing in semi-supervised SVM classification. Optimization **60**, 143–151 (2011)
2. Astorino, A., Di Francesco, M., Gaudioso, M., Gorgone, E., Manca, B.: Polyhedral separation via difference of convex (DC) programming. Soft. Comput. **25**, 12605–12613 (2021)
3. Astorino, A., Frangioni, A., Gorgone, E., Manca, B.: Ellipsoidal classification via semidefinite programming. Oper. Res. Lett. **51**, 197–203 (2023)
4. Astorino, A., Gaudioso, M.: Ellipsoidal separation for classification problems. Optim. Methods Softw. **20**, 267–276 (2005)
5. Ben-Tal, A., Nemirovski, A.: Lectures on modern convex optimization 2020. SIAM, Philadelphia (2021)
6. Benbarrad, T., Salhaoui, M., Kenitar, S.B., Arioua, M.: Intelligent machine vision model for defective product inspection based on machine learning. J. Sens. Actuator Netw. **10** (2021)
7. Boser, B.E., Guyon, I.M., Vapnik, V.N.: A training algorithm for optimal margin classifiers. In: Proceedings of the Fifth Annual Workshop on Computational Learning Theory, COLT '92, New York, NY, USA, Association for Computing Machinery, pp. 144–152 (1992)
8. Chow, C.: On optimum recognition error and reject tradeoff. IEEE Trans. Inf. Theory **16**, 41–46 (1970)
9. Cortes, C.: Support-Vector Networks. Machine Learning (1995)
10. Cristianini, N., Shawe-Taylor, J.: An Introduction to Support Vector Machines and Other Kernel-Based Learning Methods. Cambridge University Press (2000)
11. D'Orazio, P., Popoyan, L.: Fostering green investments and tackling climate-related financial risks: Which role for macroprudential policies? Ecol. Econ. **160**, 25–37 (2019)
12. Fumera, G., Roli, F.: Support vector machines with embedded reject option. In: Pattern Recognition with Support Vector Machines: First International Workshop, SVM: Niagara Falls, Canada, 10 August 2002 Proceedings, pp. 68–82. Springer (2002)
13. Gaudioso, M., Gorgone, E., Pallaschke, D.: Separation of convex sets by Clarke subdifferential. Optimization **59**, 1199–1210 (2010)
14. Glineur, F.: Pattern separation via ellipsoids and conic programming. PhD thesis, Master Dissertation, Faculty Polytechnique de Mons, Belgium (1998)
15. Grandvalet, Y., Rakotomamonjy, A., Keshet, J., Canu, S.: Support vector machines with a reject option. In: Advances in Neural Information Processing Systems, vol. 21 (2008)

16. Grzybowski, J., Pallaschke, D., Urbański, R.: A pre-classification and the separation law for closed bounded convex sets. Optim. Methods Softw. **20**, 219–229 (2005)
17. Gunjan, A., Bhattacharyya, S.: A brief review of portfolio optimization techniques. Artif. Intell. Rev. **56**, 3847–3886 (2023)
18. John, F.: Extremum problems with inequalities as subsidiary conditions, studies and essays presented to R. Courant on his 60th birthday, 8 January 1948
19. Ozsu, S., Oztuna, F., Bulbul, Y., Topbas, M., Ozlu, T., Kosucu, P., Ozsu, A.: The role of risk factors in delayed diagnosis of pulmonary embolism. Am. J. Emerg. Med. **29**, 26–32 (2011)
20. Pedregosa, F., Varoquaux, G., Gramfort, A., Michel, V., Thirion, B., Grisel, O., Blondel, M., Prettenhofer, P., Weiss, R., Dubourg, V., Vanderplas, J., Passos, A., Cournapeau, D., Brucher, M., Perrot, M., Duchesnay, E.: Scikit-learn: machine learning in Python. J. Mach. Learn. Res. **12**, 2825–2830 (2011)
21. Sabnis, S., Yadav, S.K., Salsingikar, S.: Identification of defective railway wheels from highly imbalanced wheel impact load detector sensor data. In: Neufeld, J.S., Buscher, U., Lasch, R., öst, D.M., Schönberger, J. (eds.) Operations Research Proceedings, pp. 397–403. Springer (2020)
22. Schölkopf, B., Burges, C.J.C., Smola, A.J.: Advances in Kernel Methods: Support Vector Learning. The MIT Press, Cambridge Mass (1999)
23. Tortorella, F.: Reducing the classification cost of support vector classifiers through an ROC-based reject rule. Pattern Anal. Appl. **7**, 128–143 (2004)
24. Udler, M.S., McCarthy, M.I., Florez, J.C., Mahajan, A.: Genetic risk scores for diabetes diagnosis and precision medicine. Endocr. Rev. **40**, 1500–1520 (2019)
25. Vapnik, V.: The Nature of the Statistical Learning Theory. Springer Verlag, New York (1995)
26. Vapnik, V.: The support vector method. In: Gerstner, W., Germond, A., Hasler, M., Nicoud, J-D. (eds.) Artificial Neural Networks—ICANN'97, pp. 261–271. Springer, Berlin, Heidelberg (1997)
27. Wegkamp, M., Yuan, M.: Support Vector Machines with a Reject Option. Bernoulli (2010)

Lagrangian Relaxation for the Intermediate Facilities Location Problem

Enrico Gorgone, Michele Monaci, and Daniele Vigo

Abstract The consolidation and merging of freight flows is crucial for logistics operations in order to reduce costs and keep the level of service provided to customers, especially when trucks travel with a half-full load. These kinds of operations take place in intermediate facilities or terminals located between the origins and the destinations of freight. We consider the problem of locating the intermediate facilities in a transportation network, which gives rise to a combinatorial problem. We address the problem through the Lagrangian decomposition. The Lagrangian dual is solved by efficient non-differentiable optimization methods and the sub-problem (the Lagrangian relaxation) is reduced to a binary quasi-knapsack problem. The proposed method is extensively tested on randomly generated instances showing it obtains tight lower bounds. In addition, we develop a Lagrangian-based heuristic to attain an integer feasible solution. Numerical results are reported, comparing the performance of the proposed methods with those obtained by applying a general-purpose Mixed-Integer Linear Programming solver to a mathematical formulation of the problem.

Keywords Lagrangian relaxation · Mixed-integer linear programs · Intermediate facilities · Transportation network

E. Gorgone (✉)
Dipartimento di Matematica e Informatica, Università degli Studi di Cagliari, Cagliari, Italy
e-mail: egorgone@unica.it

M. Monaci · D. Vigo
Dipartimento dell'Energia Elettrica e dell'Informazione, Università di Bologna
Alma Mater Studiorum, Bologna, Italy
e-mail: michele.monaci@unibo.it

D. Vigo
e-mail: daniele.vigo@unibo.it

M. Di Francesco et al. (eds.), *Operations Research: Closing the Gap Between Research and Practice*, AIRO Springer Series 15, https://doi.org/10.1007/978-3-031-90095-2_33

1 Introduction

Freight flow consolidation is a crucial need of modern logistics to achieve sustainable and cost-effective transportation. To this end, in most cases, intermediate facilities are used by transportation companies where goods for different destination are sorted and grouped so as to be transported together. Such a process is generally managed directly by common carriers and the shipper has not necessarily a share of the cost reduction resulting by the consolidation. Therefore, many shipper companies may be interested in playing an active role in the consolidation by creating consortia with other local companies so as to autonomously consolidate some or all their transportation flows and achieve better prices from carriers compared with those which can be obtained separately. This opportunity is particularly relevant in areas with a high density of companies operating in compatible sectors which share transportation and level of service requirements as it happens in specialized industrial districts typical of norther Italy.

To evaluate the potential benefit of these informal consortia it is necessary to identify those flows that are worth to be consolidated by means of an appropriate network flow model. A possible model to be used for such purpose turns out to be a version of an Intermediate Facilities Location (IFL) model in a transportation network similar to those used to design transportation networks in the carrier industry. In these models, the established intermediate facilities are places in which flows are aggregated to avoid vehicles traveling with half-full loads on long-distance trips. As for networks with intermediate facilities, three classes of optimization problems have been identified: (i) vehicle routing problems, (ii) transshipment problems, and (iii) service network design problems (see [7] for a survey on this class of problems). In our model we aim at finding which facilities to activate and assume that there is only one mode to reach a customer from a facility.

The major novelty of our work lies in the application of the Lagrangian Relaxation approach [4] to the IFL problem along with the use of a Bundle method [1, 2, 8], whose implementation has been described in [3]. We applied the Bundle method to the Lagrangian Relaxation of a mathematical formulation for the IFL, resulting in a non-smooth function. The usefulness of the method is the possibility of getting at each iteration a solution of the relaxed problem, from which a feasible solution of the original problem can be constructed. The proposed relaxation does not satisfy the integrality property—when splitting is not taken into account—which allows us to obtain a good lower bound, better than the one attained by using standard linear relaxation. However, we used the method with the aim at getting also a good upper bound.

The paper is organized as follows. In Sect. 2 we present the model. In Sect. 3 we propose a decomposition technique and a Lagrangian approach. In Sect. 4 we discuss the numerical results.

2 Problem Definition and Formulation

In the Intermediate Facilities Location (IFL) problem the goods are to be carried to the customer either using a direct path from the supplier to the destination or by means of an indirect path through an intermediate facility. In such a facility the goods are possibly consolidated with those from some other suppliers sharing the same destination. The decision making process concerns the choices of the facilities to be activated and the way to distribute the goods between the direct and indirect paths. For an extensive survey IFL problems we direct the reader to [7].

The IFL is defined on an appropriate transportation network in which two sets of nodes are defined, namely:

- O: the set of origins;
- D: the set of destinations.

Each destination $d \in D$ generally identifies a relatively limited area, such as a province, a town or a district, rather than a specific address, and is associated with a potential consolidation facility. Those consolidation facilities may be physically located in common infrastructure, e.g. a logistic operator site, but are kept separate in the following for modeling purposes. In addition, as we have a potential consolidation facility for each destination, set D also denotes the set of potential facilities. We consider the general case wherein the goods are characterized by a type $t \in T$ which denotes the specific handling and transportation requirements which result in specific unit transportation costs, where T is the set of goods types. For the sake of simplicity we also assume that goods of different types can be transported together in the same vehicle. We, however, note that the proposed model can be easily modified to consider incompatible goods types which must be transported separately. The amount of goods of type $t \in T$ that must flow from $o \in O$ to $d \in D$ is denoted by f_{otd}. The *direct cost* c_{otd} is paid to carry a unit of flow of type $t \in T$ from origin $o \in O$ to destination $d \in D$ through the direct path. Alternatively the goods may be sent to the destination by using an indirect path through a facility associated with their destination. Whenever a facility d, with maximum and minimum capacity U_d and L_d respectively, is used a fixed cost F_d is paid. The benefit of a facility is that of sharing vehicles to aggregate goods from many origins, thus reducing the usage of *half full-load* vehicles which may occur in the direct path. The consolidated flows from a facility to a destination are transported by a fleet of homogeneous vehicles each with capacity K. Whenever a vehicle is used to transport goods from a facility to its associated destination, a specific fixed cost p_d is paid which only depends on the destination. In addition, the *indirect cost* c'_{otd} is paid to carry a unit of flow of type t from origin o to a destination d through the indirect path through the associated facility, under the obvious assumption that $c'_{otd} \ll c_{otd}$. Finally, a minimum quantity b of goods is required to pass through all the facilities in order to ensure that the whole intermediate facilities system is sufficiently used.

The input parameters of the IFL problem can be summarized as follows:

(i) minimum (resp. maximum) facility capacity L_d (resp. U_d) for any $d \in D$;

(ii) facility activation fixed cost F_d for any $d \in D$;
(iii) vehicle capacity K, with $K < U_d$ for any $d \in D$;
(iv) fixed vehicle cost p_d for any $d \in D$;
(v) flow f_{otd} for a triplet $(o, t, d) \in (O, T, D)$;
(vi) unit direct (resp. indirect) cost c_{otd} (resp. c'_{otd}) for a triplet $(o, t, d) \in (O, T, D)$;
(vii) minimum quantity b of consolidated flow through all facilities $d \in D$.

2.1 *An Integer Programming Model for IFL*

The integer programming model of IFL makes use of the following decision variables:

- $x_{otd} \in [0, 1]$: the relative direct flow, i.e. the fraction of flow of type $t \in T$ carried from the origin $o \in O$ to the destination $d \in D$ with respect to the demand f_{otd};
- $y_{otd} \in [0, 1]$: the relative indirect flow, i.e. the fraction of flow of the type $t \in T$ carried from the origin $o \in O$ to the facility $d \in D$ with respect to the demand f_{otd};
- $w_d \in \mathbb{Z}$: the number of vehicles used at facility $d \in D$;
- $z_d = \begin{cases} 1, & \text{if the facility } d \in D \text{ is activated} \\ 0. & \text{otherwise} \end{cases}$

The problem consists in minimizing the sum of transportation costs, facility activation costs and vehicles costs while satisfying the capacities constrains. A small example of such a transportation network is illustrated in Fig. 1. The formulation of the IFL problem is the following:

$$\min \sum_{d \in D} \left[\sum_{o \in O} \sum_{t \in T} \left(\bar{c}_{otd} x_{otd} + \bar{c}'_{otd} y_{otd} \right) + F_d z_d + p_d w_d \right] \tag{1a}$$

$$x_{otd} + y_{otd} = 1, \qquad o \in O, t \in T, d \in D \tag{1b}$$

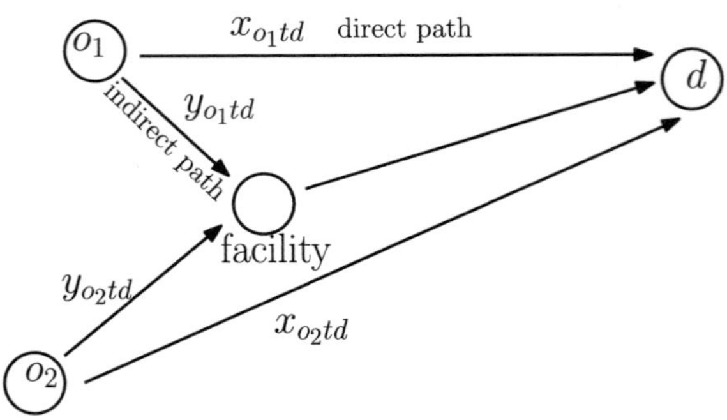

Fig. 1 An example

$$\sum_{o\in O}\sum_{t\in T} f_{otd}y_{otd} \le Kw_d, \qquad d\in D \tag{1c}$$

$$\sum_{o\in O}\sum_{t\in T} f_{otd}y_{otd} \le U_d z_d, \qquad d\in D \tag{1d}$$

$$\sum_{o\in O}\sum_{t\in T} f_{otd}y_{otd} \ge L_d z_d, \qquad d\in D \tag{1e}$$

$$\sum_{o\in O}\sum_{t\in T}\sum_{d\in D} f_{otd}y_{otd} \ge b \tag{1f}$$

$$0\le x_{otd}\le 1, \qquad o\in O, t\in T, d\in D \tag{1g}$$

$$0\le y_{otd}\le 1, \qquad o\in O, t\in T, d\in D \tag{1h}$$

$$z_d\in\{0,1\}, \qquad d\in D \tag{1i}$$

$$0\le w_d \quad \text{int}, \qquad d\in D \tag{1j}$$

where $\bar{c}_{otd} = c_{otd}f_{otd}$ and $\bar{c}'_{otd} = c'_{otd}f_{otd}$, i.e. the unit flow costs c_{otd} and c'_{otd} are scaled down by the flow f_{otd}. The constraint (1b) states that the flow f_{otd} has to reach the destination d either through the direct or the indirect path, or a combination of them. The formulation allows the flow f_{otd} to be split between the direct and indirect path. To inhibit the splitting we can add the following binary constraints:

$$x_{otd}\in\{0,1\} \qquad o\in O, t\in T, d\in D \tag{2}$$

$$y_{otd}\in\{0,1\} \qquad o\in O, t\in T, d\in D \tag{3}$$

By using constraints (1b) and removing x the variables, problem (1) can be reformulated as follows:

$$\begin{aligned}\min \sum_{d\in D}\left[\sum_{o\in O}\sum_{t\in T}\left(\bar{c}_{otd}(1-y_{otd}) + \bar{c}'_{otd}y_{otd}\right) + F_d z_d + p_d w_d\right]\\ = \sum_{d\in D}\left[\sum_{o\in O}\sum_{t\in T}\left(\bar{c}_{otd} + \bar{c}''_{otd}y_{otd}\right) + F_d z_d + p_d w_d\right]\end{aligned} \tag{4}$$

$$(1c), (1d), (1e), (1f), (1h), (1i), 1j)$$

where $\bar{c}''_{otd} = \bar{c}'_{otd} - \bar{c}_{otd}$

3 Lagrangian Relaxation

The formulation (4) is hard to be solved because of constraint (1f), which links all the facilities. In order to obtain a lower bound of (4) and decomposing the problem by facilities, we aim at finding a solution to its Lagrangian dual by relaxing the constraint (1f), whose Lagrangian multiplier is called $\mu \geq 0$. Then, the Lagrangian relaxation is the following:

$$
\begin{aligned}
z(\mu) = b\mu + \min \sum_{d\in D} & \left[\sum_{o\in O}\sum_{t\in T} \left(\bar{c}_{otd} + \left(\bar{c}''_{otd} - f_{otd}\mu\right) y_{otd}\right) + F_d z_d + p_d w_d \right] \\
& \sum_{o\in O}\sum_{t\in T} f_{otd}\, y_{otd} \leq K w_d \\
& \sum_{o\in O}\sum_{t\in T} f_{otd}\, y_{otd} \leq U_d z_d \\
& \sum_{o\in O}\sum_{t\in T} f_{otd}\, y_{otd} \geq L_d z_d \\
& 0 \leq y_{otd} \leq 1 \qquad\qquad o \in O, t \in T \\
& z_d \in \{0, 1\} \\
& 0 \leq w_d \quad \text{int}
\end{aligned}
$$

Hence, the problem is decomposed as the sum of a constant term and $|D|$ subproblems. More precisely, the constant term is

$$
z_0(\mu) = \sum_{o\in O}\sum_{t\in T}\sum_{d\in D} \bar{c}_{otd} + b\mu
$$

and each of the $|D|$ subproblems has the following form:

$$
\begin{aligned}
z_d(\mu) = \min & \sum_{o\in O}\sum_{t\in T} \left(\bar{c}''_{otd} - f_{otd}\mu\right) y_{otd} + F_d z_d + p_d w_d \\
& \sum_{o\in O}\sum_{t\in T} f_{otd}\, y_{otd} \leq K w_d \\
& \sum_{o\in O}\sum_{t\in T} f_{otd}\, y_{otd} \leq U_d z_d \\
& \sum_{o\in O}\sum_{t\in T} f_{otd}\, y_{otd} \geq L_d z_d \\
& 0 \leq y_{otd} \leq 1 \qquad\qquad o \in O, t \in T \\
& z_d \in \{0, 1\} \\
& 0 \leq w_d \quad \text{int}
\end{aligned}
$$

Because we assume that $K < U_d$, for any $d \in D$, the rounded-up number $\bar{\alpha} = \lceil U_d/K \rceil$ is greater than one. Therefore, for any integer number $\alpha \in \mathcal{B} = \{1, \ldots, \bar{\alpha} : K\alpha \geq L_d\}$, we define the following quasi-knapsack problem:

$$z_d^\alpha(\mu) = F_d + p_d\alpha + \min \sum_{o\in O} \sum_{t\in T} \left(\bar{c}''_{otd} - f_{otd}\mu\right) y_{otd} \tag{5a}$$

$$\sum_{o\in O} \sum_{t\in T} f_{otd} y_{otd} \leq \min\{K\alpha, U_d\} \tag{5b}$$

$$\sum_{o\in O} \sum_{t\in T} f_{otd} y_{otd} \geq L_d \tag{5c}$$

$$0 \leq y_{otd} \leq 1, \quad o \in O, t \in T, \tag{5d}$$

Finally, the Lagrangian relaxation can be solved by the following formula:

$$z_d(\mu) = \min \left\{ \min_{\alpha\in\mathcal{B}} \left(z_d^\alpha(\mu)\right), 0 \right\},$$

where the outer minimum operation is used to handle the case in which the optimal solution to the Lagrangian subproblem has $z_d = 0$.

Problem (5) becomes a binary quasi-knapsack problem if constraint (3) is added,, and in this case the integrality property does not hold. The best lower bound can be attained by maximizing the Lagrangian function over the space $\mu \geq 0$, which leads to a problem (the Lagrangian dual) whose objective function is non-smooth. That problem can be addressed via non-differentiable convex methods such as Bundle methods and Subgradient methods [5, 6]; alternatively line search can also be used, since it is a one-dimensional problem.

3.1 Lagrangian Heuristic

To develop a simple primal heuristic we consider the case in which the Lagrangian relaxation is an integer problem, i.e. constraint (3) is added to its mathematical formulation. The Lagrangian optimal solution is utilized to construct a feasible solution to (4), which can be easily obtained by swapping the values of the y variables and ensuring that constraint (1f) is satisfied. In practice, we sort y in the descending order of f_{otd} value. For each $d \in D$, we scan y and if we have $y_{\bar{o}\bar{t}d} = 1$, $y_{\hat{o}\hat{t}d} = 0$ with $f_{\bar{o}\bar{t}d} < f_{\hat{o}\hat{t}d}$, then we set $y_{\bar{o}\bar{t}d} = 0$, $y_{\hat{o}\hat{t}d} = 1$. This procedure is repeated till (1f) holds. The construction of such a feasible solution can be performed whenever the Lagrangian relaxation is solved during the execution of the non-differentiable method.

4 Computational Experiments

We have solved the Lagrangian dual problem by means of a *bundle method*, i.e., a stabilized version of the Cutting Plane algorithm that is exactly the naïve version of the Dantzig-Wolfe decomposition. The implementation of the bundle method we have used is in C++ (see, e.g., [3]), within the `NDOSolver/FiOracle` open-source project available at

https://gitlab.com/frangio68/ndosolver_fioracle_project

The code was compiled with GNU `g++ 4.4.5` (with `-O3` optimization option) and ran on an 2.2 GHz Intel Core i-7 processor with 16 GB of RAM, under a macOS operating system.

In our tests, we have compared the Lagrangian approach with the state-of-the-art traditional ones developed to solve mixed-integer and linear programs. To this end, we have used IBM-ILOG `CPLEX 20.1` to solve (4) and its linear relaxation. We note that we have considered the case where constraints (2)–(3) are included and the integrality property does not hold, consequently Lagrangian dual gap is tighter than the linear relaxation one. The other case, i.e. the one in which splitting is allowed, was not examined, since it would not be advantageous to use the Lagrangian approach. In fact, the bound obtained would be as good as the one attained by using linear relaxation.

4.1 Instances

The numerical experiments have been performed on 6 randomly generated instances, using the following parameters:

- unit traveling costs: $\bar{c}_{otd} \leq 1$, $\bar{c}'_{otd} \leq 0.01\bar{c}_{otd}$;
- facility activation fixed costs: $10 \leq F_d \leq 400$, $10 \leq p_d \leq 200$;
- facility capacities: $0.5\bar{f}_d \leq U_d \leq 0.9\bar{f}_d$, $0.1U_d \leq L_d \leq 0.4U_d$;
- vehicle capacities: $K \leq \bar{k} \max_{d\in D} \bar{f}_d$;
- demand: $1 \leq f_{otd} \leq 100$;
- minimum quantity of consolidated flow through all facilities: $b = 0.6\sum_{d\in D} \bar{f}_d$;

where $\bar{f}_d = \sum_{o\in O}\sum_{t\in T} f_{otd}$.

The instances characteristics are summarized in the following table.

group	P_1	P_2	P_3	P_4	P_5	P_6
$\lvert O\rvert$	100	100	500	1000	1000	2000
$\lvert T\rvert$	5	5	5	5	10	10
$\lvert D\rvert$	10	20	100	100	400	400

For each P_i, with $i = 1, \ldots, 6$, we have considered two values of $\bar{k}$ and we set P_i^a with $\bar{k} = 0.4$ and P_i^b with $\bar{k} = 0.1$.

4.2 Computational Results

For each approach, we have reported in the tables of the numerical experiments the computational time and the gap. As for the linear relaxation (LP), gap_l represents the relative gap between the linear relaxation solution and the best MIP solution provided by CPLEX. As for the Mixed Integer Problem (MIP), gap_m represents the relative gap between the best bound and the best integer solution at the end of the Branch & Bound execution. Instead, we report two gaps for the Lagrangian approach, gap_h^{LB} is the relative gap between the Lagrangian dual solution provided by the Bundle and the best MIP solution provided by CPLEX and gap_h^{UB} is the relative gap between the Lagrangian heuristic solution and the best MIP solution provided by CPLEX.

In Table 1 we fixed the time limit to 1 h and the relative gap tolerance (RT) to 1e-6. In Table 2 we have used RT = 1e-6 and RT = 1e-4. Because for instance P_6 the B&B by CPLEX is not able to find any solution, gap_h^{UB} represents the relative gap between the Lagrangian heuristic solution and the Lagrangian dual solution provided by the Bundle itself.

In Table 2 we have reported the largest instances and we have shown that in most cases, even by requesting a lower accuracy, CPLEX fails to terminate within 1 h. Hence, in Table 2 we have considered two different values of RT.

Table 1 Time limit (TL) = 3600 s, Relative tolerance (RT) = 1e-6

#	LP		MIP		Lagrangian Heuristic		
	time	gap_l	*time*	gap_m	*time*	gap_h^{LB}	gap_h^{UB}
P_1^a	0.05	9e-3	7.11	8e-7	63.76	1e-4	3e-5
P_1^b	0.09	4e-4	16.42	3e-7	491.01	1e-4	2e-4
P_2^a	0.30	7e-3	3.01	1e-6	18.01	7e-4	4e-4
P_2^b	167.21	5e-3	TL	6e-5	158.50	1e-4	8e-5
P_3^a	10.78	2e-2	TL	2e-6	16.45	6e-6	3e-4
P_3^b	13.95	1e-2	660.41	2e-8	60.88	5e-6	2e-4
P_4^a	171.91	7e-3	1139.16	1e-8	32.38	2e-6	2e-4
P_4^b	171.10	6e-3	795.10	1e-9	153.87	2e-6	1e-4

Table 2 Time limit (TL) = 3600 s

#	RT = 1e-6				RT = 1e-4			
	MIP		Lag. Heur.		MIP		Lag. Heur.	
	time	gap_m	*time*	gap_h^{UB}	*time*	gap_m	*time*	gap_h^{UB}
P_5^a	TL	1e-5	1347	1e-4	392	5e-4	707	1e-4
P_5^b	TL	1e-5	2673	1e-4	203	9e-4	1422	2e-4
P_6^a	TL	–	353	7e-5*	TL	–	241	8e-5*
P_6^b	TL	–	504	7e-5*	TL	–	314	8e-5*

For small instances CPLEX outperforms the Lagrangian approach, conversely for large instances the Lagrangian approach typically provides better results in terms of computational time and solution accuracy. We have used the Bundle method for solving the Lagrangian dual, instead of the Subgradient one, because it terminates for all instances in a few iterations. For the sake of brevity, this information was not reported in the tables.

Acknowledgements This study was carried out within the MOST—Sustainable Mobility National Research Center and received funding from the European Union Next-GenerationEU (PIANO NAZIONALE DI RIPRESA E RESILIENZA (PNRR)—MISSIONE 4 COMPONENTE 2, INVESTIMENTO 1.4—D.D. 1033 17/06/2022, CN00000023). This manuscript reflects only the authors' views and opinions, neither the European Union nor the European Commission can be considered responsible for them.

References

1. Astorino, A., Frangioni, A., Gaudioso, M., Gorgone, E.: Piecewise quadratic approximations in convex numerical optimization. SIAM J. Optim. **21**(4), 1418–1438 (2011)
2. Astorino, A., Frangioni, A., Fuduli, A., Gorgone, E.: A nonmonotone proximal bundle method with (potentially) continuous step decisions. SIAM J. Optim. **23**(3), 1784–1809 (2013)
3. Frangioni, A.: Generalized bundle methods. SIAM J. Optim. **13**(1), 117–156 (2002)
4. Frangioni, A.: About Lagrangian methods in integer optimization. Ann. Oper. Res. **139**, 163–193 (2005)
5. Frangioni, A., Gorgone, E.: Generalized bundle methods for sum-functions with "easy" components: applications to multicommodity network design. Math. Progr. **145**(1), 133–161 (2014)
6. Frangioni, A., Gendron, B., Gorgone, E.: On the computational efficiency of subgradient methods: a case study with Lagrangian bounds. Math. Progr. Comput. **9**, 573–604 (2017)
7. Guastaroba, G., Speranza, M.G., Vigo, D.: Intermediate facilities in freight transportation planning: a survey. Transp. Sci. **50**(3), 763–789 (2015)
8. Hiriart-Urruty, J.B., Lemaréchal, C.: Convex Analysis and Minimization Algorithms, vol. II. Springer-Verlag (1993)

MIX
Papier aus verantwortungsvollen Quellen
Paper from responsible sources
FSC® C105338
FSC
www.fsc.org

If you have any concerns about our products,
you can contact us on
ProductSafety@springernature.com

In case Publisher is established outside the EU,
the EU authorized representative is:
Springer Nature Customer Service Center GmbH
Europaplatz 3, 69115 Heidelberg, Germany

Printed by Libri Plureos GmbH
in Hamburg, Germany